Geographic Information Systems

Geographic Information Systems

An Introduction

Second Edition

Tor Bernhardsen

Asplan Viak
Arendal, Norway

John Wiley & Sons, Inc.
New York • Chichester • Weinheim • Brisbane • Singapore • Toronto

This book is printed on acid-free paper. ∞

Published simultaneously in Canada.

Library of Congress Cataloging-in-Publication Data:

Bernhardsen, Tor.
Geographic information systems: an introduction/Tor Bernhardsen.—2nd ed.
p. cm.
ISBN 0-471-32192-3 (cloth: alk. paper)
1. Geographic information systems. I. Title.
G70.212.B473 1999
910'.285—dc21 98-38195

Printed in the United States of America.

10 9 8 7 6 5 4 3 2 1

CONTENTS

FOREWORD

In approaching global problems, the people of the world are benefiting from fifty years of major developments in the science and technology used for studying the earth, its human population, and the natural and cultural resources that make up the global environment. Especially important among these have been the rapid developments in remote sensing of the earth environment, in computer science and technology, in global electronic communications, and in the gathering of information about the earth and its inhabitants into vast new electronic storehouses. In sum, as we leave this last decade of the twentieth century, we are in the fortunate position of having both the technical and the political means of bringing the people of the world together to deal with the world's problems; we must seize this opportunity.

One of these newly developed technologies and one of the most important components of any approach to global problem solving is geographic information systems (GIS) technology. Developed in just the last 30 years or so, GIS technology already represents a billion dollar industry worldwide, growing at perhaps 25% per year, and serving some 50,000 to 100,000 users in more than 100 countries. If projections are correct, GIS will be used by more than a million persons by about the turn of the century.

GIS has been used on problems at a wide range of scales and for geographic areas from a few hectares up to the global databases just now beginning to become available for wide use. GIS has been applied by a wide range of disciplines to a correspondingly wide range of problems. Governments, nongovernmental organizations, businesses, and educational institutions all now use GIS technology. Members of the general public will become GIS users in the years just ahead. That GIS technology has proven its value to its users is indicated by the rapid growth in its use, the rapid growth in expenditures for GIS technology, and by the very large amounts of resources now being devoted throughout the world to creating digital data for use in GISs of various kinds.

Another indication of the value of GIS technology is that the number of GIS educational programs is growing even faster than is the use of the technology. For this important educational enterprise to succeed, sound textbooks in GIS technology are needed. This volume is just such a textbook. Until just a few years ago there were few introductory textbooks dealing with GIS technology. Those who learned the technology learned chiefly or exclusively through experience. Fortunately that is now changing, and new GIS textbooks are published nearly every month. Nevertheless, because the field is so new, the contents appropriate to such a textbook are not yet widely agreed upon. To a degree, then, each new textbook is still an experiment, a creation not unlike those experimental vessels that Vikings of an earlier age built so assiduously and successfully, and by means of which they extended themselves over the seas of much of the known world. I think this book is also well designed, solidly constructed, and finely crafted; those who depend on it as they set out to explore our spatial world will be well served.

In my own work in the GIS field over the last thirty years or so, I have become convinced of the importance of providing GIS technology that works for its users. Many persons, especially those new to the field, can be overwhelmed by the rapid growth of the GIS literature and by the hyperbole that is often associated with the announcement of new products and new developments, many of which will not advance the field and some proportion of which will never, in fact, actually come into existence. Students and others new to the GIS field need the guidance of textbook authors with extensive experience in the field, discernment, and careful judgment; authors who can separate the sound and essential from that which will be ephemeral; authors who can indicate what actually works. That experience is reflected in these pages.

It is also important to realize that a GIS is considerably more than just technology; most of the problems with real GIS now have more to do with the people and the procedures they use (system design, applications programming, effective use of GIS in solving real problems, funding new systems, and so on) than with computer hardware and software. The reality is that while one can now buy a reliable turnkey GIS hardware/software system and can often obtain a great deal of GIS data through purchase or conversion, there is no similarly reliable way to create the organizational structures and experienced staff necessary to support an effective GIS. These things must be learned, and unfortunately, their important nuances probably cannot be taught.

Some years ago, in thinking about how GIS use could best be developed throughout the world, I made a list of the traits that I thought GIS technical advisers would require if they were to be successful in spreading GIS use into new disciplines, new organizations, and new areas in the world. The list of intellectual and personal traits was a formidable one, chiefly because I believe GIS technology can be applied to so many kinds of problems and in so many situations. The list, in turn, led me to reflect on how one could possibly find, recruit, and train such people. How does one go about producing people with boundless enthusiasm, extraordinary patience, coupled with appropriate technical, managerial, and political skills; those who are technicians as well as humanitarians; persons sympathetic to the ills of humankind and not afraid to wade in and try to improve things even when, often, they will not be successful?

In the end, I have concluded that such persons will, naturally and inevitably, be drawn to GIS technology, bringing with them the many skills and abilities that they have already acquired elsewhere. When they understand just what GIS technology is capable of doing for many kinds of problem solving, they will embrace it and take it with them wherever they go. Books like this one will foster that process, and so contribute to the improvement of the world in which we live.

I hope that the readers of this book will profit from what it has to teach them. If we are to solve many of the problems facing us—in the cities, in the wild areas of the earth, in the atmosphere and the oceans, problems of the earth as a whole—we shall need the help of skilled users of GIS technology. If readers can master what is in this volume, they will be well started on this enterprise.

Jack Dangermond

PREFACE

The first English edition of the book came out in 1992. The positive response that we received from all parts of the world inspired us to start work on a second edition. For this edition, an extensive revision of the text has been carried out. However, we have chosen to maintain our original philosophy that the book should be suitable for a reader with little or no knowledge of GIS. The second edition has more but shorter chapters and has been supplemented by additional material on the newest technological developments, such as Internet and multimedia. We have also incorporated information drawn from recent experience in the practical application and introduction of GIS.

Since this is an introductory book, some aspects have been dealt with relatively superficially and others have only been mentioned in passing. For example, there is relatively little discussion of different mathematical methods and formulas for handling the geometry. However, questions related to the introduction of GIS into organizations are treated in reasonable detail. Addressing these issues correctly is linked directly with the successful functioning of the systems being introduced.

In this respect I would like to express my particular appreciation of the assistance given by Professor Michael F. Goodchild, University of California, Santa Barbara. He took a critical look at the first edition and made comments, both positive and negative, and presented constructive proposals for change. We have to a great extent followed his guidelines in restructuring the material. I would also like to thank Professor Goodchild for his substantial contribution to Chapter 1. Further, I would like to thank Professor Inge Revhaug at the Agricultural University of Norway for his valuable assistance in supplying the correct statistical terminology in Chapter 11, and Professor Jan Terje Bjørke at NTNU, the Norwegian University of Technology and Science, for his evaluation of, and contribution to, Chapter 16.

I would like to express my gratitude for the fine cooperation of Mahala Mathiassen of Asplan Viak for her English translation of all the new material in the second edition. I would also like to thank my employer, Asplan Viak, for making it possible for me to use a great deal of my working time in preparing this book.

Tor Bernhardsen
Arendal, Norway

PREFACE TO THE FIRST EDITION

This book is an attempt to meet the need for a comprehensive presentation of the various fields currently associated with the term "geographic information systems" (GIS). Very few limitations have here been set on interpretations of the term GIS. By looking at the table of contents, one might find that the book could just as well have been entitled "Geographic Information Technology" (GIT). In allowing for a wide range of definitions we have taken into consideration the fact that most users of geographic data are addressing specific tasks, and that they will always choose the technology most applicable to those tasks, regardless of whether or not the technology falls within a particular definition of GIS.

Questions related to the introduction of GIS into organizations are in this book treated in relatively great depth. This is because correctly addressing these issues is directly linked with the successful functioning of the systems being introduced. This is the reasoning behind the sub-title "the beneficial use of geographic information technology."

The idea for this book arose out of my work with the Nordic joint project named "Community Benefits of Digital Spatial Information—Nordic KVANTIF I & II," 1985–1990.

The first edition of the book came out in Norwegian in 1989. During this first preparation I was given invaluable assistance, in particular by Professor Øystein Andersen at the Agricultural University of Norway, Harald Danielsen of the Nidar District Administration, and Arild Reite of the Norwegian Cartographic Union.

Prior to this second edition, which is now being published in English, an extensive revision has been carried out. With regard to this I would like to express my particular appreciation of the assistance given again by Professor Øystein Andersen and by the Director of GRID-Arendal, Olav Hesjedal, both of whom have provided valuable suggestions as to how the book might be improved. Svein Tveitdal, the International Director of Viak I.T., has also made important contributions to Chapter 12. Knut Heggen of Asplan Viak Sør has provided practical examples for Chapter 9, and Robert Sandvik of the Norwegian Naval Cartography Administration has supplied material for Chapter 13.

I would like to express my gratitude for the fine cooperation and teamwork of Michael Brady Consultants who have been responsible for the English translation. They have also made contributions in other respects, thereby improving the overall quality of the book.

Dr. William Warner at the Agricultural University of Norway has made a major contribution by reviewing the technical and scientific terminology of this English edition. I would also like to thank Jürg Keller of AB-Trykk for the fine layout and illustrations.

Finally, I would like to express my deepest thanks to my wife and two children for their patience and forbearance while so much of my free time has been spent working on the book.

Tor Bernhardsen
Arendal, May 1992

Geographic Information Systems

CHAPTER 1

Geographical Information Systems and Geographical Information

1.1 Basic Concepts

Society is now so dependent on computers and computerized information that we scarcely notice when an action or activity makes use of them. Over the past few decades we have developed extremely complex systems for handling and processing data represented in the only form acceptable to computers: strings of zeros and ones, or bits (binary digits). Yet is has proved possible to represent not only numbers and letters, but sound, images, and even the contents of maps in this simple, universal form. Indeed, it might be impossible to tell whether the bits passing at high speed down a phone line, or stored in minute detail on a CD-ROM (compact disk—read-only memory) represent a concerto by Mozart or the latest share prices. Unlike most of its predecessors, computer technology for processing information succeeds in part because of its ability to store, transmit, and process an extremely wide range of information types in a generalized way.

Computerization has opened a vast new potential in the way we communicate, analyze our surroundings, and make decisions. Data representing the real world can be stored and processed so that they can be presented later in simplified forms to suit specific needs. Many of our decisions depend on the details of our immediate surroundings, and require information about specific places on the Earth's surface. Such information is called *geographical* because it helps us to distinguish one place from another and to make decisions for one place that are appropriate for that location. Geographical information allows us to apply general principles to the specific conditions of each location, allows us to track what is happening at any place, and helps us to understand how one place differs from another (Figure 1.1). Geographical information, then, is essential for effective planning and decision making.

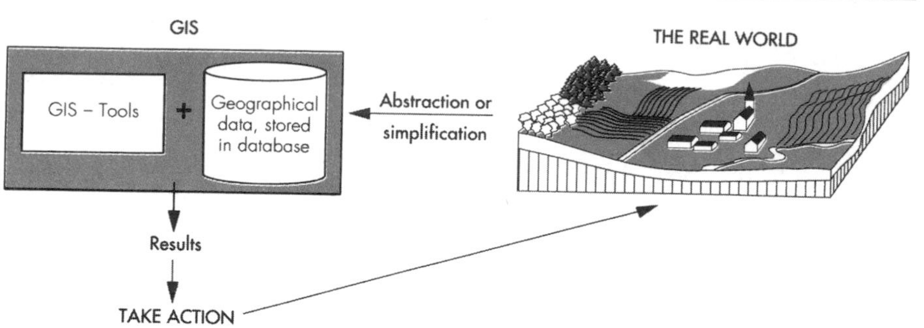

Figure 1.1
GIS is intended to be a means of improving everyday life. It is therefore important that the information that results from data processing be applied to guide the real world in the right direction.

We are used to thinking about geographical information in the form of maps, photos taken from aircraft, and images collected from satellites, so it may be difficult at first to understand how such information can be represented in digital form as strings of zeros and ones. That problem is one of the central issues of this book, and the fact that many alternatives exist is one of the reasons why this book is as long as it is. If we can express the contents of a map or image in digital form, the power of the computer opens an enormous range of possibilities for communication, analysis, modeling, and accurate decision making (Figure 1.2). At the same time, we must constantly be aware of the fact that the digital representation of geography is not equal to the geography itself—any digital representation involves some degree of approximation.

Since the mid-1970s, specialized computer systems have been developed to process geographical information in various ways. These include:

- Techniques to input geographical information, converting the information to digital form
- Techniques for storing such information in compact format on computer disks, compact disks (CDs), and other digital storage media
- Methods for automated analysis of geographical data, to search for patterns, combine different kinds of data, make measurements, find optimum sites or routes, and a host of other tasks
- Methods to predict the outcome of various scenarios, such as the effects of climate change on vegetation
- Techniques for display of data in the form of maps, images, and other kinds of displays
- Capabilities for output of results in the form of numbers and tables.

The collective name for such systems is *geographical information systems,* (GISs). The acronym GIS has come to signify much more than a software system that processes, stores, and analyzes geographical

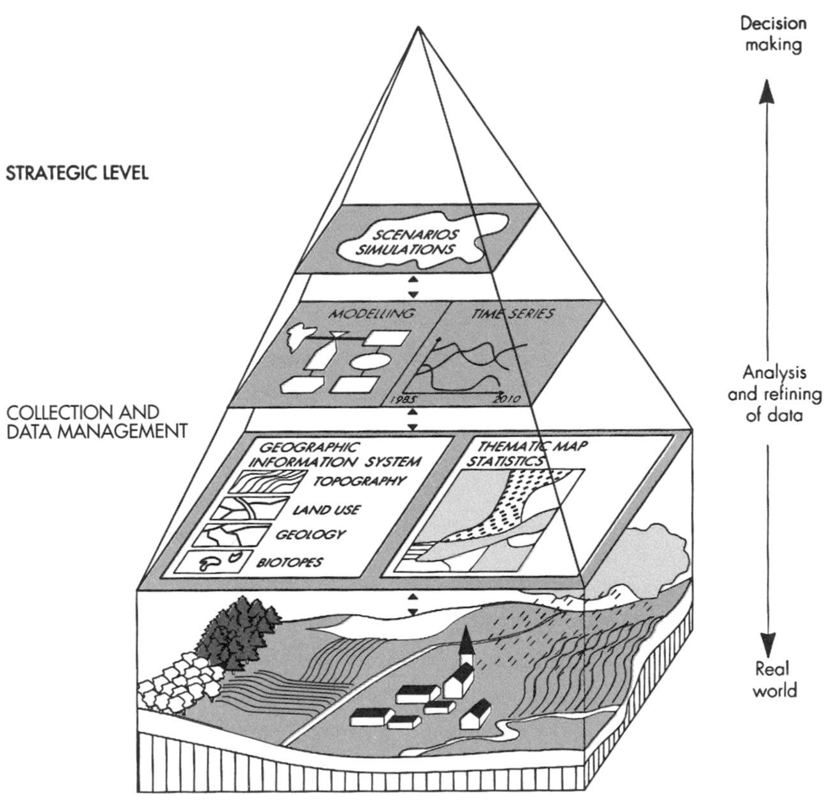

Figure 1.2
By use of geographical information systems, a simplified world can be brought into the computer. Different techniques can be applied to analyze and simplify the data, and the foundation is laid for the decision-making pyramid. Today, geographical information systems are in the process of filling the upper half of the pyramid. (Figure freely adapted from Grossman 1983.)

data. GIS is a "hot" application area for digital technology. Its software industry has been growing at more than 20% a year for many years, and recent figures for total annual sales of GIS software exceed $500 million. The term 'GIS' has come to be associated with any activity involving digital geographical data; we now talk about GIS data, GIS decisions, and even GIS systems.

Although it is very easy to purchase the constituent parts of a GIS (the computer hardware and basic software), the system functions only when the requisite expertise is available, the data are compiled, the necessary routines are organized, and the programs are modified to suit the application. A computer system can function at what may appear to be lightning speed, yet the entire time span of a GIS project can stretch to months and even years. These facets of an overall GIS are interlinked (Figure 1.3). In general, procurement of the computer hardware and software is vital but straightforward. The expertise required is often underestimated, the compilation of data is ex-

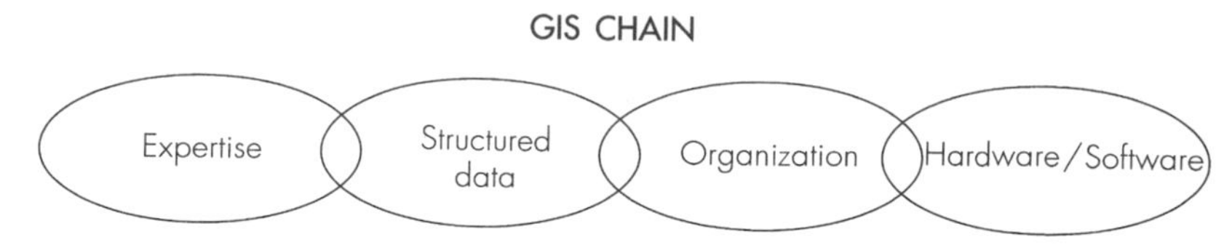

Figure 1.3
A GIS system cannot be bought off the shelf. The system has to be built up within an organization. When planning to introduce GIS, it is important that equal attention be given to all four links in the GIS chain.

pensive and time consuming, and the organizational problems can be most vexing. These facets of an overall GIS are discussed in detail later.

Traditionally, geographical data are presented on maps using symbols, lines, and colors. Most maps have a legend in which these elements are listed and explained—a thick black line for main roads, a thin black line for other roads, and so on. Dissimilar data can be superimposed on a common coordinate system. Consequently, a map is both an effective medium for presentation and a bank for storing geographical data. But herein lies a limitation. The stored information is processed and presented in a particular way, usually for a particular purpose. Altering the presentation is seldom easy. A map provides a static picture of geography that is almost always a compromise between many differing user needs. Nevertheless, maps are a substantial public asset. Surveys conducted in Norway indicate that the benefit accrued from the use of maps is three times the total cost of their production.

Compared to maps, GIS has the inherent advantage that data storage and data presentation are separate. As a result, data may be presented and viewed in various ways. Once they are stored in a computer, we can zoom into or out of a map, display selected areas, make calculations of the distance between places, present tables showing details of features shown on the map, superimpose the map on other information, even search for the best locations for retail stores! In effect, we can produce many useful products from a single data source (Figure 1.4).

GIS defined

The term *geographical information system* (GIS) is now used generically for any computer-based capability for the manipulation of geographical data. A GIS computer-based capability for the manipulation of geographical data. A GIS includes not only hardware and software, but also the special devices used to input maps and to create map products, together with the communication systems needed to link various elements. The hardware and software functions of a GIS include:

- Acquistion and verification
- Compilation

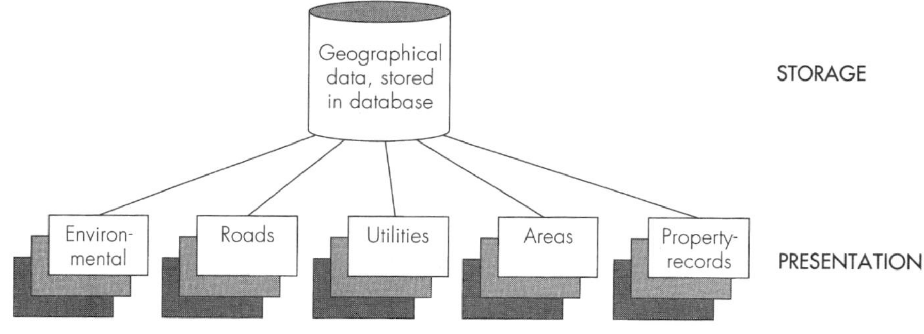

Figure 1.4
A map can be both a presentation medium and a storage medium, with resulting limitations. With GIS, storage and presentation are separated, thereby enabling a wide variety of products to be created from the same basic data.

- Storage
- Updating and changing
- Management and exchange
- Manipulation
- Retrieval and presentation
- Analysis and combination

All of these actions and operations are applied by a GIS to the geographical data that form its database.

All of the data in a GIS are georeferenced, that is, linked to a specific location on the surface of the Earth through a system of coordinates. One of the commonest coordinate systems is that of latitude and longitude; in this system location is specified relative to the equator and the line of zero longitude through Greenwich, England. But many other systems exist, and any GIS must be capable of transforming its georeferences from one system to another.

Geographical information attaches a variety of qualities and characteristics to geographical locations (Figure 1.5). These qualities may be physical parameters such as ground elevation, soil moisture level, or atmospheric temperature, as well as classifications according to the type of vegetation, ownership of land, zoning, and so on. Such occurrences as accidents, floods, or landslides may also be included. We use the general term *attributes* to refer to the qualities or characteristics of places, and think of them as one of the two basic elements of geographical information, along with locations.

In some cases, qualities are attached to points, but in other cases they refer to more complex features, either lines or areas, located on the Earth's surface; in such cases the GIS must store the entire mapped shape of the feature rather than a simple coordinate location. Examples of commonly mapped features are lakes, cities, counties, rivers, and streets, each with its set of useful attributes. When a feature is used as a reporting zone for statistical purposes, a vast amount of in-

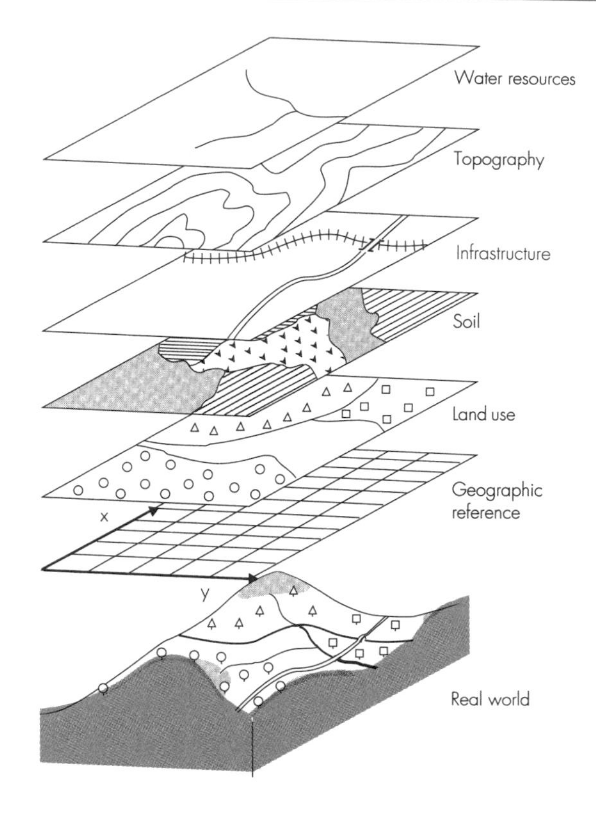

Figure 1.5
One can visualize the data stored as theme layers in the computer, with each layer linked to a common georeferencing system.

formation may be available to be used as attributes for the zone in GIS. In market research, for example, it is common for postal codes to be used as the basis for reports on demographics, purchasing habits, and housing markets.

The relationships between geographical features often provide vital information. For example, the connections of a water supply pipe network may be critical for firefighters, who need to know which valves to close in order to increase water pressure while extinguishing a fire. The details of properties bordering a road are necessary if all property owners affected by roadwork are to be properly notified. Connections between streets are important in using a GIS to assist drivers in navigating around an unfamiliar city. The ability of a GIS to store relationships between features in addition to feature locations and attributes is one of the most important sources of the power and flexibility of this technology. Some GISs can even store flows and other measures of interaction between features, to support applications in transportation, demography, communication, and hydrology, among other areas.

Stored data may be processed in a GIS for presentation in the form of maps, tables, or special formats (Figure 1.6). One major GIS strength

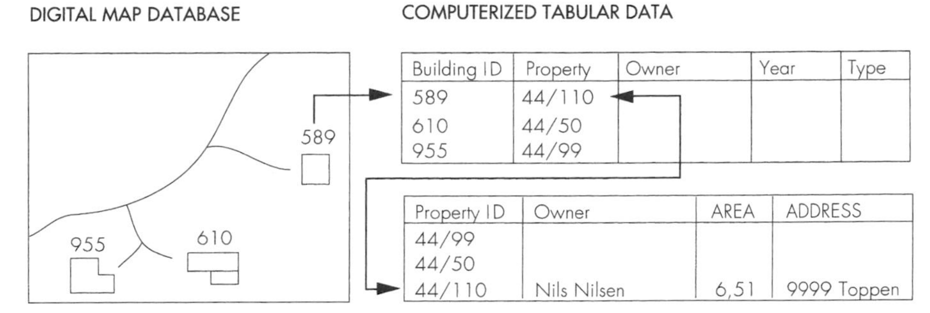

Figure 1.6
How GIS functions, based on the interaction between a digital map database and computerized tabular data.

is that geographical location can be used to link information from widely scattered sources. Because the geographical location of every item of information in a GIS database is known, GIS technology makes it possible to relate the quality of groundwater at a site with the health of its inhabitants, to predict how the vegetation in an area will change as the climate warms, or to compare development proposals with restrictions on land use. This ability to overlay gives GIS unique power in helping us to make decisions about places and to predict the outcomes of those decisions. The only requirement is that the geographical information from each source be expressed in compatible georeferencing systems.

A GIS can process georeferenced data and provide answers to questions involving, say, the particulars of a given location, the distribution of selected phenomena, the changes that have occurred since a previous analysis, the impact of a specific event, or the relationships and systematic patterns of a region. It can perform analyses of georeferenced data to determine the quickest driving route between two points and help resolve conflicts in planning by calculating the suitability of land for particular uses.

Many recent GISs can process data from a wide range of sources, including data obtained from maps, images of the Earth obtained from space satellites, videofilm of the Earth taken from low-flying aircraft, statistical data from published tables, photographs, data from computer-assisted design (CAD) systems, and data obtained from archives by electronic transmission over the Internet and other networks. Data integration is one of the most valuable functions of a GIS, and the data that are integrated are more and more likely to be obtained from several distinct media—multimedia is an active area for research and development in GISs (Figure 1.7).

Technically, a GIS organizes and exploits digital geographical data stored in databases. As we have already seen, the data include information on locations, attributes, and relationships between features.

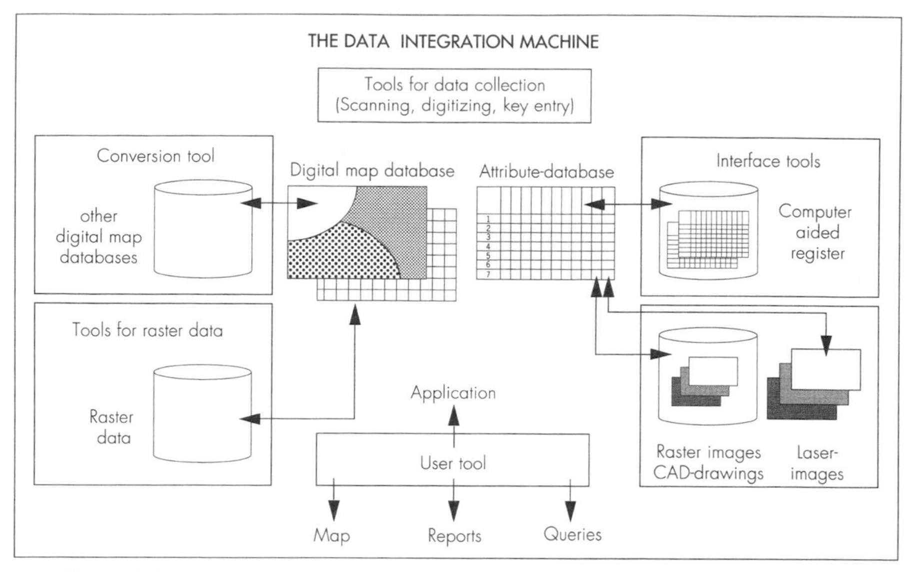

Figure 1.7
GIS is in the process of becoming a typical data integration machine. Some modern systems can receive, process, and transmit data of widely varying origin. (Figure freely adapted from ESRI.)

But a database can only approximate the real world, since the storage capacity of a database is minuscule in comparison with the complexity of the real world, and the cost of building a database is directly related to its complexity. The contents of a book of 100,000 words can be stored in digital form in roughly 1 million bytes (the common unit of computer storage is a byte, defined as 8 bits; 1 megabyte is slightly more than 1 million bytes). The information on a topographic map is comparatively dense, and it commonly takes 100 megabytes to capture it in digital form. A single scene from an Earth observing satellite might contain 300 megabytes, the information content of 300 books. Thus even crude approximations to the complexity of real-world geography can rapidly overtake the capacity of our digital storage devices.

Although we often think of the contents of a GIS database as equivalent to a map, there are important differences. On a map, a geographical feature such as a road or a power line is shown as a symbol, using a graphic that will readily be understood by the map reader (Figure 1.8). In a geographical database a road or power line will be represented by a single sequence of points connected by straight lines, and its symbolization will be reattached when it is displayed. A windmill will be represented by a single point, with the attribute "windmill", and will be replaced by a symbol when displayed. This approach is economical, since the geometric form of the windmill symbol will be

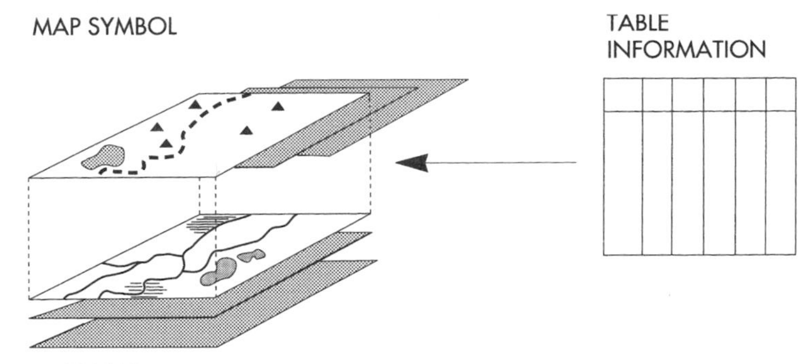

Figure 1.8
For presentation purposes, the table information is translated into map symbols, which are then superimposed on existing map data.

stored only once rather than repeated at each windmill location, and it also allows analysis to be more effective.

Databases are vital in all geographical information systems, since they allow us to store geographical data in a structured manner that can serve many purposes. Many GISs impose further structure by using a database management system (DBMS) to store and manage part or all of the data in a largely independent subsystem under the GIS itself (Figure 1.9). A DBMS is a general-purpose software product, and GISs that use this approach are often able to function in conjunction with a wide range of DBMS products.

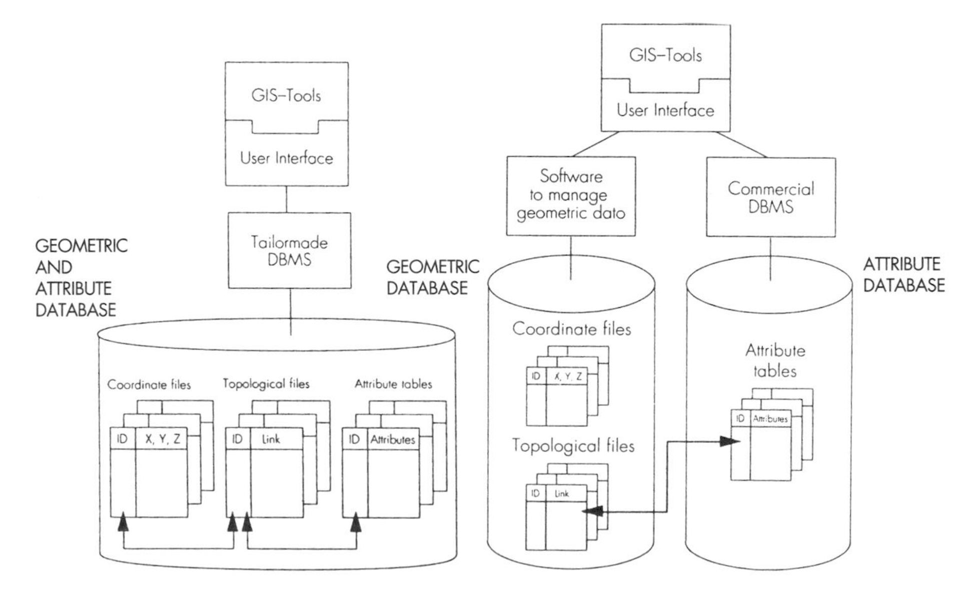

Figure 1.9
Alternative DBMS solution for GIS.

The database underlying a GIS achieves many objectives. It ensures that data are:

- Stored and maintained in one place
- Stored in a uniform, structured, and controlled manner than can be documented
- Accessible to many users at once, each of whom has the same understanding of the database's contents
- Easily updated with new data

This contrasts with the traditional way of organizing and storing data on paper in filing cabinets, in which data are often:

- Stored in ways that are understandable to one person only
- Easily corrupted by use, or edited in ways that are meaningful only to the editor
- Inaccessible to anyone other than the creator of the system
- Stored in formats and at scales that are so diverse that they cannot be compared or collated
- Difficult to update

The constituent parts and modes of operation of GISs are discussed in detail in Chapter 3.

GIS diversity

Although the general definition of GIS given here is quite valid, in practice the diversity of GIS has spawned various definitions. First, users have contrived working definitions suited to their own specific uses. Thus they may vary according to whether operators are planners, water-supply and sewage engineers, support service personnel, or perhaps professional and public administrators or Earth scientists. Second, those with a more theoretical approach, such as research workers, software developers or sales and training staff, may use definitions that are different from those used in practical applications. Systems can be tailor-made by assembling them from available software tool kits of semi-independent modules, assorted computer hardware components, and other interoperable devices. Many applications can be addressed by acquiring a single, generic GIS product and a standard configuration of hardware. There are many views of GISs, including:

- A data processing system designed for map production or visualization
- A data analysis system for examining conflicts over plans or optimizing the design of transport systems
- An information system for responding to queries about land ownership or soil type
- A management system to support the operations of a utility company, helping it to maintain its distribution network of pipes or cables

- A planning system to aid the design of road systems, excavations, or forest harvest operations
- An electronic navigation system for use in land or sea transport.

GISs are often designated according to application. When used to manage land records they are often called land information systems (LISs); in municipal and natural resource applications they are important components of urban information systems (UISs) and natural resource information systems (NRISs) respectively. The terms *spatial* and *geospatial* are often used almost interchangeably with *geographical,* although *spatial* is also used to refer more generally to any two- or three-dimensional data whether or not it relates directly to the surface of the Earth. The term *automatic mapping/facility management* (AM/FM) is frequently used by utility companies, transportation agencies, and local governments for systems dedicated to the operation and maintenance of networks. Nonetheless, *GIS* is now accepted internationally as an umbrella term for all digital systems designed to process geographical data, and is used in that sense in this book.

As befits this breadth of applications, the field of GIS involves many disciplines, applications, types of data, and end users, examples of which are:

- *Disciplines*: computer sciences, cartography, photogrammetry, surveying, remote sensing, geography, hydrography, statistics, information sciences, planning
- *Applications*: operation and maintenance of networks and other facilities, management of natural resources, real estate management, road planning, map production
- *Data*: digital maps, digital imaging of scanned maps and photos, satellite data, ground truth data, video images, tabular data, text data
- *Users*: water supply and sewage engineers, planners, biologists and cartographers, surveyors

The software capabilities required for a GIS often overlap those needed by other computer applications, particularly image processing and computer-assisted design (CAD). Image processing systems are designed to perform a wide range of operations on the images captured by videocameras, still cameras, and remote-sensing satellites. Today, the distinction between image processing and GIS is becoming increasingly blurred as images become more and more important sources of GIS data. Broadly, though, it is convenient to think about image processing systems as concerned primarily with the extraction of information from images, and GIS as concerned with the analysis of that information.

CAD systems have been developed to support design applications in engineering, architecture, and related fields. Broadly, CAD systems emphasize design over analysis and often lack the capabilities needed to process the complex attributes and information of georeferenced data or to integrate georeferenced data from many sources. Nevertheless, the distinction between CAD and GIS has become increasingly blurred in recent years; by adding appropriate features, many former vendors of CAD systems are now able to compete effectively in the GIS market.

1.2 Socioeconomic Challenges

Various surveys have been conducted to determine the importance of geographical data by assessing the proportion of all data that are geographical, as defined above. In local government bodies and in utility companies the proportion is often over 80%; in mapping agencies it may come very close to 100%. These surveys have also shown that major users of geographical data—the construction sector, public administration, agriculture, forestry and other resource management, telecommunications, electricity supply, transportation—spend 1.5% to 2% of their annual budgets acquiring geographical data. In relation to gross national product (GNP), annual expenditures average 0.5% in industrialized countries and 0.1% in developing countries. In the United States, a 1993 survey by the Office of Management and Budget found annual expenditures on geographical information in federal agencies totaling over $4 billion, suggesting that the total in all areas of U.S. society is well over $10 billion. The major challenges to system developers and users alike are now very different, and related to the comparative ease of use of the technology, the problems of finding and accessing suitable data, and the lack of trained personnel able to exploit the technology's potential to its full.

Our complex society

Modern societies are now so complex, and their activities so interwoven, that no problem can be considered in isolation or without regard for the full range of its interconnections. For example, a new housing development will affect the local school system. Altered age distribution in a village will affect health and social expenditure. The volume of city traffic will put constraints on the maintenance of buried pipe networks, affecting health. Street excavations may drastically reduce the turnover of local retail shops. Traffic noise from a new road or motorway may well drive people from their homes. The actions needed to solve such problems are best taken on the basis of standardized information that can be combined in many ways to serve many users (Figure 1.10). GISs have this capability.

Populations are now extremely mobile; changing jobs and moving to another location have become commonplace. When key personnel leave a company, they take their expertise with them; if that expertise

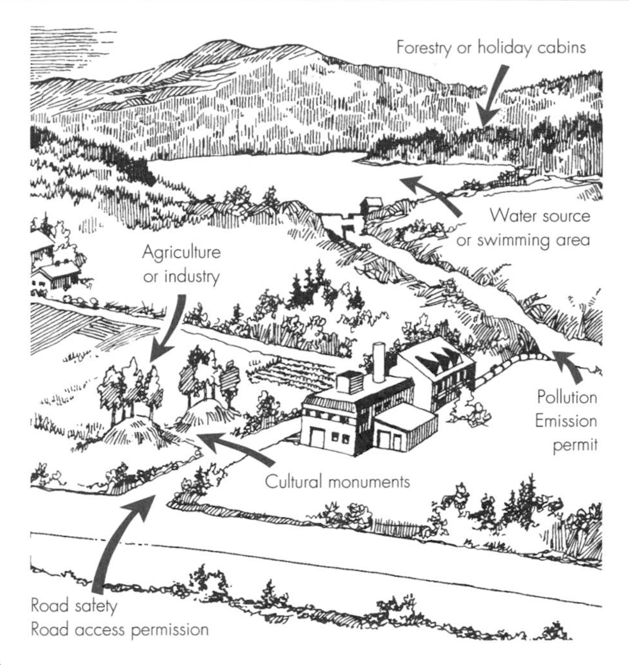

Figure 1.10
Today's society has a growing need for information. Conflicts of interest have to be brought out into the open and information has to be made available to the parties concerned. It has been shown that at least 50% of the public administration sector needs geographical data in one form or other. (Courtesy of the Norwegian Cartographers Union et al.)

involves specific knowledge of, say, the water supply and sewage network of a community, the loss can be serious if the information is otherwise inadequately documented. Here, too, GIS has an advantage in that it can act as an effective filing system for dissimilar sectors of a complex society.

Operation and maintenance

Another aspect of developed industrial societies is that emphasis has shifted from the planning and development of infrastructure to its operation and maintenance, particularly for such facilities as buried pipes and cables, and transportation systems. As we approach the twenty-first century the infrastructure of many industrial societies is deteriorating and the costs of maintenance are increasing accordingly (Figure 1.11). It is estimated that the average age of the water supply pipes of many cities of eastern North America is now over 100 years, and some 20% to 30% of the water supplied to these systems is now lost through leakage. The costs of replacing this aging infrastructure over the next decades will be truly astronomical. In Sweden, too, costs of operation and maintenance of the water supply system in the 1990s are double those of the 1980s.

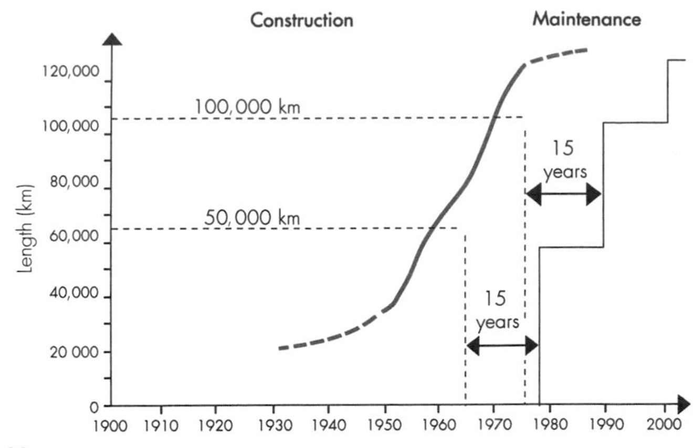

Figure 1.11
Total length of the water and sewerage system in Sweden from 1935 to date. Maintenance is normally carried out about 15 years after the pipes have been laid. This means that maintenance costs in the 1990s are double those of the 1980s.

In Norway in 1990, municipal budgets for operation and maintenance were nine times those allocated to building new facilities. The large sums involved indicate that even minor gains in efficiency in this area can result in considerable savings. Yet more efficient operation and maintenance is possible only when full information is available on a facility's location, condition, age, construction material(s), and service record.

As many as 35 different types of piping and cable may be buried beneath the surface of a major city street. These include telecommunications and optical cables, high-voltage, local power, signal, heating, and TV cables, as well as piping for water supplies, sewerage, gas, and remote heating (Figure 1.12). Indeed, such is the confusion of pipes and cables that it is impossible to include all the relevant information on an ordinary map. Since the mid-1970s, therefore, piping and cable data have increasingly been computerized.

Computerization is the only realistic way of systematizing and standardizing the enormous amounts of data involved. For instance, even in a small city (population 500,000) there may be several million meters of buried pipes and cables. Even though the cities of industrialized countries have long had such problems, the overall picture seems equally serious in the cities of developing countries. In some cases, construction workers have been electrocuted when high-voltage cables whose locations were not pinpointed have been cut accidentally. The situation with water supply and sewage networks is equally adverse. Far too many human tragedies may be ascribed to an inadequate, leaky water supply, deplorable sewerage, lack of maintenance, and lack of knowledge of the networks involved. The prime requirement for the information involved is, of course, that it must be manageable. This means that now and in the future, it must be computer based.

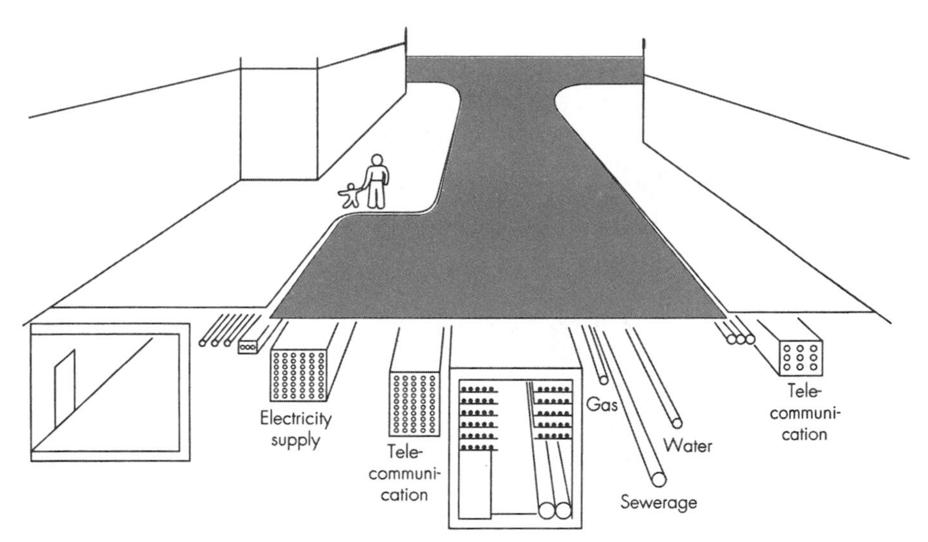

Figure 1.12
Complexity of the systems under a city street. There can be up to 35 different utility systems under a single street (standard specifications for piping cartography).

Environmental and resource management

Decision making is becoming increasingly complex as dwindling natural resources and more demanding economic priorities diminish the chances of today's decision being right tomorrow. Furthermore, environmental awareness is constantly increasing among the general public, particularly among the younger generation. The pressing global challenges are now:

- Uncontrolled desertification
- Erosion, particularly of productive agricultural soils
- Monoculture, particularly in areas devoted to cereal production such as the American Midwest, and its associated pollution and soil deterioration
- Loss of endangered animal and plant species
- Acid rain and associated deterioration of forest and aquatic environments
- Pollution of rivers, lakes, and oceans
- Environmentally related illnesses
- Contaminated groundwater and other environmentally related stresses
- The greenhouse effect and resultant climatic change
- Reduction in the ozone layer
- Repeated environmental catastrophes, such as oil spills, poison leaks, and radioactive releases

Despite many comprehensive studies, the global environment is still not well understood, because nature is complex and most effects are

Figure 1.13
Today the world's natural resources are under great pressure and it is important to ensure sustainable development built on a strong foundation. GIS can be instrumental in initiating such a process.

interrelated. For example: a small decrease in the atmospheric ozone layer permits more ultraviolet radiation to reach the surface of the Earth; this kills the marine algae on which fish feed. Fish populations dwindle, which reduces catches, and threatens the economy of fishing villages in major coastal areas. To help us map and monitor such changes, and plan appropriate responses that take account of the complex interactions of the Earth system, many countries now have comprehensive programs to capture and archive information on existing natural resources and known sources of pollution, using technologies such as satellite remote sensing and GIS (Figure 1.13). Worldwide databases have been established by agencies such as the United Nations Environmental Program, whose GRID program makes such data available to a wide range of users in GIS format.

Environmental data may be used both to expose conflicts and to examine environmental impacts. Impact analyses and simulated alternatives will probably become increasingly important. GIS is playing a key role in:

- Monitoring and documenting natural conditions and detecting change
- Documenting the suitability of resources and land areas for various uses
- Exposing conflicts and conflicting interests
- Revealing cause–effect relationships
- Modelling the interaction of various components of the Earth's environment to predict the effects of changes: for example, the ef-

fects of continued burning of fossil fuels on atmospheric carbon dioxide, or the effects of global climate change on forests and agriculture

Addressing environmental problems is complicated by a sparsity of information. Even the data that have been compiled are of limited use in decision making because of non uniform storage and filing, lack of verification, lack of a systematic approach to data acquisition, and even obsolescence. GIS techniques and databases are ideally suited for the optimal manipulation of such environmental data.

Planning and development

As discussed above, the planning and development of new housing, roads, and industrial facilities requires data on the terrain and other geographical information. Development often involves building on marginal terrain, increasing the density of building in areas already built up, or both. Yet the new structures must fit within the existing technical infrastructure; here computerization is a great aid. For example, when Stockholm's Central Railway Station was expanded, between 1986 and 1989, all technical infrastructure facilities above and below ground were registered, and the entire project managed with the use of computers and GIS. In the United States, GIS has been used to plan and manage the massive Inner Relief Project in Boston, Massachusetts. One of the benefits GIS holds for such projects is a minimalization of disruption to the existing infrastructure (Figure 1.14).

Escalating construction costs have made the optimizing of building and road location extremely important. Minimizing blasting and earthmoving are significant aspects of minimizing costs. Flexibility is vital: plans should be amenable to rapid change as decisions are made. The influence of special-interest groups and of individual citizens require that initial plans be presented efficiently and in a manner that is easily understood. Simplified, visualized plans are instrumental in conveying both the content of a scheme and the nature of any likely impact on those concerned.

Management and public services

In modern societies, decisions should be made quickly, using reliable data, even though there may be many differing viewpoints to consider and a large amount of information to process. Today, the impact of development decisions is ever greater, involving conflicts between society and individuals, or between development and preservation. Information must therefore be readily available to decision makers; the majority of such information is likely to be geographical in nature, and best handled using GIS.

Figure 1.14
Today's requirements for physical planning are stringent. Projects that involve encroachments on nature have to be carried out in the most environmentally friendly, yet economical way. (Courtesy of Asplan Viak.)

Overviews of administrative units and properties are crucial in the development of both virgin terrain and built-up areas, in both developing and developed nations. In many countries, property registration is extensive: even in smaller states, 2 or 3 million properties may be involved. Moreover, property is also an economic factor in taxation and as security for loans, so comprehensive overviews are essential to a well-ordered society. Computerized registers based on GIS technology are now well established in many countries. They are playing a vital role in restructuring the system of land ownership of former communist countries and in modernizing land records in developing countries.

Safety at sea

In many waters, the volume of shipping has increased considerably. Offshore oil rigs and other fixed facilities have been erected in customary shipping channels. As alterations along channels often affect navigation, there is an urgent need in shipping circles for up-to-date information. Today, a ship sailing in international waters may carry as many as 2,000 navigational charts, each of which may need updating as often as weekly (Figure 1.15).

Technology now permits the sending of information to mariners via public telecommunications networks, while electronic charts may

Figure 1.15
A prerequisite for safe navigation at sea is rapid access to relevant navigational data. The risk of collision at sea is enhanced by the increasing speed of vessels and complexity of shipping lanes. Electronic sea chart systems function in the same way as GIS at sea and improve safety.

be combined with positioning and radar displays on a ship's bridge. This means that ships can now sail more safely and at greater speeds in hazardous waters. Computers are now widely used for the production of traditional paper charts as well as for the production of updates.

Land transportation

In many countries, the greater part of transportation has shifted from rail and water to roads; at the same time, the use of private cars has greatly increased. These developments have created traffic problems, which cause loss of time and money. Large (and sometimes hazardous) goods are now transported by road. In most countries the annual costs of traffic accidents have become extremely high.

The transportation sector has always been a major consumer of maps and geographical data, so new technologies may realize considerable savings. The automobile industry is now investing heavily in the development of driver information systems (Figure 1.16), and several systems are now on the market. In principle, all of them involve

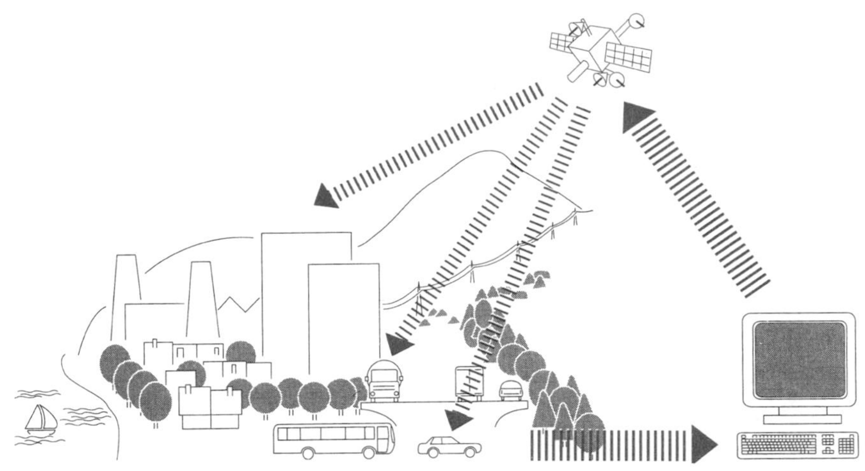

Figure 1.16
The transport sector has always made extensive use of geographical data. The introduction of new technology has opened a new range of applications in the fields of transport planning and traffic information.

simple GIS functions with digital maps and supplementary information. In many urban areas, where traffic is complex, increased driver information can ease congestion and improve safety. Such innovative programs to address problems of traffic congestion, known collectively as *intelligent transportation systems* (ITSs), are the subject of much research, particularly in the United States and Europe. They rely heavily on database representations of road networks and related information of importance to drivers and travelers, and on GIS.

Military uses

Modern military planning, training, and warfare are extensively computerized. In many ways, military computer systems include GIS functions. Digital data representing the form of the land surface are used in flight simulators and as the basis of automatic navigation and targeting in many weapons systems, such as the Cruise missile. Digital information on shorelines, water depth, land cover, and roads is now widely used by the military, and is processed using military forms of GIS technology.

The *global positioning system* (GPS) is another increasingly useful form of geographical information technology, initially developed by the U.S. military but now widely used for both military and civilian applications around the world, often in conjunction with GIS. It consists of a constellation of 24 satellites whose signals can be analyzed by a very simple hand-held device to determine position on the Earth's surface. A single receiver, available at a retail price of less than $300, can provide positioning in civilian applications to about 100 m, and in military applications to about 40 m. By using a combination of a roving receiver and a base station it is possible to achieve much higher accuracy, down to the centimetre level. An investment of $10,000 in a base station and receiver will yield positional accuracy relative to the base station of better than 1 m.

The importance of geographical data and information technologies in the Gulf War was such that it was described on more than one occasion as the "first GIS war". The implications for the GIS community of such statements, and of the role of GIS in warfare in general, have been the subject of comment. As a rule, military operations are relatively independent of civilian activities. However, the increasingly demanding military specifications pertaining to equipment and data access will undoubtedly benefit civilian GIS in due course, and trends in modern warfare indicate that military GIS activities can be expected to increase.

Conclusions

This is the age of information, and information technologies such as GIS and GPS will have increasing impact on our lives. GIS technology can be used to build detailed information databases on individual citizens, and thus threaten individual privacy, and it can be used to improve our monitoring of the global environment and the decisions we make about the future of the planet. It can be used to make either reliable decisions based on good data, or unreliable decisions based on bad data or inappropriate methods of analysis. In such issues the computer itself is no more than a mechanical processor. The responsibility for wise use lies entirely with the user and with those who provide its data, develop its software, and build its hardware. The technology has enormous potential, both for use and abuse, both conscious and unconscious. There is sometimes a tendency to give computer results more credibility than they deserve, and the wise use of a powerful technology like GIS often requires more responsibility on the part of its users than did the earlier techniques of map use and analysis.

1.3 Benefits of Computerizing Information

Almost all aspects of modern society use digital information, and the total amount that flows through our communication networks daily is truly staggering. GIS offers its users the ability to process quantities of data far beyond the capacities of manual systems (Figure 1.17). Data in GIS are stored in a uniform, structured manner, as opposed to manual systems in which data are stored in archives and files, in agencies, on file cards, on maps, or in long reports. Data may be retrieved from GIS databases and manipulated far more rapidly and reliably than data in manual systems. In addition, data are quickly compiled into docu-

Figure 1.17
There are considerable gains to be made by converting to computerized information. However, unlike with slot machines, the winnings are not based on chance.

ments using techniques that include automatic mapmaking and direct report printouts. The potential gains from switching from manually prepared maps and ordinary files to computerized GIS are considerable, in both the public and private sectors.

A joint Nordic research project (Nordic KVANTIF) has evaluated the gains realized by implementing GIS through studies of its use in 50 to 60 organizations in the United States, Canada, Italy, and the five Nordic countries—Norway, Sweden, Denmark, Finland, and Iceland. The study showed that considerable benefits may be achieved, provided that the strategy used to implement GIS is suitably chosen. The study also showed that benefits are often related to objectives and that the following benefit/cost ratios may be attained by introducing GIS (Figure 1.18):

1. If computerized GIS is used for automated production and maintenance of maps, the benefit/cost ratio is 1:1.
2. If the system is also used for other internal tasks such as work manipulation and planning, the benefit/cost ratio may be 2:1.
3. The full benefit of the system is first realized when information is shared among various users. The benefit/cost ratio may then be 4:1.

The benefit/cost ratios quoted above refer to municipal services. Studies have shown that corresponding ratios for nationwide uses are somewhat lower, up to 2:1 to 3:1. Nonetheless, it is obvious that investment in GIS is at least as productive as investment in other sectors. These benefits are not automatic. They depend largely on proper choice of an acquisition and implementation strategy, following care-

Objective level	Map production	Map production and internal use of data	Map production, internal use of data and shared use of data
Tasks	• storage • manipulation • maintenance • presentation	• map production • planning • facility maintenance • project management	• map production • project • planning • facility maintenance • coordination • general service • facility management • economic planning • service and information
Benefit/cost ratio:	1:1	2:1	4:1

Figure 1.18
The benefit/cost ratio of transferring to GIS depends on how the system and information are applied. The ratio will normally be 1:1 for pure map production; 2:1 for other applciations such as planning, and as much as 4:1 for a multiple-user information system.

ful study of the objectives and requirements of GIS investment, and careful selection of the appropriate system. Without these safeguards, many GIS projects eventually fail to deliver the promised benefits and may eventually fail entirely, at considerable cost to the institution. Even with a carefully selected strategy it is difficult to estimate benefits precisely. The ratios discussed above are average over many projects varying widely in scale and scope. Some figures, however, are impressive, with benefit/cost ratios of up to 8 to 10:1 or more. Detroit Edison, a public/electric company in the United States, cut production time for a particular type of map overview from 75 hours to 1½ minutes. The city of Burnaby in Canada experienced costs that exceeded benefits by $1 million for the first three years of the project, but after seven years achieved an annual benefit of $400,000 dollars.

Some benefit/cost ratios for various sectors and activities are (Tveitdal 1987).

National data	
Forestry	2.0:1
Transportation	4.0:1
Environmental data	1.8:1
Documentation	2.0:1
Municipal data	
Map production	1:1
Agency uses	2:1
Joint uses	4:1

A number of GIS projects have been terminated because benefits did not meet expectations; in other cases, a project might not have been terminated had benefits been evaluated in a rigorous, objective manner. The Nordic KVANTIF study showed that GIS projects almost always involve high financial risk but that successful projects often results in benefits greater than anticipated.

Benefits are a function of many factors, including the goals and objectives of the project, the strategy adopted in its implementation, and the structure of the system built to serve the objectives (Figure 1.19). Systematic planning and implementation often set profitable GIS projects apart from those that are unprofitable. Projects based on carefully estimated cost and benefit calculations are often more profitable than projects driven by pure technology. Profitable projects are user oriented rather than production oriented. Profitable projects start by being defined so clearly and convincingly that they are funded outside the ordinary operating budget.

The measurable benefits of GIS are usually expressed as gains in efficiency in terms of time saved, but there are also many cases of direct increases in income and reductions in costs. Measurable benefits may include:

- Improved efficiency due to more work being performed by the same staff, or the same work performed by a smaller staff
- Reduction in direct operating costs through better bases for financial management, less costly maintenance of facilities, and joint uses of available data
- Increases in income due to increased sales, or sales of new products and services

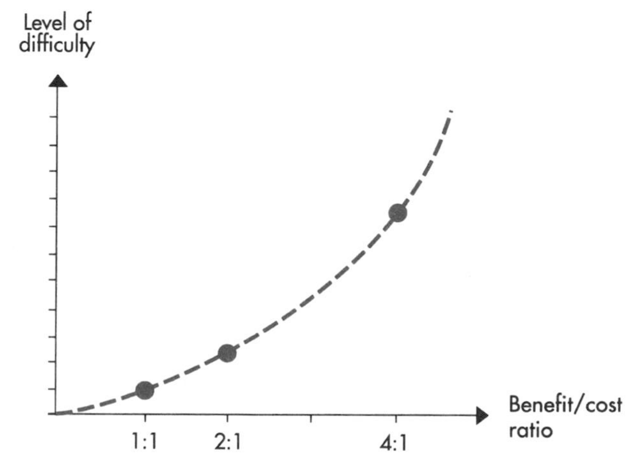

Figure 1.19
High benefit/cost ratio figures are not easy to achieve, and the level of difficulty of the project tends to cause the ratio to increase proportionately.

Experience indicates that when GIS makes some traditional jobs superfluous, staff are not made redundant but instead put to tasks in the GIS environment that create more value.

Intangible benefits may also accrue. They cannot be expressed directly in monetary terms, but attempts should always be made to include them when benefits are evaluated. Intangible benefits may include:

- Improved public and private decision making in administration, planning, and operations
- Improved information and service to the public
- Increased safety, and reduction in the impact of disasters through better planned evacuation and more efficient management of emergency services
- An improved environment for future generations
- Better presentation of plans and their associated effects
- Improved decisions regarding new development, and better analysis of market and site conditions

The greatest long-range global benefits of GISs are probably in the sectors where decisions have an environmental impact. The environment and the natural relationships within it are complex and not yet fully understood. It is, however, widely known that environmental degradation is implicated in the causes of many modern illnesses, such as asthma and cancer, the annual costs of which are enormous.

Increases in safety at sea are associated with the introduction of electronic navigation systems, such as the *electronic chart display and information system* (ECDIS). The benefits realized can best be appreciated by considering what the ECDIS might help prevent. Among the more serious environmental disasters are oil spills from tankers that run aground. Subsequent cleanups have cost from $40 per kilogram of oil to totals of up to $2 billion. If improved navigation using ECDIS can prevent one such disaster per year, the benefits are considerable, albeit difficult to assess directly.

1.4 Users of GIS

In this chapter we have reviewed some of the basic concepts of GIS and some of the issues that arise it its use. Today, the widespread acquisition of digital computers by businesses, schools, and households has allowed technologies such as GIS to penetrate many aspects of our lives. Nevertheless, computer processing of geographical data remains problematic, and GIS are widely regarded as difficult to learn about and to use. We hope that subsequent chapters of this book will provide a conceptual and technical understanding of GIS that will allow readers to make effective use of its capabilities in one or more of the many areas of its application.

Users of GIS naturally fall into two groups. Some are professional operators of GIS, who spend much of their lives working with the technology in their jobs. They are well trained in the particular software they use and are well aware of its capabilities. In many cases they do not use the results of their work themselves, but pass them to end users. The results may be maps, designed and produced by the GIS operator, results of analysis to be used in planning harvesting of trees, or work orders for maintenance staff in a major utility company.

The second group of users spend a relatively small proportion of their lives using GIS. They may maintain a GIS capability on their personal workstation in order to produce an occasional map, to find a restaurant in an unfamiliar city, to plan a driving route for a vacation, or to carry out analysis of map data in connection with a research project. In these cases the opportunities for lengthy training are much less, so the GIS must be simple and easy to use.

This second group also comprises end users and primary users who make professional decisions based on GIS products. The group includes:

- Operation and maintenance engineers; a typical decision may be whether to replace or repair a damaged water main.
- Regional planners; characteristic tasks involve presentations of plans to municipal authorities in a realistic, varied, visual manner.
- Building authority functionaries; representative jobs include processing building permit applications involving access roads, water supply, or sewage.
- Revenue officials, typically dealing with tax assessment and taxpayer addresses.
- Road engineers, whose responsibilities include locating new roads to minimize cut-and-fill operations.
- Information officers; information produced may include complete packages to newly established firms with details on industrial areas, schools, and transportation.
- Local officials, who may require updated overviews on the effects of effluents on water quality at public beaches or the effects of zoning on school capacities.
- Fire brigades, for whom rapid, reliable information on the locations of fires and the presence of hazards such as explosives would be invaluable.
- Forest managers planning harvest operations, computing volumes of annual growths, estimating road costs and identifying sensitive wildlife areas.

- Bank officials, perhaps wishing to verify ownership of properties offered as collateral.
- Oil tanker captains maneuvring a ship in hazardous waters.
- Truck drivers seeking to minimize the problems of transporting an extrawide load between two points.

Such a list could be endless; only imagination sets a limit. Indeed, GIS may spawn information technologies applicable to completely new sectors, such as optimum warehousing or, even, brain mapping.

CHAPTER 2

Historical Development: Geographical Data and GIS

2.1 Early Developments

Geographical information systems evolved from centuries of map-making and the compilation of registers. The earliest known maps were drawn on parchment to show the gold mines at Coptes during the reign (1292–1225 B.C.) of Rameses II of Egypt. Perhaps earlier still are Babylonian cuneiform tablets that describe the world as it was then known (Figure 2.1). At a later date, the Greeks acquired cartographic skills and compiled the first realistic maps. They began using a rectangular coordinate system for making maps around 300 B.C. About 100 years later, the Greek mathematician, astronomer, and geographer Eratosthenes (ca. 276–194 B.C.) laid the foundations of scientific cartography. One of the earliest known maps of the world was constructed by Claudius Ptolemaeus of Alexandria (ca. A.D. 90–168).

The Romans were more concerned with tabulations and registers. The terms *cadastre* (an official property register) and *cadastral* (of a map or survey that shows property or other boundaries) originate from the late Greek *katà* stíkon, which means "by line." But it was the Romans who first employed the concept to record properties, in the *capitum registra,* literally, "land register." In many countries, the term *cadastre* designates map and property registers. In France it also applies to the administrative agency in charge of such documents.

Throughout history, as societies organized, it became necessary to meet the expense of this. Some of the better known earlier examples include taxation levied by emperors and kings to meet military expenses. These direct levies are the foundations of today's complicated revenue systems involving the taxation of income, property, and goods. Since both the ancient Egyptians and the Romans taxed property, property registration was early systematized to assure tax revenues.

The earliest maps were drawn almost exclusively to facilitate commercial sea voyages (Figure 2.2). On them, coasts were meticulously detailed and harbors were plumbed, while interiors remained unknown, apart from details of important trade and caravan routes.

Figure 2.1
One of the oldest known maps in the world, a Babylonian cuneiform tablet dating from about 1400 B.C.

The Arabs were the leading cartographers of the Middle Ages. European cartography degenerated as the Roman Empire fell. But in the fifteenth century, old skills were revived and Claudius Ptolemaeus's *Geographia* was translated into Latin to become the then existent view of the world. Although cartography was neglected, in many countries property registry thrived. The best known example is the Domesday Book, the record of the lands of England compiled in 1086 for the first Norman king, William the Conqueror (1027–87). The data included specifications of properties and their value, and a count of inhabitants and livestock, as well as incomes earned and taxes paid.

The travels and explorations of Marco Polo, Christopher Columbus, Vasco da Gama, and others resulted in increased trade. In turn, maps were needed of previously unmapped seas and coasts. As the European countries and the newly discovered regions evolved to more organized societies, the need for geographical information increased.

Figure 2.2
One of the oldest known maps of the northern European seas, dating from 1570. The cartographer was obviously more familiar with the southern than with the northern tracts.

Ordnance developments, such as the introduction of artillery, made maps important in military operations, and military agencies became the leading mapmakers. In many countries, the military mapmakers became responsible for both topographic land maps and navigational charts (Figure 2.3). Vestiges of this trend remain: map agencies, partic-

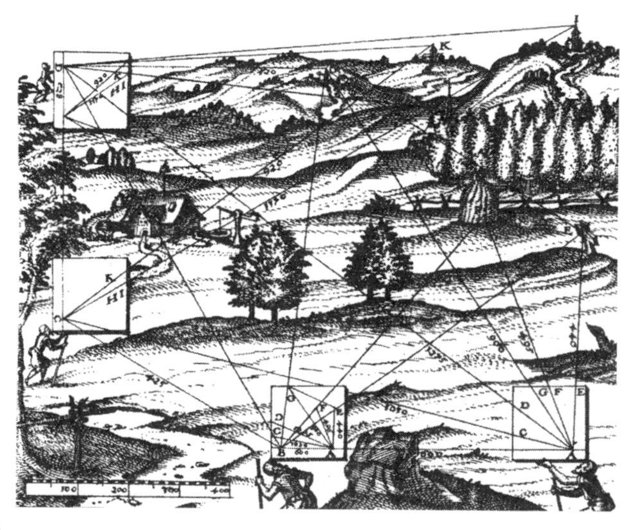

Figure 2.3
In the eighteenth century, vast land areas were surveyed using relatively exact methods. This task was often carried out by military mapmakers.

ularly nautical chart agencies, seem characteristically military. For example, the official mapmaking agency of Great Britain is the Ordnance Survey. The introduction of mass printing techniques enabled maps to be produced as consumer articles rather than as works of art, as was often the case earlier when maps were drawn by hand.

Until the nineteenth century, geographical information was used mostly for trade and exploration by land and sea and for tax collection and military operations. New needs arose in step with evolving infrastructures, such as roads, railways, telegraph and telephone lines, and gas and water supplies. Planning these facilities required information about the terrain beyond that commonly available. The accurate location of towns and cities, lakes and rivers, mountains and valleys became increasingly important. Detailed topographic information was needed to lay out railway and road gradients and curve radii. Then, as now, foundations were a major challenge, so maps showing the type of soil and the quality, location, and properties of bedrock were required. As planning advanced, specialized maps became more common. The first geological map of Paris was compiled in 1811. In 1838, the Irish government compiled a series of maps for the use of railway engineers, which may be regarded as the first manual geographical information system.

Development became increasingly dependent on socioeconomic factors. The rights of property owners entered the picture because the construction of roads and railways often necessitated the expropriation of private lands. New applications arose for property registers and maps as builders needed to compile overviews of affected properties in order that their owners might be justly compensated.

As cities grew larger and more complex, accurate urban planning became a necessity. Many countries began compiling statistical information relating to urban planning in the early nineteenth century. By 1837 the British Registrar General's Office had amassed extensive population statistics.

On the Continent, traditional village property ownership became a hindrance to effective farming. Many properties had become fragmented over the years, owing to inheritance settlements. In some cases a single property might comprise several hundred dispersed parcels of land. Sometimes the ownership of, or rights to, a parcel were divided: one owner could have timber rights, another grazing rights, a third hunting rights, and so on. Therefore, property mapping in the late nineteenth century aimed to wrest order from chaos. With reference to available land registers, the various parcels were assembled into properties that were easier to work. Borders were consolidated, clarifying ownership and facilitating the taxation of property.

Aerial photography accelerated the progress of mapmaking. The first aerial photograph was used for mapmaking, and the first mapmaking instrument devised, in 1909. Photogrammetry, the technique of making measurements from photographs, developed rapidly in the 1920s and 1930s, and the two world wars also hastened developments. After World War II, photogrammetry became widely used in mapmaking, mostly for maps in scales from 1:500 to 1:50,000. Aerial photographs themselves became important sources of quantitative information in evaluating such features as vegetation and geological formation.

2.2 First Automatic Processing of Geographical Information

Although Blaise Pascal is credited with devising the first true calculating machine in 1647, large amounts of data were first processed automatically in 1890, when a tabulating device conceived by Hermann Hollerith was used in compiling the U.S. census. In Hollerith's first apparatus, census data were punched on cards which were then read electromechanically to compile data in separate registers. In the first half of the twentieth century, Hollerith's various mechanisms were developed further. Data processing using punch cards became an industry.

During World War II, data processing again advanced, primarily to meet the military need for predicting ballistic trajectories. One of the most famous computers developed for that purpose was ENIAC, an acronym for *electronic numerical integrator and calculator.* After the war, computer development continued. In 1953, IBM launched the model 650, which became the "Model T of the Computer Age" by virtue of being the first electronic computer not to be hand-made. More than 100 were produced—in those days an amazing quantity. In today's computer terms, ENIAC, Whirlwind, the IBM 650, and other early electronic computers are referred to as *first generation.* All first-generation computers suffered from a common drawback: they used vacuum tubes, which, like light bulbs, gave off heat and had limited lifetimes. That alone limited their application. One 25,000-tube computer of the period was continuously manned by a staff of 10, of which two were technicians assigned to continuous replacement of burned-out tubes. Nonetheless, computerization was the established technology for processing large amounts of data. By 1952, all U.S. governmental statistical data were processed by electronic computers.

By the late 1950s and early 1960s, second-generation computers using transistors became available, outperforming their vacuum-tube predecessors. Suddenly, computers became affordable in disciplines other than those of major governmental agencies. Meteorologists, geologists, and other geophysicists began using electronic mapmaking devices. Initially, the quality was poor, not least because automatic drawing machines had yet to be developed.

As the uses of second-generation computers spread, theoretical models were evolved to use statistical data. Then, as now, public and private decision making was often based on analyses of various classes of geographical data. These included demographic trends, cost-of-living variations, the distribution of natural resources, wealth and social benefits, and the demography of employment. The first geographical information system was constructed by the government of Canada in the late 1960s, and by modern standards was both unbelievably crude and expensive. It required a large mainframe computer, and its output was entirely in the form of tables. This was, in part, because no computer-controlled devices were available at that time to draw maps and in part because of the system's emphasis on analysis. Later, in the United States, a similar system, MIDAS, began processing data on natural resources.

The need for reliable geographical data multiplies with the expansion of road, rail, telecommunications and sewage networks, airports, electricity and water supplies, and other essential services vital to the infrastructure of urban areas. Terrain information on maps is now a vital planning tool, from the first conceptual stage to the final, legally binding plan. Burgeoning road networks have mandated extensive analyses of transport patterns. Indeed, since the mid-1950s, computers have been used in the United States to simulate traffic flows in relation to population distribution.

Often, conflicts of interest arise between developers and conservationists, or between municipal and regional planners and individuals. Most countries now have laws intended to resolve such conflicts—planning and building statutes, environmental and historical preservation laws, and the like. Enforcement of these laws requires an improved overall view of property ownership. Many countries began computerizing property registers in the mid-1970s; such systems may be viewed as the initial systems of land information systems (LISs).

2.3 The Microprocessor

In the 1960s and early 1970s, integrated circuits were developed and computer programs refined. The result: third-generation computers which brought computerization to virtually all professional disciplines, especially those processing large amounts of data.

The next major breakthrough came in 1971–1972 with the development of the microprocessor. In 1974, a microprocessor was used to build the first fourth-generation desktop computer. Seven years later, the first microprocessor-based desktop computer was launched as a personal computer (PC). By the mid-1980s, the computer field was divided into three categories according to size of computer: mainframes,

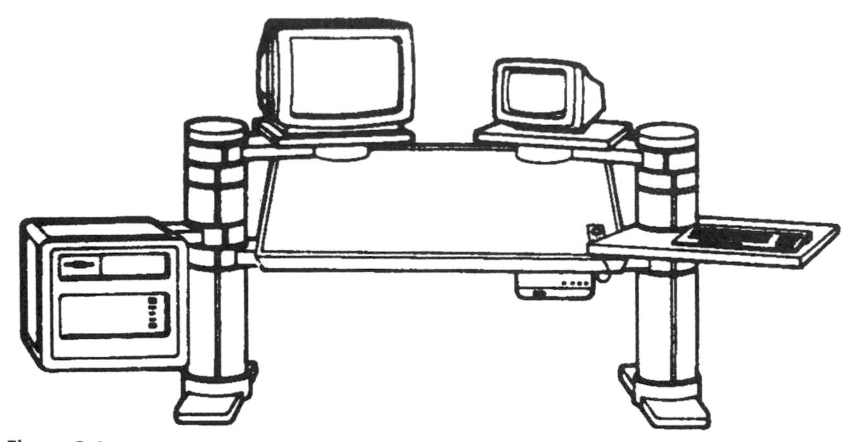

Figure 2.4
Early workstation, with graphics display, digitizer, secondary storage, and data processor. (Courtesy of SysScan.)

the descendants of the original large computers, intended for major data processing and computational tasks; PCs, the increasingly ubiquitous desktop computers; and minicomputers/workstations (Figure 2.4), which were smaller than mainframes but larger than desktop PCs. By the early 1990s, mainframes had become physically smaller and computationally more capable. That trend was reflected strongly in PCs, which by 1990 were outperforming minicomputers built only a few years earlier. This development signaled the demise of the minicomputer, an event that was further hastened by the introduction of PC networks, in which processing and storage capacity may be shared and distributed. The development of powerful workstations in the mid-1980s, however, led to an increasing acceleration in the use of GIS. The overall trend is best illustrated in terms of the costs of computing: a computer's processing and storage capabilities which, in 1960, cost $100,000, could be purchased for $10 in 1984 and $0.005 in 1997. In other words, cost efficiency increased by a factor of 10 every two to three years.

In the 1970s and 1980s, various systems were evolved to replace manual cartographic computations. Workable production systems became available in the late 1970s and system development continued through the 1980s. Nonetheless, by the mid-1990s, elegant approaches to some cartographic tasks have yet to be found, and computerized cartographic research and development remains a continual challenge. The spread of PCs spurred user-friendly operations and programs capable of processing in ways previously not possible, for example, by considering the *logical* connections in geographical data.

Increases in microprocessor computing capacity also made the processing of digital and satellite images and other types of raster images commercially available in the mid-1980s. Software systems have developed apace. Relational database systems, such as dBase and Oracle which first appeared in the late 1980s, are particularly useful in processing geographical data. Commercially available relational databases are now used routinely in GIS systems.

In the late 1980s, computing capability became widely accessible as microprocessors were used for a multitude of devices, from household appliances and automobiles to an extensive range of specialized instruments, including those used in GIS. For GIS users, microprocessors have improved such devices as:

- Surveying instruments
- GPS (global positioning system)
- Digitizing table
- Scanners
- Environmental monitoring satellites and data presentation systems, including graphic displays, electrostatic plotters, and laser printers

2.4 Recent Developments

The 1990s have produced even faster and more powerful computer equipment and peripherals. However, new developments in the field of data networks and communications are of equal importance, specifically local area network (LAN), wide area network (WAN), and last, but certainly not least, Internet and World Wide Web (WWW). The development of the Internet was initiated by the U.S. Department of Defense as long ago as the late 1960s. World Wide Web was developed at CERN (European Organization for Nuclear Research) in Switzerland in 1990.

Data networks have opened up a whole new range of opportunities for geographical data search and distribution, thereby considerably increasing the value of GIS, particularly since common data have become more easily accessible. The most spectacular development in the GIS arena has occurred in the field known as *multimedia.* Multimedia techniques are based on the combination of elements such as figures, text, graphics, pictures, animation, sound, and video (Figure 2.5). Multimedia brings geospatial information into living maps and makes complex information understandable to those who are not technically sophisticated. Multimedia technology is available on the Internet and has proved to be eminently suitable as an information tool in planning (city, roads), tourism, and the distribution of environmental information.

Flight simulators are perhaps the best-known example of the application of data technology to create near real-life situations, thus making them ideal for use in training. The concept behind flight simula-

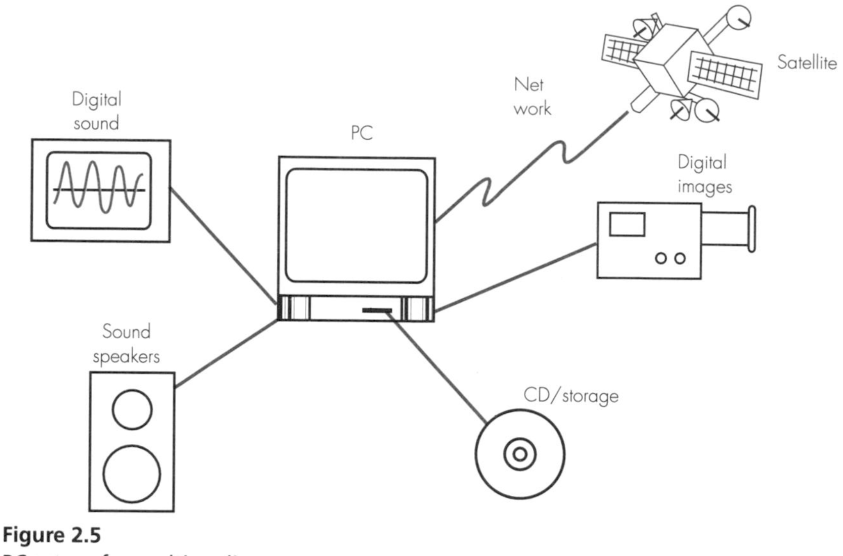

Figure 2.5
PC set up for multimedia.

tors has now been adopted for other activities and is known as *virtual reality.* Virtual reality is a term used in multimedia; it was marketed as fiction in 1984 and became commercial technology in 1992. The most extreme virtual reality experience is attained by "dressing up" in a computer and moving into a world where almost all external impulses are artificial. Virtual reality and GIS have many features in common and are becoming more and more integrated.

CHAPTER 3

From the Real World to GIS

3.1 The Real World

In many ways GIS presents a simplified view of the real world. Since the processes involved are seldom straightforward because realities are irregular and constantly changing, perception of the real world depends on the observer. For example, a surveyor might see a road as two edges to be surveyed, the roadwork authority might regard it as an asphalt surface to be maintained, and the driver will see it as a highway. Moreover, the real world may be described in terms of countless phenomena, from basic subatomic particles up to the dimensions of oceans and continents. The complexity of the real world, as well as the broad spectrum of its interpretations, suggests that GIS system designs will vary according to the capabilities and preferences of their creators. This human factor can introduce an element of constraint, as data compiled for a particular application may be less useful elsewhere.

The systematic structuring of the data determines its ultimate utility and consequently the success of the relevant GIS application. This aspect is also characteristic of the data available in traditional maps and registers. The real world can be described only in terms of models that delineate the concepts and procedures needed to translate real-world observations into data that are meaningful in GIS. The process of interpreting reality by using both a real-world and a data model is called data *modeling.* The principles involved are illustrated in Figure 3.1 and Figure 3.2.

3.2 Real-World Model

The arrangement of the real-world model determines which data need to be acquired. The basic carrier of information is the *entity,* which is defined as a real-world phenomenon that is not divisible into phenomena of the same kind. An entity consists of:

- Type classification
- Attributes
- Relationships

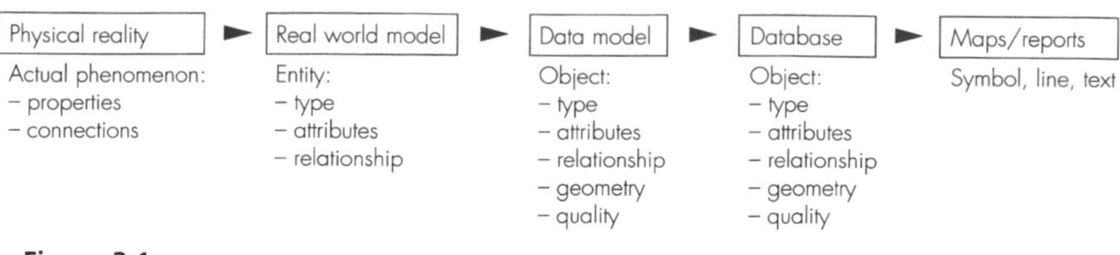

Figure 3.1
To bring the real world into GIS, one has to make use of simplified models of the real world. Uniform phenomena can be classified and described in the real-world model which is converted into a data model by applying elements of geometry and quality. The data model is transferred to a database that can handle digital data, from which the data can be presented.

Entity types

The concept of *entity types* assumes that uniform phenomena can be classified as such. During the classification process, each entity type must be uniquely defined to preclude ambiguity. For example, "house" must be defined in such a way that "detached house at No. 10 Church Road" is classified under "house" and not under "industrial building."

Some user organizations may need to classify entity types into categories as well as according to type. For example, national highways, country roads, urban roads, and private roads might come under the "roadways" category; alternatively, all entities within a specific geographical area might belong to a unique category of that area. In geographical data an entity type is also known as the nominal scale or qualitative data (Figure 3.3).

Entity attributes

Each entity type may incorporate one or more attributes that describe the fundamental characteristics of the phenomena involved (Figure 3.4). For example, entities classified as "buildings" may have a "material" attribute, with legitimate entries "frame" and "masonry" and a "number of stories" attribute with legitimate values of 1 to 10, and so on.

In principle an entity may have any number of attributes. For example, a lake may be described in terms of its name, depth, water quality, or fish population as well as its chemical composition, biological activity, water color, algae density, potability, or ownership. Attributes may also describe quantitative data, which may be ranked in three levels of accuracy: ratio, interval, and ordinal. The most accurate are *ratio* or *proportional* attributes, such as length and area, which are measured with respect to an origin or starting point and on a continuous scale. *Interval* data, such as age and income category, comprise numerical data in groups and are thus less accurate. The least accurate are *ordinal* data of rank, such as "good," "better," and "best," which describe qualitative data in text form. These could also be characterized as quality data.

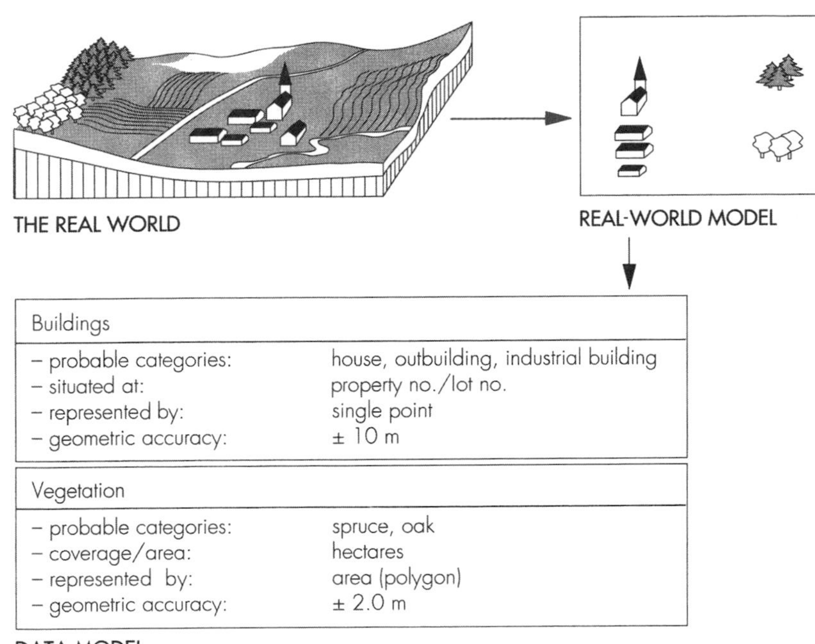

ID	Type	Property No.	X	Y	Accuracy
1	House	44 113	350	575	± 10.0
2	Outbuilding	45 6	375	600	± 10.0
3	Industrial	45 11	345	630	± 10.0

ID	Type	Area	Coordinates	Accuracy
10	Spruce	100	250,420 250,455 370,475 360,420 250,420	± 2.0
20	Oak	50	360,420 370,475 425,395 425,420 360,420	± 2.0

DATA BASE

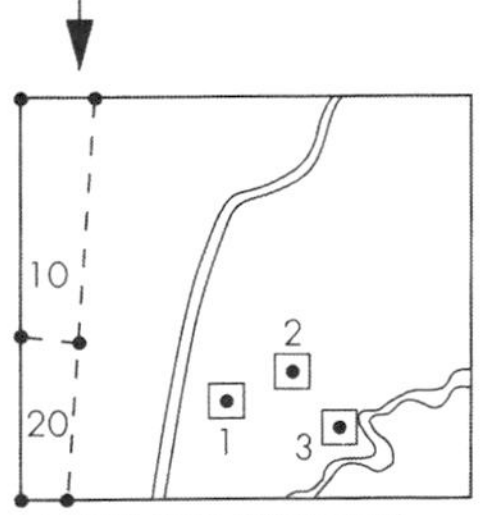

MAP WITH SYMBOLS

Figure 3.2
Modeling process. The transformation of the real world into GIS products is achieved by means of simplification and models in the form of maps and reports.

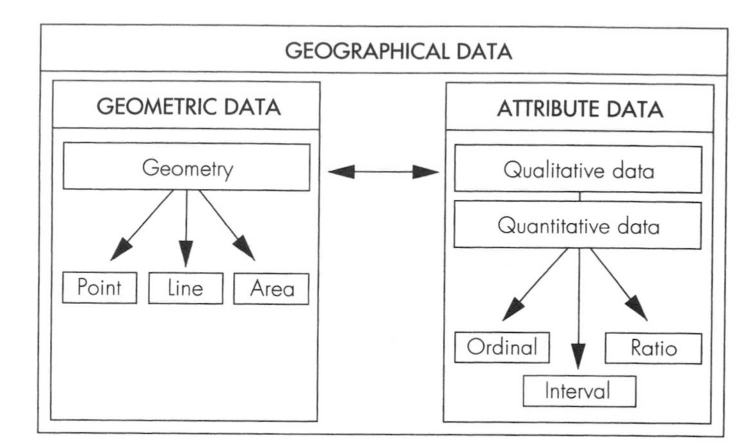

Figure 3.3
Geographical data can be divided into geometric data and attribute data. Attribute data can in turn be subdivided into qualitative data and quantitative data.

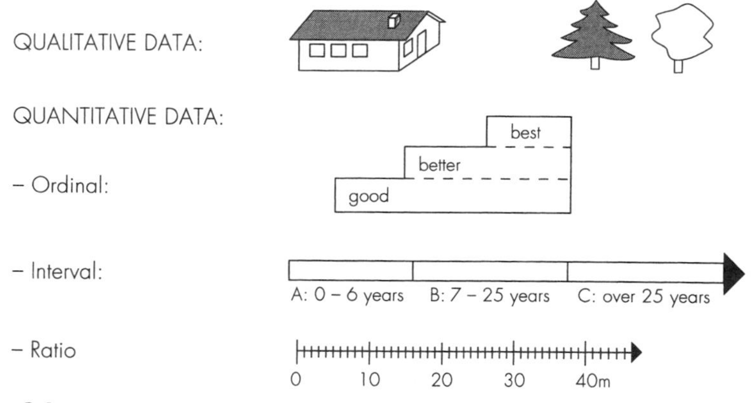

Figure 3.4
Attribute data consist of qualitative or quantitative data. Qualitative data specify the type of object, while quantitative data can be categorized into ratio data, data measured in relation to a zero starting point; interval data, data arranged into classes; and ordinal data, which specify quality by the use of text.

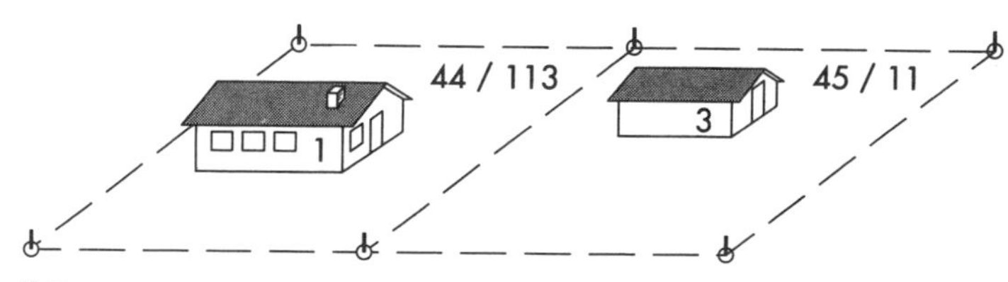

Figure 3.5
The computer cannot see the real world so it is necessary to specify the various relations between entities, such as *belong to, comprise, are located in/on,* and *border on.*

Entity relations

Relations often exist between entities. Typically, these include (Figure 3.5):

Relation	Typical example(s)
Pertains/belongs	A depth figure *pertains* to a specific shoal, or a pipe *belongs* to a larger network of contiguous pipes.
Comprises	A country or a state *comprises* counties, which in turn *comprise* townships.
Located in/on	A particular building is *located on* a specific property.
Borders on	Two properties have a common *border.*

Although such relations are intuitively obvious on ordinary maps, computers have no intuition. The computer processing or relations therefore requires either further descriptive information or instructions on how it may be compiled. This aspect of map reading—the human ability to see what a computer needs to be "told" to "see"—highlights pivotal differences between human and computer processing. As a sign in one international computer research laboratory states: Computers: Fast, Accurate, Stupid. Humans: Slow, Slovenly, Smart.

Some relations may be unwieldy. One cause may be complexity, as in networks where the states of switches or contacts, open or closed, determine which parts of the network may be viewed as comprising a logical entity. In such situations, one may differentiate between actual and potential relations.

3.3 Data Model

A real-world model facilitates the study of a selected area of application by reducing the number of complexities considered. Those outside the selected area are considered insignificant. However, if a real-world model is to be of any use, it must be realized in a database. A data model makes that possible.

Unlike humans, computers cannot "learn" the essentials of manholes, property lines, lakes, or other types of objects. What they can do is to manipulate geometric objects such as points, lines, and areas, which are used in data models. The carriers of information in data models are known as *objects* (Figure 3.7).* These correspond to entities

* The terms *entity, feature,* and *object* are often interchangeable. In this book we have chosen to use *entity* as a term for a real-world thing or phenomenon. The term *feature* is used very little in the book but is synonymous with the term *object* and is defined as a group of spatial elements which together represent a real-world entity.

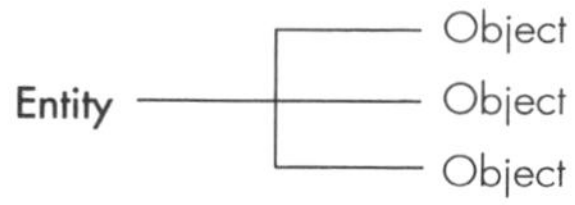

Entity — Object, Object, Object

Figure 3.6
A single entity can be described by several objects (i.e., there are many relationships between entities).

in real-world models and are therefore regarded as database descriptions of real-world phenomena.

Objects are characterized by:

- Type
- Attributes
- Relations
- Geometry
- Quality

Real-world models and entities cannot be realized directly in databases, partly because a single entity may comprise several objects. For instance, the entity 'Church Road' may be represented as a compilation of all the roadway sections between intersections, with each of the sections carrying object information. Multiple representations produced by such divisions may promote the efficient use of GIS data (Figure 3.6). This means that information-carrying units and their magnitudes must be selected before the information is entered in a database. For example, the criteria for dividing a roadway in sections must be selected before the roadway can be described.

Objects

Objects in a GIS data model are described in terms of identity type, geometric elements, attributes, relations, and qualities. *Identities,* which may be designated by numbers, are unique: no two objects have the same identity. Type codes are based on object classifications, which can usually be transferred from entity classifications. An object may be classified under one type code only.

Data models may be designed to include:

- Physical objects, such as roads, water mains, and properties
- Classified objects, such as types of vegetation, climatic zones, or age groups
- Events, such as accidents or water leaks
- Continuously changing objects, such as temperature limits
- Artificial objects, such as elevation contours and population density
- Artificial objects for a selected representation and database (raster)

3.3.1 Graphical representation of objects

Graphical information on objects may be entered in terms of (Figure 3.7):

- Points (no dimensions)
- Lines (one dimension)
- Areas (two dimensions)

Points

A *point* is the simplest graphical representation of an object. Points have no dimensions but may be indicated on maps or displayed on screens by using symbols. The corner of a property boundary is a typical point, as is the representative coordinate of a building. It is, of course, the scale of viewing that determines whether an object is defined as a point or an area. In a large-scale representation a building may be shown as an area, whereas it may only be a point (symbol) if the scale is reduced.

Lines

Lines connect at least two points and are used to represent objects that may be defined in one dimension. Property boundaries are typical lines, as are electric power lines and telecommunications cables. Road and rivers, on the other hand, may be either lines or areas, depending on the scale.

Areas

Areas are used to represent objects defined in two dimensions. A lake, an area of woodland, or a township may typically be represented by an area. Again, physical size in relation to the scale determines whether an object is represented by an area or by a point. An area is

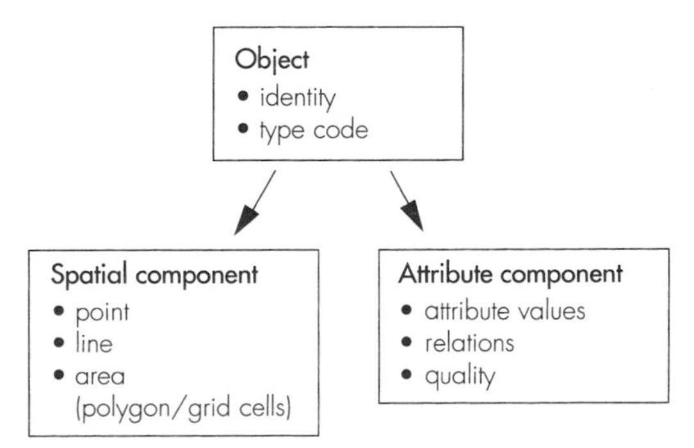

Figure 3.7
In a data model, objects are categorized as object classifications, geometric elements (point, line, area), attributes, relations between the entities, and quality definitions of these descriptive elements.

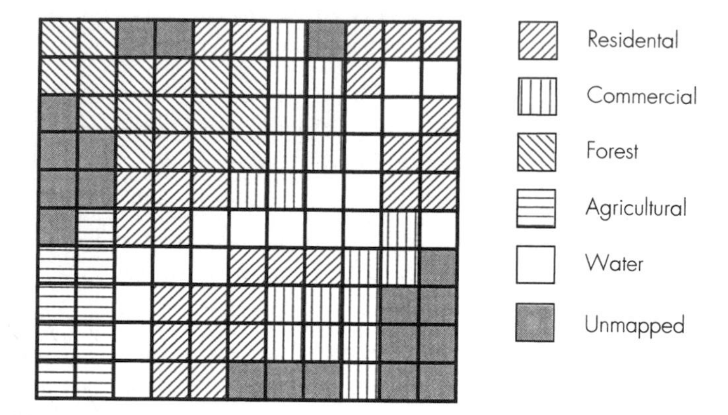

Figure 3.8
Land use in a town, in the form of a raster map. The land use is registered in a raster system with cells of varying size. Each category is given its own symbol on the map.

delineated by at least three connecting lines, each of which comprises points. In databases, areas are represented by polygons (i.e., plane figures enclosed by at least three straight lines intersecting at a like number of points). Therefore, the term *polygon* is often used instead of *area*.

Physical reality is often described by dividing it into regular squares or rectangles so that all objects are described in terms of areas (Figure 3.8). This entire data structure is called a *grid*. Population density is well suited to grid representation; each square or rectangle is known as a *cell* and represents a uniform density or value. The result is a generalization of physical reality. All cells of a grid in a data model or a database are of uniform size and shape but have no physical limits in the form of geometric lines.

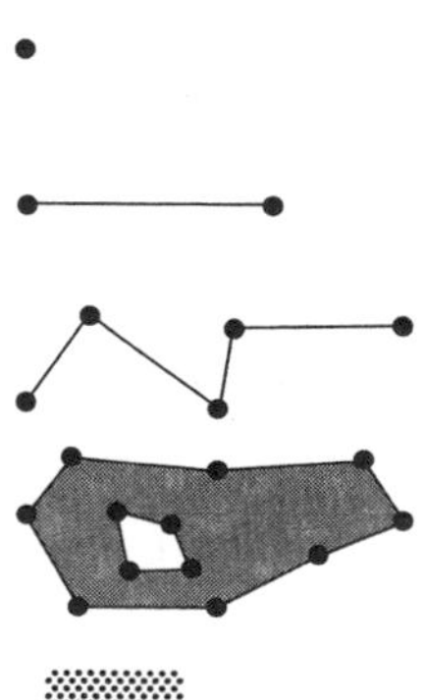

Point: A zero-dimensional object that specifies geometric location specified through a set of coordinates.

Line segment (vector): A one-dimensional object that is a direct line between two endpoints.

String: A sequence of line segments.

Area/polygon: A two-dimensional object bounded by at least three one-dimensional line segments.

Raster cell/pixel: A two-dimensional object (area) that represents an element of a regular tesselation of a surface.

Figure 3.9
Point, line, area.

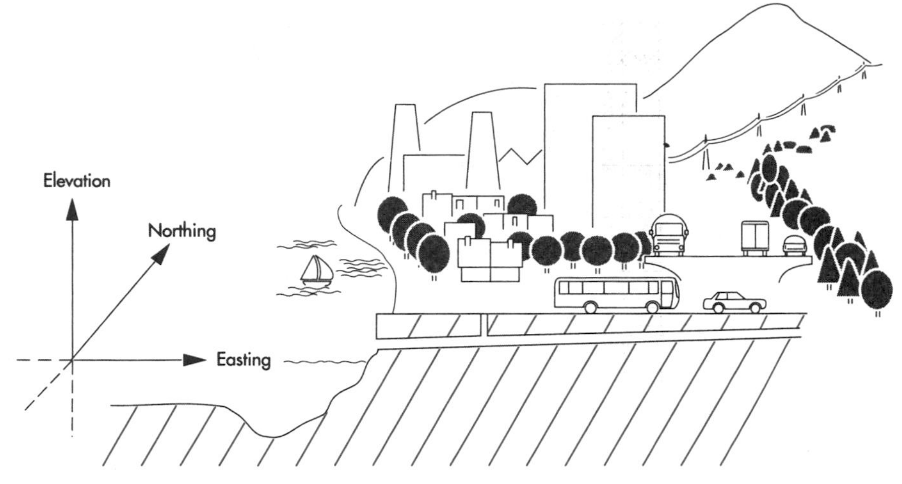

Figure 3.10
The world is three-dimensional with phenomena having a location and surface area in both elevation and ground plane.

In the traditional layer-based data model heights are treated as attributes to the objects, not as a part of the geometry. But the real world is three-dimensional (Figure 3.10), and three-dimensional data models are discussed in Chapter 5.

3.3.2 Object attributes

Object attributes are the same as the entity attributes of the real-world model. Attributes describe an object's features and may thus be regarded as a computer's "knowledge" of the object. In practice, object attributes are stored in tables (Figure 1.6), with objects on lines and attributes in columns. Theoretically, attribute values connected to grid data can be presented in the same way. Each grid cell corresponds to an object (Figure 3.11).

Cell	Attributes				
no.	A	B	C		X
01					
02					
03					
11					
12					
–					
nn					

Figure 3.11
In principle, the difference between vector data and raster data is not that great. Raster data could well be arranged in tabular form with each cell number representing a line and each attribute (layer or raster values) a column.

3.3.3 Object relations

Object relations are the same as entity relations in the real-world model. Differentiation is made between:

1. Relations that may be calculated from:
   a. The coordinates of an object: for example, which lines intersect or which areas overlap
   b. Object structure (relation), such as the beginning and end points of a line, the lines that form a polygon, or the locations of polygons on either side of a line
2. Relations that must be entered as attributes, such as the division of a county into townships or the levels of crossing roads that do not intersect

3.3.4 Quality

The true value of any description of reality depends on the quality of all the data it contains: graphics, attributes, and relations. Graphical data accurate to ± 0.1 m obviously describe reality more faithfully than data accurate to ± 100 m. Similarly, recently updated data are preferable to five-year-old data (which bring in temporal factors).

In the initial data modeling stage, the assessment of the data quality should include:

- Graphical accuracy (such as ± 1.0 m accuracy)
- Updating (when and how data should be updated)
- Resolution/detailing (e.g., whether roads should be represented by lines or by both road edges)
- Extent of geographical coverage, attributes included, and so on
- Logical consistency between geometry and attributes
- Representation: discrete versus continuous
- Relevance (e.g., where input may be surrogate for original data that are unobtainable)

Information on the quality of data is important to users of the database. Requirements for data quality are discussed in greater detail in Chapter 11.

3.4 From Database to GIS to Map

Once a data model is specified, the task of realizing it in a computer is technical and the task of entering data is simple and straightforward, albeit time consuming. A database need seldom be made to suit a data model, as many databases compatible with GIS applications are now on the market. The problem at hand is more one of selecting a suitable database with regard to:

- Acquisition and control
- Structure
- Storage
- Updating and changing
- Managing and exporting/importing
- Processing
- Retrieval and presentation
- Analyses and combinations

Needless to say, a well-prepared data model is vital in determining the ultimate success of the GIS application involved. Users view reality using GIS products in the form of maps with symbols, tables, and text reports. The dissemination of information via maps is discussed further in Chapter 16.

3.5 Shortcomings of the Traditional GIS Data Model

3.5.1 Entities and fields

In the real world, one specific area or field may have many different characteristics; one area will in reality represent a number of entities or object types, such as coniferous forest, protected area, property no. 44/133, and so on. We experience on a daily basis that it is the area as an entity that carries the information. However, in our real-world model we split phenomena into entities (entity: a real-world phenomenon that is not divisible into phenomena of the same kind) and allow the entities to be bearers of information. This model will allow an entity to represent only one phenomenon (e.g., only coniferous forest or only protected area). To adapt the model to reality, overlapping phenomena (entities/objects) are separated into different layers (Figure 3.12). Reality is thus adapted to fit into a layer system, which is also traditionally used in map presentation. In the real world, areas are not

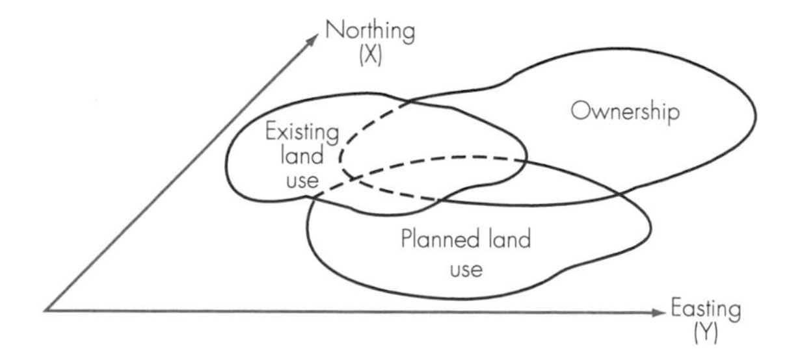

Figure 3.12
In the field model, the bearers of information consist of areas that often overlap. There are systems available today that can handle overlapping areas in the same layer.

divided into any form of horizontal two-dimensional physical layer—not even geological layers (strata) are presented in this way.

We can say that geometry—where coordinates define points, lines, and areas—is in many ways an artificial concept and an unnatural way to describe reality. Coordinates are not tangible and are never used in our everyday description of reality. Instead, we define a phenomenon's location in relation to other phenomena with which the recipient of the information is familiar. We can therefore establish that our model of reality is not perfect. During the 1990s, new models have been developed, known as *object-oriented models,* which to a certain extent can allow for the fact that the entity bearing the information can represent many phenomena. Object-oriented database systems are currently little used in commercial GIS but would appear to have many advantages over traditional database systems.

3.5.2 Uncertainty

To regard the real world as consisting of geometric constructs (points, lines, areas) means viewing objects as discrete data model representations. That is, all objects have clearly defined physical limits. These limitations are most obvious on maps, where lines imply sharp demarcations with no smooth, continuous transitions.

A discrete data model does not always suit reality. Difficulties arise in depicting phenomena that lack clear physical demarcation, such as soil types, population densities, or prevailing temperatures. There can also be uncertainty in the attribute values to be retained. In the traditional discrete model, entities or objects are defined as being either within specific classes or outside them and thus operate only with areas that are homogeneous with respect to limitation and classification. In reality, phenomena will often vary even within small, limited areas. For example, coniferous forest often contains deciduous trees, population density is variable, and terrain surface changes continuously. Once again, we have established that our real-world model is not perfect and that it is closely linked to traditional mapping concepts. Some of these problems can be partly solved by using the *fuzzy set theory,* which allows an object to belong only partially to a class. The fuzzy theory has as yet been little used in commercial GIS software; thus the significance of this type of deficiency in the data model is left to the person interpreting the results (maps and reports) of the GIS process.

3.5.3 Conceptual generalization

When points, lines, and polygons are selected as the geometric representation of objects, this very often results in a *generalization* of the real world; a town can be represented by a point rather than a polygon, and a road will frequently be represented by a center line and not two road verges. The need to divide objects into classes also results in a generalization. For example, an area of forest that is mainly coniferous, with some deciduous, will often be generalized and classified as coniferous, not as a combination. Thus conceptual generalization is also a method for handling uncertain elements.

It will always be necessary to make choices about such generalizations in relation to the real world when making data models. This may be seen as a problem, but generalization is also a technique that makes it possible to obtain an overview of our complex reality. It can also be difficult to create data models that have a uniform and clear definition of the objects' classes. For example, does a pedestrian area that is accessible to emergency vehicles classify as a road?

3.5.4 Role of maps in data modeling

Maps are, in general, good sources for describing objects and their attributes. However, maps always represent particular models of the real world, and GIS should represent the real world, not the maps that depict it. For instance, ferry routes are often shown by dotted lines on maps, whereas in transport planning data models should form integral parts of a contiguous road network. As a rule of thumb, therefore, always look at a map as a data source, not as a data model.

3.6. Extension of the Reality Concept

The traditional model for transformation from the real world to GIS, as described above, has its obvious faults. In addition, it only describes flat and unchanging reality. Models for describing objects in three-dimensional space and terrain have not yet been discussed, nor has the fourth dimension—time—and its inroads into a geographical data model. The same applies to models for dealing with objects (traffic) moving along defined networks.

Here it is also most practical to use the same basic concept: a geometry consisting of points, lines, and polygons, and attributes that describe the objects or phenomena. Elevation values can be linked to points, lines, and polygons and thereby give the objects a position in space. The surface of the terrain can be described with the help of sloping areas or with the help of horizontal surfaces with an elevation value linked as an attribute. Elevation values can also be linked to objects such as towers, wells, and buildings as attributes. The time factor can be accommodated by storing *all* historical data for the objects, such as changes in the geometry or attribute values. The movement of objects (traffic) along a road network can be simulated by assigning attribute values to elements in the network. These should be values that are relevant to the transfer speeds, and the sum of attribute values for different routes will be the measurement of passage in time or distance. A technical description of how this can be realized in GIS is given in Chapter 5, while the basic data models are described in Chapter 4.

Undoubtedly the traditional data model concept has definite drawbacks when describing these new real-world elements. We must accept that the real world is too complex to be described in full at present, although researchers are continuously engaged in developing improved models.

CHAPTER 4

Basic Data Models

4.1 Introduction

As discussed in Chapter 3, GIS depicts the real world through models involving geometry, attributes, relations, and data quality. In this chapter, the realization of models is described, with the emphasis on geometric spatial information, attributes, and relations. Data quality is discussed further in Chapter 11.

Spatial information is presented in two ways: as vector data in the form of points, lines, and areas (polygons); or as grid data in the form of uniform, systematically organized cells. Geometric presentations are commonly called *digital maps.* Strictly speaking, a digital map would be peculiar because it would comprise only numbers (digits). By their very nature, maps are analog, whether they are drawn by hand or machine, or whether they appear on paper or displayed on a screen. Technically speaking, GIS does not produce digital maps—it produces analog maps from digital map data. Nonetheless, the term *digital map* is now so widely used that the distinction is well understood.

4.2 Vector Data Model

The basis of the vector model is the assumption that the real world can be divided into clearly defined elements where each element consists of an identifiable object with its own geometry of points, lines, or areas (Figure 4.1). In principle, every point on a map and every point in the terrain it represents is uniquely located using two or three numbers in a coordinate system, such as in the northing, easting, and elevation Cartesian coordinate system. On maps, coordinate systems are commonly displayed in grids with location numbers along the map edges (Figure 4.2). On the ground, coordinate systems are imaginary, yet marked out by survey control stations. Data usually may be transformed from one coordinate system to another. (See Chapter 6 for further details on coordinate systems.)

With few exceptions, digital representations of spatial information in a vector model are based on individual points and their coordinates. The exceptions include cases where lines or parts of lines (e.g.,

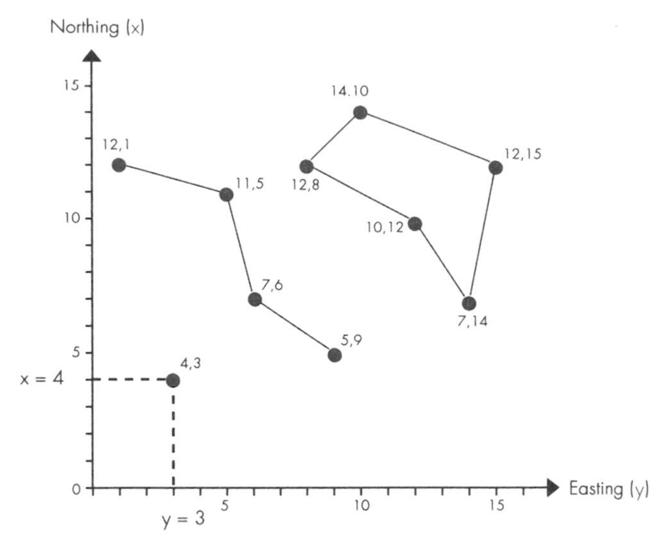

Figure 4.1
In the vector data model, the objects are described as points, lines, or areas (polygons). These three geometric phenomena may be described individually by a single point in a coordinate system and with connected lines (for lines and areas). Therefore, each object comprises one or more sets of coordinates.

those representing roads or property boundaries) may be described by mathematical functions, such as those for circles or parabolas. In these cases, GIS data include equation parameters: for example, the radii of the circles used to describe parts of lines. Together with the coordinate data, instructions are entered as to which points in a line are uncon-

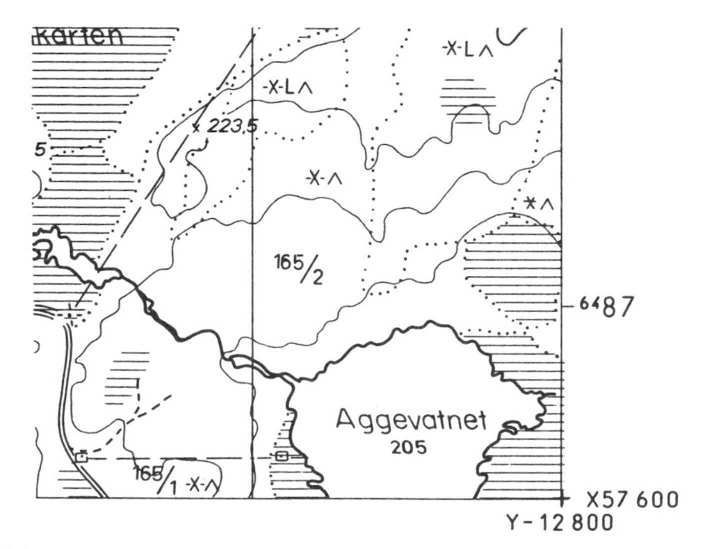

Figure 4.2
On maps, coordinate systems in an *x*- and *y*-direction format are generally displayed in grids, with location numbers along the map edges.

nected and which are connected. These instructions can subsequently be used to create lines and polygons and to trigger "pen up" and "pen down" functions in drawing.

Coordinate systems are usually structured so that surveys along an axis register objects in a scale of 1:1; that is, 1 m along the axis corresponds to 1 m along the ground. In principle, the degree of accuracy of measurements along an axis is decided by the type of measuring method applied, while the required degree of precision will naturally influence the amount of work required to gather the data.

Mathematically, a vector is a straight line, having both magnitude and direction. Therefore, a straight line between two data coordinate points on a digital map is a vector—hence the concept of vector data used in GIS and the designation of vector-based systems. In a vector model, points, lines, and areas (polygons) are the homogeneous and discrete units that carry information. As discussed above, these three types of object may be represented graphically using coordinate data. However, as we shall see, the objects may also carry attributes that can be digitized, and all digital information can be stored.

4.2.1 Coding digital data for map production

Anyone familiar with maps knows that map data are traditionally coded. Roads, contour lines, property boundaries, and other data indicated by lines are usually shown in lines of various widths and colors. Symbols designate the locations of churches, airports, and other buildings and facilities. In other words, coordinates and coding information identify all objects shown on a map.

Not surprisingly, then, the digital data used to produce maps are also coded, usually by the assignment of numerical codes used throughout the production process—from the initial data to computer manipulation and on to the drawing of the final map. Each numerical code series contains specific codes assigned to the objects in the group. For example, the codes for boundaries may be as illustrated in Figures 4.3 and 4.4.

Numerical code series	Object group
1 000	Survey control stations
2 000	Terrain formation
3 000	Hydrography
4 000	Boundaries
5 000	Built-up areas
6 000	Buildings and facilities
7 000	Communications
8 000	Technical facilities

Figure 4.3
Digital map data often use numerical coding, in the form of different numerical series, to identify object groups.

Numerical code	Object type
4 001	National border
4 002	County boundary
4 003	Township boundary
4 011	Property boundary
4 022	National park border
	etc.

Figure 4.4
By using a numerical coding system, codes can be assigned to all levels of detailed information on the objects.

Digital data for map production comprise sequences of integers, such as

– 53144011123456789123406780 – 53144011123336788123306700

Use of the format permits the numerical sequence to be divided into groups and read

– 5/314/4011/12345/6789/12340/6780
– 5/315/4011/12333/6788/12330/6700

The figures designations are as follows:

Figure	**Designates**
– 5	Start of a continuous sequence of data (i.e., if there are several coordinates, they are to be connected in a line: *pen down*)
314	Serial number of data sequence (such as of a unique line)
4011	Property boundary (such as might produce a final line width of 0.3 mm)
12345	First easting coordinate
6789	First northing coordinate (pen moves to next coordinate set).
12340	Last easting coordinate
6780	Last northing coordinate
– 5	End of data sequence, start of next sequence (*pen up*—moved and set down for a sequence of new coordinates, etc.)

In thematic coding, which may be compared to the overlay separation of conventional map production (Figure 4.5), data are divided into single-topic groups, such as all property boundaries. Information on symbol types, line widths, colors, and so on, may be appended to

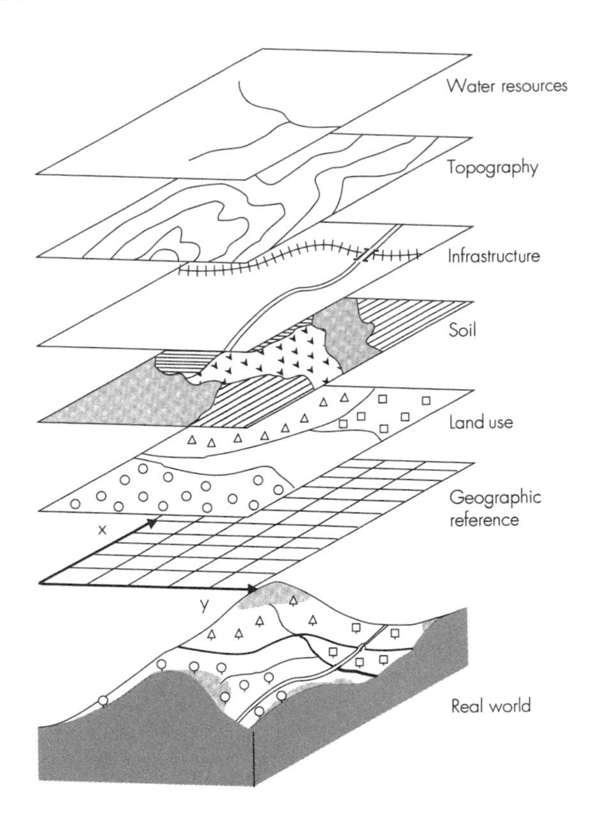

Figure 4.5
Theme codes in the digital map data can be used to separate data into thematic layers.

each thematic code, and various combinations of themes may be drawn. Data may be presented jointly in this way only if all objects are registered, using a common coordinate system.

Coding digital data for GIS

Point objects may easily be realized in a database because a given number of attributes and coordinates is associated with each point (Figure 4.6). Line and polygon objects are more difficult to realize in a database because of the variation in the number of points composing them. A line or a polygon may comprise two points or 2000 or more points, depending on the extent of the line and the complexity of the area, which is delineated by a boundary line that begins and ends at the same point.

Object spatial information and object attributes are often stored in different databases to ease the manipulation of lines and areas, but in some systems they are stored together. As pivotal attributes are often available in existing computer memory files, dividing the databases conserves memory by precluding duplicate storage of the same data.

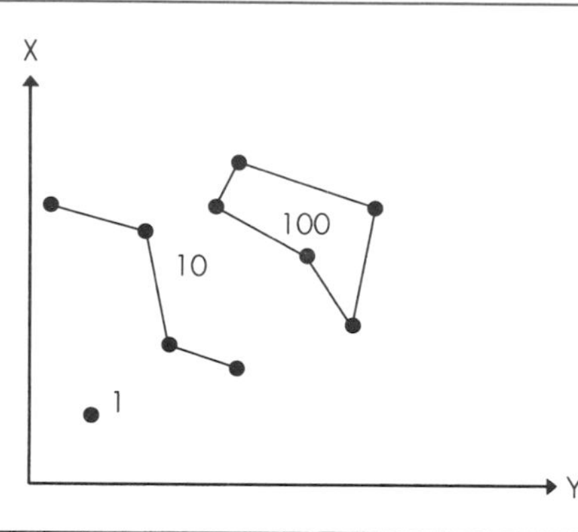

Geometry	ID	Attributes		Coordinates
Point	1	A	B	4,3 (single point)
Line	10	C	D	2,1 11,5 · · 5,9 (string)
Area	100	C	E	14,10 12,14 · · · · 14,10 (closed polygon)

Figure 4.6
Each object is assigned attributes and coordinates. However, the number of coordinates for lines and polygons will vary considerably, depending on the length of the line and circumference of the polygon. This may make it inappropriate to store attributes and geometry together and is one of the reasons why many systems store them separately.

The separate storage of attribute and spatial information data requires that all objects in the attribute tables be associated with the corresponding spatial information. This association is achieved by inserting spatially stable and relevant attribute data or codes from the attribute table into the spatial information, or vice versa. In other words, identical objects have the same identities in both databases.

The identity (ID) codes used to label and connect spatial information and attribute table data are most often numerical, but may be alphanumerical. Typical identity codes include building numbers, property numbers, and street numbers and addresses. If the data are ordered in a manuscript map, each object may be assigned a serial number used in both the spatial information and the attribute databases. Polygons for vegetation mapping can, for example, be numbered from 1 onward, while pipes, manholes, and so on, are usually numbered according to an administrative system. ID codes allow differentiation between objects, whereas theme codes allow for differentiation between different groups of objects. In theory, identity codes and thematic codes are both attribute data. However, they are very closely tied to geometry and are therefore often treated as such, as described above.

Spatially defined objects without attributes need no identifiers, but they are required for all objects that are listed in attribute tables, and manipulated spatially. Identifiers are normally entered together with the relevant data, but they may also be entered later, using an interac-

ZERO-DIMENSIONAL OBJECTS:

Point: A zero-dimensional object that specifies geometric location specified througth a set of coordinates.

Node:.A zero-dimensional object that is a topological junction and may specify geometric location.

ONE-DIMENSIONAL OBJECTS

Line segment (vector): A one-dimensional object that is a direct line between two end points.

Link: A one-dimensional object that is a direct connection between two nodes.

Directed link: A link between two nodes with one direction specified.

String: A sequence of line segments.

Chain: A directed sequence of nonintersecting line segments with nodes at each end.

Arc: A locus of points that forms a curve that is defined by mathematical function. Also defined as a string or chain.

Ring: A sequence of any line segments with closure.

TWO-DIMENSIONAL OBJECTS

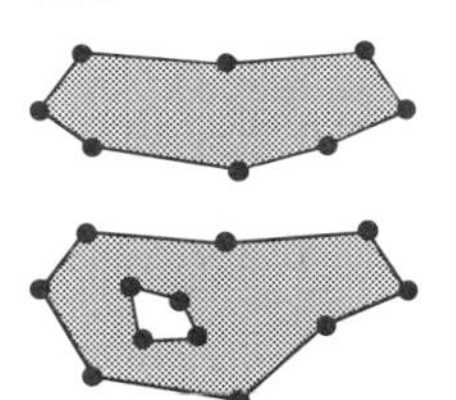

Simple area/polygon: An area defined by an outer ring that may not have inner rings (holes).

Complex area/polygon: An area defined by an outer ring with optional inner rings defining holes.

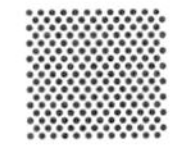

Raster cell/pixel

Figure 4.7
Geometric objects. (Adapted from Agnar Renolen.)

tive human–machine process such as keying in identifiers for objects pointed out on the screen.

Some systems tie a polygon's ID code to a characteristic point in the polygon, known as the *label point.* Label points may be computed or identified interactively on screen, and codes may be entered manually for the relevant polygons. The attribute values of the polygon are then linked to this label point. Today, systems are available which treat polygons as independent objects. Typical digital geometric data for GIS are illustrated in Figure 4.8.

Plotting may be controlled by appending drawing instructions to the thematic code, to the individual identifiers, or to other object at-

I.D.	Thematic code	X-coordinates	Y-coordinates
11	30	23 999.80	10 008.55
	30	23 990.50	10 015.10
34	40	24 876.30	11 122.86
	40	24 890.10	11 150.30
–	–	–	–
	–	–	–
122	20	24 870.25	11 130.23

Code list	
20	Topography
30	Infrastructure
40	Soil

Figure 4.8
Typical section of digital map data with relevant code list.

tribute values. In a finished map, tabular data appear on a foreground map against the background of a base map derived from the remaining map data. Look-up tables are usually used to translate tabular data to map symbols (Figure 4.9). Map presentations are discussed further in Chapter 16.

4.2.2 Spaghetti model

As illustrated in Section 4.2.1, digital map data comprise lines of contiguous numerals pertaining to spatially referenced points. *Spaghetti data* are a collection of points and line segments with no real connection (Figure 4.10). What appears as a long, continuous line on the map

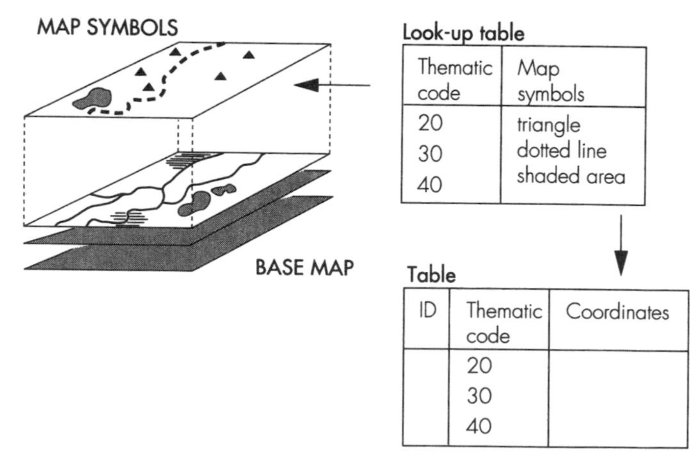

Figure 4.9
Drawing instructions are often designated in look-up tables. Thematic code values or attribute values are often input values in the tables, whereas output values can be symbol types, colors, line thickness, and so on.

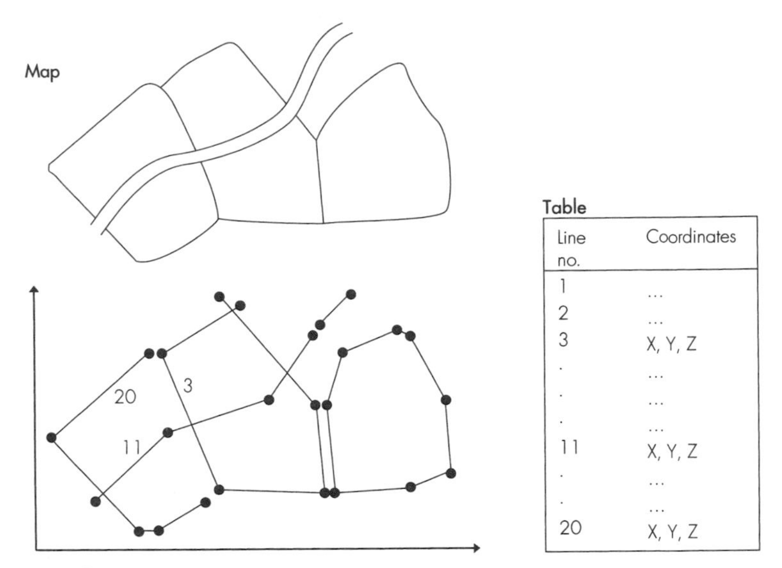

Figure 4.10
Spaghetti data is a term often used to describe digital map data with crossing lines, loose ends, double digitalization of common boundaries between adjacent polygons, and so on. These data lie in a pile, just like spaghetti. Several line segments are to be found in odd places in the data file.

or in the terrain may consist of several line segments which are to be found in odd places in the data file. There are no specific points that designate where lines might cross, nor are there any details of logical relationships between objects. Polygons are represented by their circumscribing boundaries, as a string of coordinates so that common boundaries between adjacent polygons are registered twice (often with slightly differing coordinates). The lines of data are unlinked and together are a confusion of crossings.

Unlinked (spaghetti) data usually include data derived either from the manual digitizing of maps or from digital photogrammetric registration. Consequently, spaghetti data are often viewed as raw digital data. These data are amenable to graphic presentation—the delineation of borders, for example—even though they may not form completely closed polygons. Otherwise, their usefulness in GIS applications is severely limited.

One drawback is that both data storage and data searches are sequential. Hence search times are often unduly long for such routine operations as finding commonality between two polygons, determining line intersection points, or identifying points within a given geographical area. Other operations vital in GIS, such as overlaying and network analysis, are intractable. Furthermore, unlinked data require an inordinate amount of storage memory because all polygons are stored as independent coordinate sequences, which means that all

Map scale	Map sheet	MBytes
1 : 250,000	50 x 60 cm	25 – 50
1 : 50,000	50 x 60 cm	15 – 25
1 : 5000	48 x 64 cm	2,5 – 10
1 : 1000	60 x 80 cm	1 – 3

Figure 4.11
Digital map data require an inordinate amount of storage memory. Normally, elevation data comprises 25 to 90% of the total data volume.

lines common to two neighboring polygons are stored twice. The typical memory required for unlinked data is illustrated in Figure 4.11.

4.2.3 Topology model

Topology is the branch of mathematics that deals with geometric properties which remain invariable under certain transformations, such as stretching or bending. The *topology model* is one in which the connections and relationships between objects are described independent of their coordinates; their topology remains fixed as geometry is stretched and bent. Hence the topology model overcomes the major weakness of the spaghetti model, which lacks the relationships requisite to many GIS manipulations and presentations.

The topology model is based on mathematical graph theory and employs nodes and links. A *node* can be a point where two lines intersect, an endpoint on a line, or a given point on a line. For example, in a road network the intersection of two roads, the end of a cul-de-sac, or a tunnel adit may generate a node. A *link* is a segment of a line between two nodes. Links connect to each other only at nodes. A closed polygon consisting of alternating nodes and links forms an area. Single points can be looked upon as a degenerate node and as a link with zero length (Laurini and Thompson 1992). Theme codes should be taken into consideration when creating nodes to ensure that they are created only between relevant themes (e.g., at the junction between a national highway and a county road, not between roads and property boundaries).

Unique identities are assigned to all links, nodes, and polygons, and attribute data describing connections are associated with all identities. Topology can therefore be described in three tables (Figure 4.12):

1. The polygon topology table lists the links comprising all polygons, each of which is identified by a number.
2. The node topology table lists the links that meet at each node.
3. The link topology table lists the nodes on which each link terminates and the polygons on the right and left of each link, with right and left defined in the direction from a designated start node to a finish node. The system creates these tables automatically.

A table with point coordinates to the links ties these features to the real world and permits computations of distances, areas, intersections, and other numerical parameters. The geometry of the objects is stored in its own subordinate table (see Figure 4.12). Numerous spatial analyses may then be performed, including:

- Overlaying
- Network analyses
- Contiguity analyses
- Connectivity analyses

Topological attribute data may be used directly in contiguity analyses and other manipulations with no intervening, time-consuming geometric operations.

Once the topology has been created, a map can be plotted with solid colors. This is not possible with spaghetti data. Thematic layers of topological data can also be used to steer the plotting sequence. The sequence influences what becomes visible on the map. For example, a green area superimposed on a white house will render the house invisible on the map (unless the house creates a window in the area).

Topology requires that all lines should be connected, all polygons closed, and all loose ends removed. Even gaps as small as 0.001 mm may be excessive, so errors should be removed either prior to or during the compilation of topological tables.

A function known as *snap* can also be used in digitalization. Using the snap function with a defined tolerance of, say, 1 mm, a search can be carried out around the end of a line or around an existing point which is assumed to have the same coordinates as the last point registered. When this point is found, the two points will be snapped together to form a common node, thereby closing the polygon. The same procedure can be carried out automatically on existing data. A node can also be created in existing data by calculating the point of intersection between lines. Meaningless loose ends can be removed by testing with a given minimum length.

Topological information permits automatic verification of data consistency to detect such errors as the incomplete closing of polygons during the encoding process. The graph theory contains formulas for the calculation of such data errors. There has to be a fixed relationship between the number of nodes, lines, and polygons in one data set. A run-through of the data in positive and negative directions will produce the same result.

The topological model has a few drawbacks. The computational time required to identify all nodes may be relatively long. Uncertainties and errors may easily arise in connection with the closing of polygons and formation of nodes in complex networks (such as in road interchanges). Operators must solve such problems. When new data are

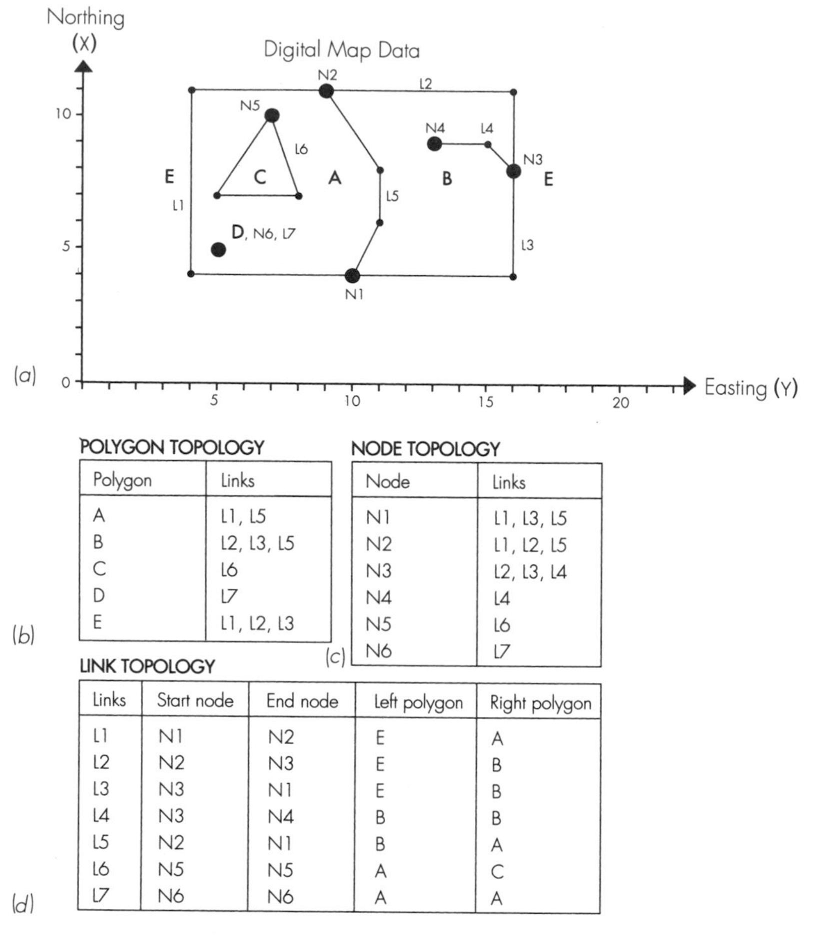

(b)

POLYGON TOPOLOGY

Polygon	Links
A	L1, L5
B	L2, L3, L5
C	L6
D	L7
E	L1, L2, L3

(c)

NODE TOPOLOGY

Node	Links
N1	L1, L3, L5
N2	L1, L2, L5
N3	L2, L3, L4
N4	L4
N5	L6
N6	L7

(d)

LINK TOPOLOGY

Links	Start node	End node	Left polygon	Right polygon
L1	N1	N2	E	A
L2	N2	N3	E	B
L3	N3	N1	E	B
L4	N3	N4	B	B
L5	N2	N1	B	A
L6	N5	N5	A	C
L7	N6	N6	A	A

(e)

LINK COORDINATES

Link	Coordinates			
L1	4,10	4,4	11,4	11,9
L2	11,9	11,16	8,16	
L3	8,16	4,16	4,10	
L4	8,16	9,15	9,13	
L5	11,9	8,11	6,11	4,10
L6	10,7	7,8	7,5	10,7
L7	5,5			

Figure 4.12
The idea behind the topology model is that all geometric objects (*a*) digital map data can be represented by nodes and links. The objects' attributes and relationships can be described by storing nodes and links in three tables: (*b*) a polygon table, (*c*) a node topology table, and (*d*) a link topology table. (*e*) An additional table gives the objects' geographical coordinates.

entered and existing data updated, new nodes must be computed and the topology tables brought up to date.

Topological data may require a longer plotting time than spaghetti data because of the separation of lines into nodes and links. However, the overall advantages of the topology model over the spaghetti model make it the prime choice in most GISs. Today, efficient software and faster computers enable topology to be established on-the-fly; thus the disadvantages of topological data as compared to spaghetti data have become less important.

Computer storage techniques and problems are discussed further in Chapters 8 and 12. Here it suffices to say that usually, map data are not stored in a contiguous unit, but rather, divided into lesser units that are stored according to a selected structure. This structure may be completely invisible to the user, but its effects, such as rapid screen presentation of a magnified portion of a map, are readily observable.

4.2.4 Data compression

The amount of memory needed can be reduced by using data compression techniques. Most of these automatic techniques are based on removing points from continuous lines (contour lines, etc.). Good data compression techniques, therefore, are those that preserve the highest possible degree of geometric accuracy. The most basic technique involves the elimination of repetitive characters: for example, the first character of all coordinates along a particular axis. The repetitive character needs to be entered only once; subsequently, it may be added to each set of coordinates. This particular technique has no effect on the geometry. Savings in characters stored are illustrated in Figure 4.13.

There are other automatic methods of removing points. One simple means is to keep only every *n*th point on a line. The lower the value of *n*, the greater the number of points that will be removed. This method does not take into account geometric accuracy; however, this can be compensated for by testing the curvature of the line. One method is to

Original data	
Northing	Easting
10,234	70,565
10,245	70,599
10,167	70,423
–	–
–	–
etc.	etc.

Compacted data	
10,000	70,000
234	565
245	599
167	423
–	–
–	–
etc.	etc.

Figure 4.13
Simple data compression. The volume of data to be stored can be reduced to a single entry, giving the value common to all coordinate values.

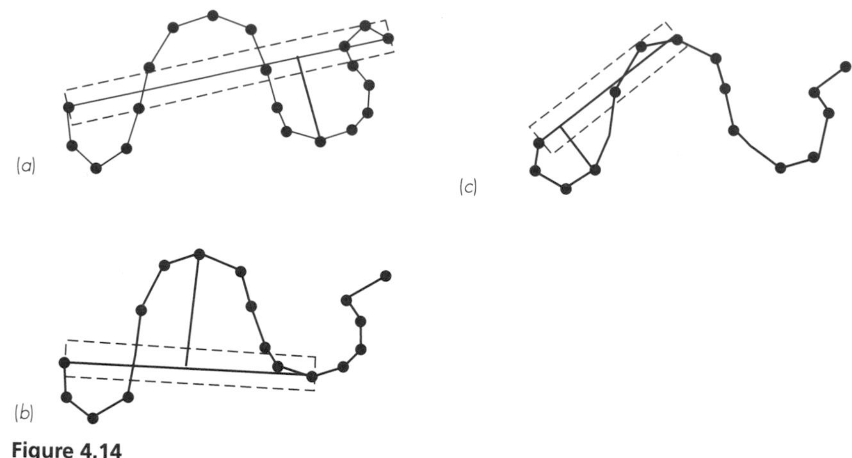

Figure 4.14
Douglas–Peucker method, to save storage space.

draw a straight line between the first and last points on a curved stretch of line and to calculate the orthogonal distance from each point on the curved line below the straight one. Points that are closer than a given distance from the straight line will be removed. The endpoint of the straight line is then moved to the point with the greatest distance and the same procedure for removing points is repeated. This continues until all the relevant points are removed. This method is known as the *Douglas–Peucker algorithm* (Figure 4.14).

Points of little or no value in describing a line may be eliminated by moving a corridor step by step along a line and deleting points that are closer to the neighboring point than a given value or where the vectors create an angle that is smaller than the given value (Figure 4.15). Contours and other lines can also be replaced with mathematical

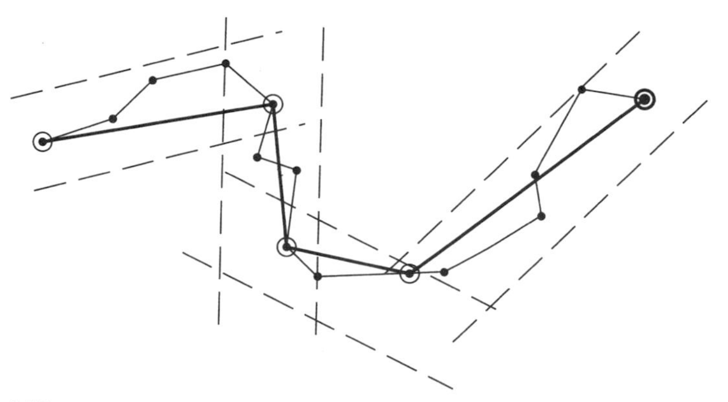

Figure 4.15
Reducing lines by use of a corridor. The number of points needed to describe a line may be reduced by moving forward a corridor of a given width until it touches the digitized line. All the points on the line in the corridor, apart from the first and last, are thereby deleted. This process is repeated until the entire line has been trimmed.

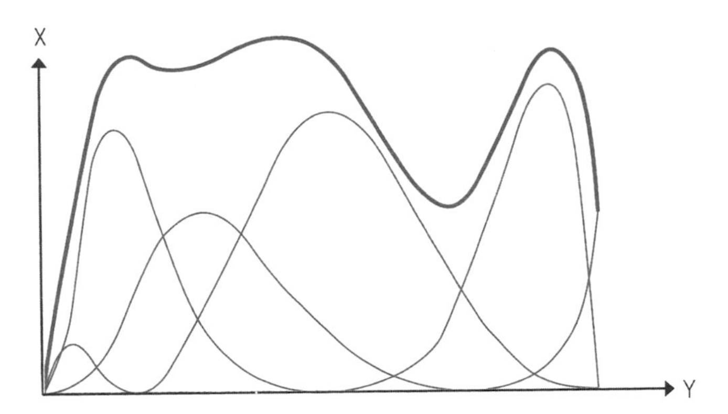

Figure 4.16
A complicated line course can be replaced by a few simple polynomials (spline). This reduces considerably the volume of data emanating from line courses.

functions, such as straight lines, parabolas, and polynomials. A spline function comprises segments of polynomials joined smoothly at a finite number of points so as to approximate a line. As shown in Figure 4.16, a spline function can involve several polynomials to build a complex shape. It has been reported that a spline function representing nautical chart data has reduced data volume by 95%.

The amount of memory required to store a given amount of data often depends on the format in which data are entered. Some formats contain more administrative routines than others, some have vacant space. Thus, the gross volumes stored are frequently related to format.

4.2.5 Storing vector data

The manner in which digital map data are stored in a record is determined by a format, a set of instructions specifying how data are arranged in fields. The latter are groups of characters or words, which, in turn, are treated as units of data. The format stipulates how the computer will read data into the fields: total number of fields specified, number of characters permissible in each field, number of spaces between fields, which fields are numeric and which are text, and so on. An example is shown in Figure 4.17.

The information content of the data is designated not in the format but ancillary to it,—for example, in a heading. Typical specifications for information content might include field assignments, such as the point number in the first field, the thematic code in the second, easting in the third, northing in the fourth, and elevation in the fifth. The meanings of the numeric codes used must also be given. The spaghetti data are stored in a simple file structure and in the order in which the data have been registered.

Users of conventional maps know the frustrations of extracting information from maps produced by various agencies using differing

Record	Field 1	2	3	4	5
1	10	4003	1556.3	2303.1	131.0
2	11	4003	1590.2	2370.2	135.5
3	12	4003	1610.6	2415.4	139.7
4	13	4011	1605.3	2541.6	133.2
5	14	4011	1540.1	2623.3	139.2
6	15	4011	1510.5	2759.9	145.6
|					
|					
|					
|					

Figure 4.17
In a single data file, data are stored in records and each record is divided into fields where the values are stored. The format determines how the data are arranged in fields, according to the number of characters, whether they are text or numeric, and so on.

map sheet series, varying scales and coordinate systems, and frequently, different symbols for the same themes. Moreover, the cartographic version of Murphy's law dictates that the necessary information is all too often located in the corners where four adjoining map sheets meet.

Database storage of cartographic data can overcome these problems because it involves standardization of data through common reference systems and uniform formats. Cartographic data from various sources can, with few limitations, be combined. The results are then independent of map sheet series and scales.

Standardized storage makes the presentation of data compiled from dissimilar sources much easier. For example, uniform storage formats permit the combination of telecommunications administration network data with property survey data, or of geological information from 1:50,000 scale maps with vegetation data from 1:20,000 scale maps.

Digital map data are stored in databases, the computerized equivalent of conventional file drawers and cabinets. Although data entries in a database can be updated far more rapidly than data printed on map sheets on file, the information is found more quickly from map sheets than by searching in a database. This is because a single map sheet contains an enormous amount of information, usually equivalent to 100,000 or more sets of coordinates. A sequential computer search of 100,000 items in a database is slow even for the most powerful computers in comparison with a quick visual scan of a map sheet. Therefore, "smart" programs known as database management systems (DBMSs) have been compiled to maintain, access, and manipulate databases. The various DBMSs differ primarily in the ways in which the data are organized. Their selection and use are vital in GIS applications because they determine the speed and flexibility with which data may be accessed.

It is usual to split topological data into different thematic layers to simplify storage and to improve access to data. This division is done so that no overlap occurs between polygons within each thematic layer. For example, property boundaries are stored in one layer while other data overlapping the property, such as roads, buildings, and vegetation boundaries, are stored in another. The disadvantage of this system is that common lines between objects (e.g., roads and properties) that are stored in different layers have to be removed several times. This problem can be avoided by using object-based storage.

4.2.6 Comments on spaghetti and topology models

When digitizing lines such as those on land-use maps, the borders of surfaces are digitized both as spaghetti data and as separate objects. When creating topology, this model is converted to a layer model. The discussion of spaghetti and topology is very much based on the assumption that a class of area entities is always a tiling of the plane in which every point lies in exactly one polygon. (See the discussion on entities and fields in Section 3.5.1.) However, the problems related to spaghetti and topology have changed somewhat during recent years with the advent of new GIS software which treats polygons as independent objects that may overlap and need not fill the plane, and with systems permitting shapes. Many of the traditional arguments for area coverage/layer model and use of topology are based on the assumption of needing to avoid computation. New and more powerful computers eliminate the need for reduction in calculation time. Today, topology can easily be built on-the-fly.

4.3 Raster Data Models

Raster data are applied in at least four ways:

1. Models describing the real world
2. Digital image scans of existing maps
3. Compiling digital satellite and image data
4. Automatic drawing driven by raster output units

In the first example, raster data are associated with selected data models of the real world: in the second and third, with compilation methods, and in the fourth, with presentation methods. The respective computer manipulations may have much in common; as discussed later, satellite data may be entered in a raster model. For ease of explanation, models based on raster data are discussed in Section 4.3.2, digital raster maps are treated in Section 9.3, satellite data are discussed in Section 9.6, and output units generally are the subject of Chapter 7.

Raster model

The raster model represents reality through selected surfaces arranged in a regular pattern. Reality is thus generalized in terms of uniform, regular cells, which are usually rectangular or square but may be triangular or hexagonal. The raster model is in many ways a mathematical model, as represented by the regular cell pattern (Figure 4.18). Because squares or rectangles are often used and a pictorial view of them resembles a classic grid of squares, it is sometimes called the *grid model.* Geometric resolution of the model depends on the size of the cells. Common sizes are 10 × 10 m, 100 × 100 m, 1 × 1 km, and 10 × 10 km. Many countries have set up national digital elevation models based on 100 × 100-m cells. Within each cell, the terrain is generalized to be a flat surface of constant elevation.

The rectangular raster cells, usually of uniform size throughout a model, affect final drawings in two ways. First, lines that are continuous and smooth in a vector model will become jagged, with the jag size corresponding to cell size. Second, resolution is constant: regions with few variations are as detailed as those with major variations, and vice versa.

The cells of a model are given in a sequence determined by a hierarchy of rows and columns in a matrix, with numbering usually starting from the upper left corner (Figure 4.19). The geometric location of a cell, and hence of the object it represents, is stated in terms of its row and column numbers. This identification corresponds to the directional coordinates of the vector model. The cells are often called *pixels* (picture elements), a term borrowed from the video screen technology used in television and computer displays. A pixel is the smallest ele-

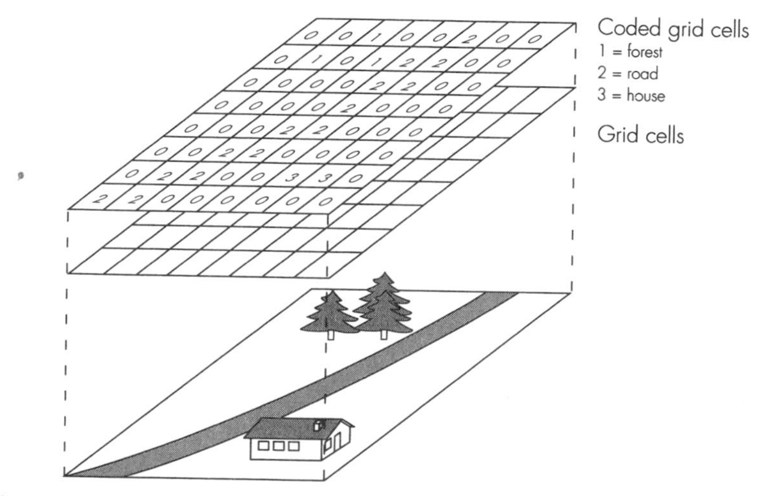

Figure 4.18
Raster data can be visualized as a grid lying over the terrain. Each grid cell has a code stored in the database describing the terrain within that particular cell.

ment of an image that can be processed and displayed individually. The raster techniques used in GIS are siblings of the rasters long used to facilitate the manipulation and display of information and consequently are suited to computerized techniques.

4.3.1 Realizing the raster model

Raster models are created by assigning real-world values to pixels (Figure 4.19). The assigned values comprise the attributes of the objects that the cells represent—and because the cells themselves are in a raster, only the assigned values are stored. Values, usually alphanumeric, should be assigned to all the pixels in a raster. Otherwise, there is little purpose in drawing empty rows and columns in a raster.

Consider a grid of cells superimposed on the ground or on a map. Assigning the values/codes of the underlying objects/features to the cells creates the model. The approach is comprehensive because everything covered by the raster is included in the model. *Draping* a ground surface in this way regards the ground or map as a plane surface.

Some GISs can manipulate both numerical values and text values (such as types of vegetation). Hence cell values may represent numerous phenomena, including:

- Physical variables, such as precipitation and topography, respectively, with amounts and elevations assigned to the cells
- Administrative regions, with codes for urban districts, statistical units, and so on
- Land use, with cell values from a classification system
- References to tables of information pertaining to the area(s) the cells cover, such as references to attribute tables
- Distances from a given object
- Emitted and/or reflected energy as a function of wavelength—satellite data

A single cell may be assigned only one value, so dissimilar objects and their values must be assigned to different raster layers, each of which deals with one thematic topic (Figure 4.20). Hence in raster models as in vector models, there are thematic layers for topography, water supply systems, land use, and soil type. However, because of the differences in the way attribute information is manipulated, raster models usually have more layers than those in vector models. In a vector model, attributes are assigned directly to objects. For instance, a pH value might be assigned directly to the object "lake." In a raster model, the equivalent assignment requires one thematic layer for the lake, in which cells are assigned to the lake in question, and a second thematic layer for the cells carrying the pH values. Raster databases may, therefore, contain hundreds of thematic layers.

	0	1	2	3	4	5	6	7
0	0	0	1	0	0	2	0	0
1	0	1	0	1	2	2	0	0
2	0	0	0	0	2	2	0	0
3	0	0	0	0	2	0	0	0
4	0	0	0	2	2	0	0	0
5	0	0	2	2	0	0	0	0
6	0	2	2	0	0	3	3	0
7	2	2	0	0	0	0	0	0

Cell table

Cell no.	Cell value
00	0
01	0
02	1
03	0
04	0
05	2
06	0
.	0
65	3
66	3
67	0
.	.
70	2
71	2
.	.
.	.

Code list	
0	unmapped
1	forest
2	road
3	house

Figure 4.19
A line number and column number define the cell's position in the raster data. The data are then stored in a table giving the number and attribute value of each cell.

In practice, a single cell may cover parts of two or more objects or values. Normally, the value assigned is that of the object taking up the greater part of the cell's area, or of the object at the middle of the cell, or that of an average computed for the whole of the cells.

Cell locations, defined in terms of rows and columns, may be transformed to rectangular ground coordinates, for example, by assigning ground coordinates to the center of the upper left cell of a raster (cell 0,0). If the raster is to be oriented north–south, the columns are aligned along the northing axis and the rows along the easting axis.

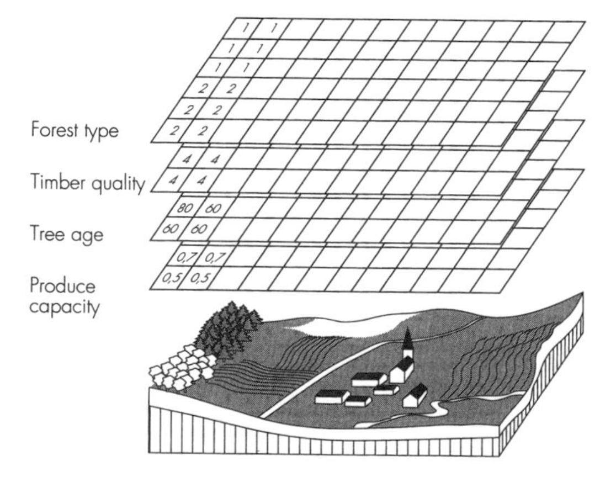

Figure 4.20
Only one attribute value may be assigned to each cell. Objects that have several attributes are therefore represented with a number of raster layers, one for each attribute.

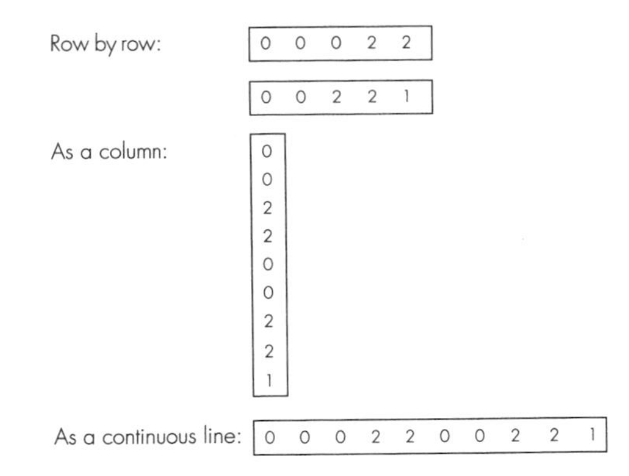

Figure 4.21
Typical cell input. Raster data may be stored in the database as a series of rows, a series of columns, or a continuous line.

The coordinates of all cell corners and centers can then be computed using the known cell shapes and sizes.

Object relations, which in the vector model are described by topology, are only partly inherent in the raster structure. When the row and column numbers of a cell are known, the locations of neighboring cells can easily be calculated. In the same way, cells contained in a given polygon may be located simply by searching with a stipulated value. It is much more difficult, however, to identify all the cells located on the border between two polygons. Polygon areas are determined merely by adding up constituent cells. Some operations, though, are more cumbersome. An example of this is computation of a polygon's perimeter length, which requires a search for, and identification of, all the cells along the polygon's border.

Overviews of phenomena in a given area are obtained from a raster model quickly and easily by searching all the thematic layers for cells with the same row and column numbers. The relevant overlay analysis is described in Section 14.11. Raster data are normally stored as a matrix, as described above. However, they can also be stored in tabular form, where each individual cell in a raster forms a line in the table (see Figure 4.21).

4.3.2 Coding raster data

Numerical codes and, in some systems, text codes may be assigned to cells. Cell values are entered from word processing files, databases, or other sources in the same sequence as they are registered (Figure 4.21). The way in which the figures are read is dictated by format. For instance, it is essential to know the number of columns per row.

Raster data may be available from a variety of dissimilar sources, ranging from satellite data and data entered manually to digital elevation data. Their collocation requires that cells from differing sources and thematic layers correspond with each other (Figure 4.22). In other

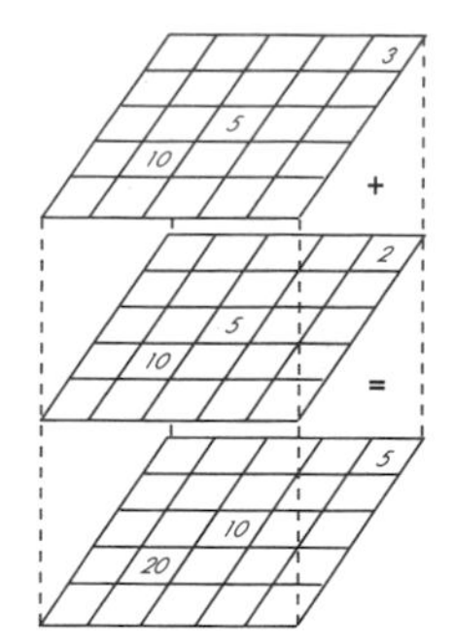

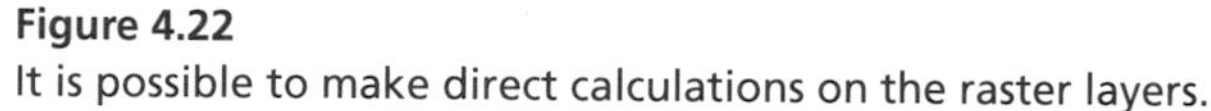

Figure 4.22
It is possible to make direct calculations on the raster layers.

words, cells having the same row and column numbers must refer to the same ground area. Various computations may be necessary to accommodate any differences in cell shape and size.

Cells may contain values referenced to attribute tables. The cells of a thematic layer may be coded so that their values correspond to identities in a given attribute table. Attribute data or tabular data may be coded independently, irrespective of whether the geometry is represented using vector data or raster data.

4.3.3 Compression of raster data

If the cell values of a raster model are entered in fixed matrices with rows and columns identical to those of the registered data, only the cell values need to be stored; row and column numbers need not. Even when only the cell values are stored, the volumes of data can easily become unwieldy. Typical operations may involve 200 thematic layers, each containing 5000 cells. The total number of cell values stored is thus $200 \times 5000 = 1$ million. A Landsat satellite raster image contains about 7 million pixels, a Landsat TM image about 35 million pixels.

Various devices may be employed to reduce data volume and, consequently, storage memory requirements. Cells of the same value are often neighbors because they pertain to the same soil type, the same population density of an area, or other similar parameters. Thus cells of the same value in a row may be compacted by stating the value and their total. This type of compacting, called *run-length encoding*, is illustrated in Figure 4.23. Further compacting may be achieved by applying the same process recursively to subsequent lines.

4.3.4 Quad-tree model

Traditionally, the raster model is based on dividing the real world into equal-sized rectangular cells. However, in many cases, it can be more

Raster image											Compacted
X	X	X	X	W	W	R	R	R	X	=	4X 2W 3R 1X
X	X	X	W	W	W	W	R	R	Y	=	3X 4W 2R 1Y
X	X	W	W	W	W	W	W	Y	Y	=	2X 6W 2Y
W	W	W	W	W	Y	Y	Y	Y	Y	=	5W 3Y 2X

Figure 4.23
Run-length encoding is the method used to save data storage space by reducing a row of cells with the same value to a single unit having a specific value and quantity.

practical to use a model with varying cell size. Larger cells (lower resolution) may be used to represent larger homogeneous areas, while smaller cells (higher resolution) may be used for more finely detailed areas. This approach, known as the *quad-tree representation,* is illustrated in Figure 4.24. In representing a given area, the aggregate amount of data involved is proportional to the square of the resolution (into cells). Since the quad-tree model is a very practical concept, it is therefore also preferable for the storage of both small and large volumes of data. The hierarchical data structuring typical of the quad-tree has other advantages which we look at later in the book.

The quad-tree paradigm divides a geographical area into square cells of size varying from relatively large down to that of the smallest cell of the raster. Usually, the squares are then quartered into four smaller squares (Figures 4.25 to 4.28). The quartering may be continued to a suitable level, until a square is found to be so homogeneous that it no longer needs to be divided, and the data on it can be stored as a unit. A larger square may therefore comprise several raster cells having the same attribute values. However, homogeneous areas that are not square or do not coincide with the pattern of squares employed may be further divided into homogeneous squares.

The structure of the quad-tree resembles an inverted tree, whose leaves are the pointers to the attributes of homogeneous squares and whose branch forks are pointers to smaller squares—hence the name quad-tree (Figure 4.28).

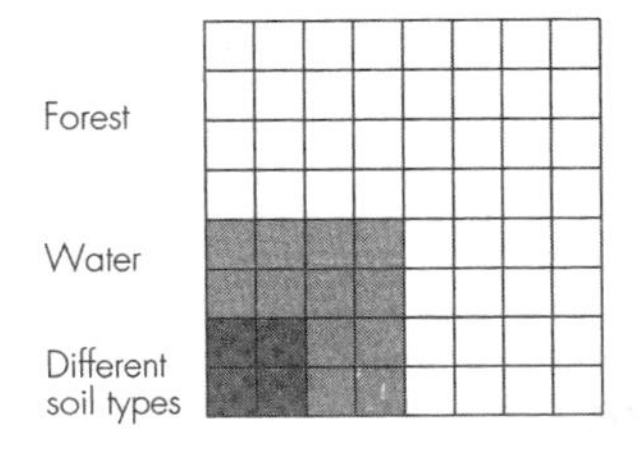

Figure 4.24
Computer memory capacity requirements can be reduced by storing homogeneous and uniform (quadratic) areas, consisting of many basis cells, as one single unit.

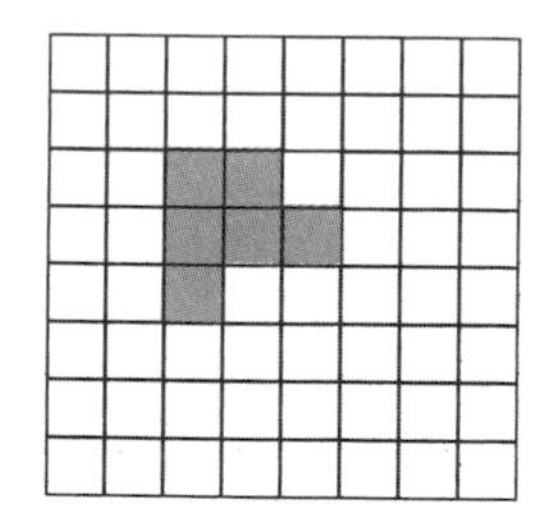

Figure 4.25
Quad-tree raster encoding recursively subdivides a map into quarters to define the boundary of an area. The shaded squares on this grid represent the region that is to be stored in a quad-tree. (Adapted from Green and Rapek 1990.)

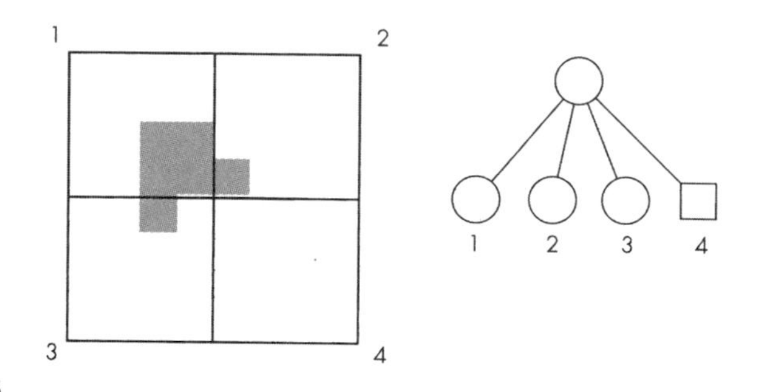

Figure 4.26
Initially, the grid is divided into four quadrants. If part of the region falls within a quadrant, it is assigned a white circle. If no part of the region falls in a quadrant, it is assigned a white square. (Adapted from Green and Rapek 1990.)

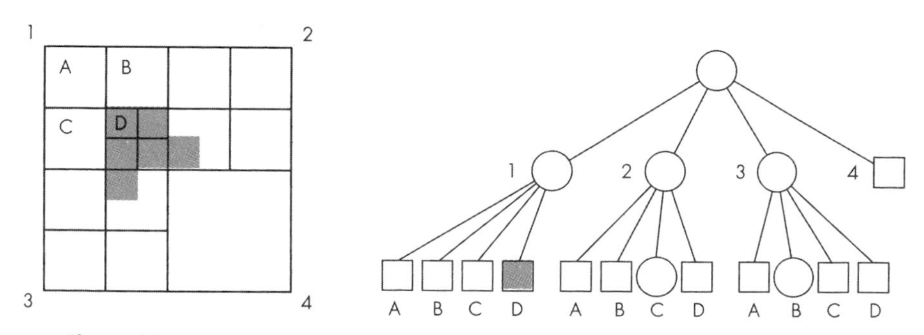

Figure 4.27
The quadrants that contain part of the region are again subdivided, which adds another level to the tree. If a quadrant is filled completely by part of the region, it is represented by a black square. (Adapted from Green and Rapek 1990.)

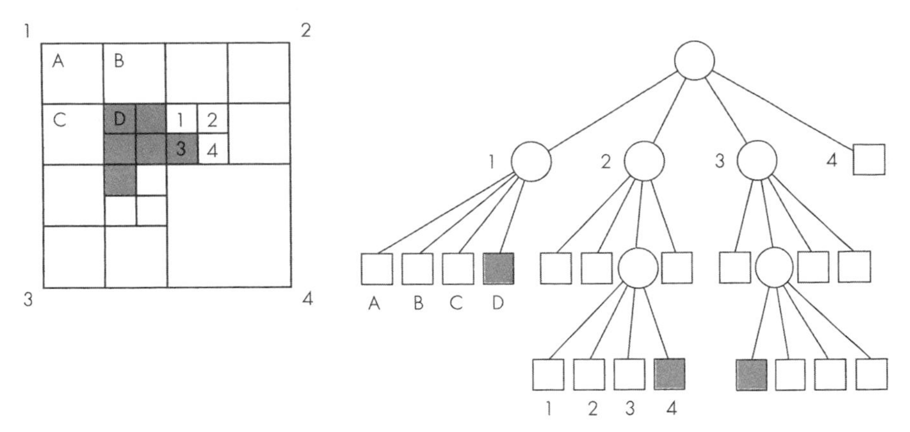

Figure 4.28
Subdivision of the grid continues until each leaf in the tree is represented by either a black or a white square. Only 19 leaves are required to store this region in a quad-tree. (Adapted from Green & Rapek 1990.)

The quad-tree principle can also be applied to vector data (Figure 4.29). Quad-tree models have been developed for both spaghetti and topological data. Even parts of lines/links will thus represent homogeneous units, right down to the smallest unit, consisting of a single point or node. Indexing of vector data is used to give more rapid data access in specific areas. This subject is dealt with in greater detail in Chapter 12.

The advantages of the quad-tree model are:

- Rapid data manipulation because homogeneous areas are not divided into the smallest cells used
- Rapid search because larger homogeneous areas are located higher up in the point structure
- Compact storage because homogeneous squares are stored as units
- Efficient storage structure for certain operations, including searching for neighboring squares or for a square containing a specific point

The disadvantages of the quad-tree model are:

- Establishing the structure requires considerable processing time.
- Protracted processing may prolong alterations and updating.
- Data entered must be relatively homogeneous.
- Complex data may require more storage capacity than ordinary raster storage.

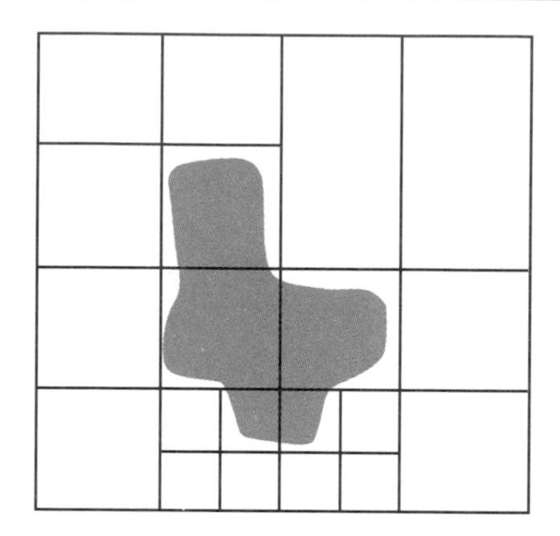

Figure 4.29
Quad-tree representation of a polygon.

4.4 Automatic Conversion between Vector and Raster Models

GIS applications sometimes require data in a form differing from that which is available. As a result, many GIS now have facilities for automatic conversion between vector and raster models. Raster data are converted to vector data through *vectorization.* The reverse process, which is just as common, is *rasterization.* In vectorization, areas containing the same cell values are converted to polygons with attribute values equivalent to the pre-conversion cell values (Figure 4.30). In the reverse process of converting polygons to cells, each cell falling within a polygon is assigned a value equal to the polygon attribute value (Figure 4.31).

Various routines are available for converting raster data to vector data, and vice versa. The former is the more complex and time consuming of the two processes, and different conversion programs can yield differing results from the same set of raster data. Normally, some information/data are lost in conversions. Consequently, converted data are less accurate than original data. These conversion processes apply specifically to data, not to the conversion of raster data from scanned maps into vector form, which is discussed further in Chapter 9.

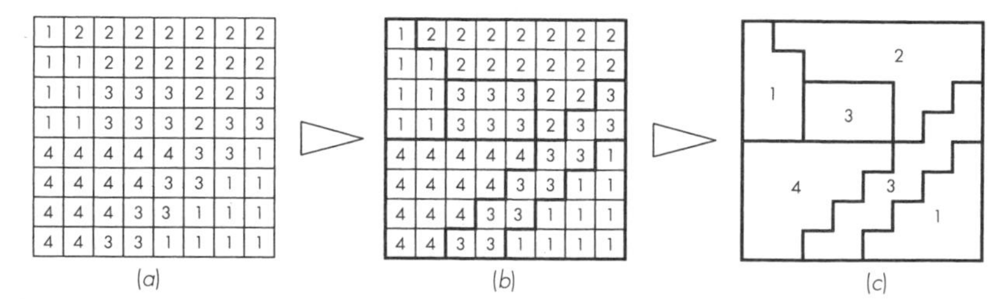

Figure 4.30
Conversion of raster data to vector data: (*a*) each raster cell is assigned an attribute value; (*b*) boundaries are set up between different attribute classes; (*c*) a polygon is created by storing *x* and *y* coordinates for the points adjacent to the boundaries.

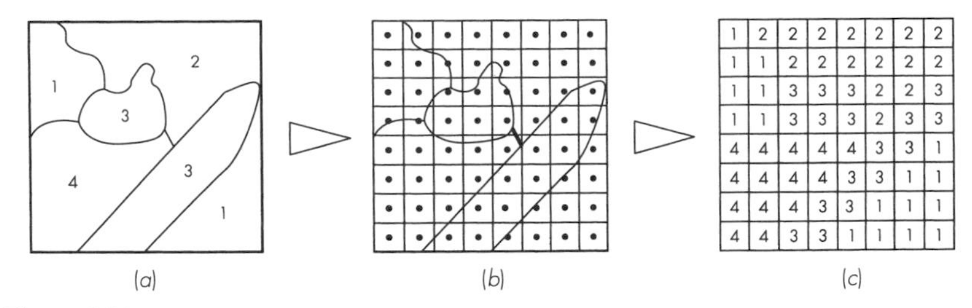

Figure 4.31
Conversion of vector data to raster data: (*a*) coded polygons; (*b*) a grid with the right cell size overlays the polygons (the polygons that contain the center of the individual cells are identified); (*c*) each cell is assigned the attribute code of the polygon to which it belongs.

4.5 Vector versus Raster Models

One of the basic decisions in GIS design involves the choice between vector and raster models, each of which has advantages and disadvantages. In the vector model, the observation units are end points and/or variable line or polygon magnitudes, whereas the raster model presupposes fixed observation areas in a grid. Otherwise, the models are identical.

As discussed previously, vector and raster data have varying ability to represent reality. It is not always easy to recognize vector data's discrete objects out in the terrain. This applies especially to phenomena with diffuse borders, such as vegetation and population density. However, many real phenomena are related to locations. Measurements are often made at points, infrastructures are often related to lines, and administrative units are frequently described in terms of defined areas of various shapes and sizes.

Raster GIS emphasizes properties: here, the basic units of observation are regular cells in a raster. Not all phenomena are related directly to such grid patterns. At the time of writing, satellite data, digital ortophoto, and digital elevation data account for the bulk of data available in raster form. In many countries, national elevation data models have been established based on a fixed grid (e.g., 100×100 m). Other types of data usually have to be reworked to a greater or lesser degree to suit rasters. The accessibility of raster data may thus be a major problem and perhaps the greatest drawback of a raster GIS in comparison with a vector GIS. However, as we have seen, there are methods for converting data from vector to raster.

A vector model, on the other hand, often requires the time-consuming and costly compilation of digital map data, while maps are integral parts of the data compiled for a raster model. Maps may be drawn for all cells as soon as they are assigned values.

	Raster	Vector
Data collection	Rapid	Slow
Data volume	Large	Small
Graphic treatment	Average	Good
Data structure	Simple	Complex
Geometrical accuracy	Low	High
Analysis in network	Poor	Good
Area analysis	Good	Average
Generalization	Simple	Complex

Figure 4.32
Some of the characteristic differences between raster and vector data.

Despite oversimplification from a functional viewpoint, vector data may be considered best suited for documentation, while raster data are more adept at showing the geographical variation of phenomena. Another simplification might be that vector data are preferable for line presentations, while raster data are superior for area presentations. Selected characteristics of the vector and raster models are compared in Figure 4.32.

To date, the vector model has been dominant in commercial GIS implementations. The raster model, on the other hand, has been used more frequently in natural resource planning and management and also in teaching because it is more easily explained and used. Many newer GIS can manipulate both vector and raster models. With dual capability, a GIS can exploit the respective advantages of both: vector data might be converted to raster data to perform overlaying or other operations more easily performed using rasters, then converted back to vector data.

4.6 Attribute Data and Computer Registers

With the advantages of easy updating, rapid search, and the flexible superimposition of data, the computerized filing of information has become commonplace in administrative work. Frequently, inaccessible, massive quantities of traditional records and files have been replaced by workstations from which very large amounts of information are rapidly accessible. Physical separation by rooms, buildings, national borders, or intervening distances is no longer a barrier to ready availability of information.

In the days when all records were kept on paper, each agency, organization, or user structured its own manual files. The result was a proliferation of parallel files, often containing nearly identical material. Computerization permits a simplification and coordination of registration efforts and can eliminate duplication and rationalize the overall filing process. In the public sector, central registers have been established as a common resource for numerous users.

Georeferencing	Attributes
ID	ID
Geometry/ Coordinates	Location
Topology	Description
	Dating

Figure 4.33
Geometric content is often limited to identifications geometry/coordinates and topology. Attribute content often comprises location (address), various descriptions of the object, and dating (age, etc.).

Most countries have various computerized registers, including registers of:

- Customers/clients
- Motor vehicles
- Properties
- Literature
- Movables
- Buildings
- Addresses
- Citizens
- Pleasure boats
- Deeds
- Environment
- Companies
- Road and signs
- Personal data

Some of these registers are important in GIS applications. Others are of less interest. In many countries, though, work is under way to make public registers available to GIS users. Upon entry, register data are selected (structured) so that registers contain uniform and limited data. As for digital map data, register data are stored using formats. There is no general pattern for register content, but usually the items registered will have identities, locations, descriptive details, time and date notations, and sometimes references to other registers (Figure 4.33). For example, typical entries in a building register may contain all the details listed in Figure 4.34.

GIS users may be obliged to compile registers when they are unavailable in computerized form, either by using GIS techniques directly or by using other systems and subsequently exporting data to GIS. In all cases, data models should be specified.

4.6.1 Coding and entering attribute data

Attribute data may be coded for several reasons in order to:

- Establish an ID code between geometry and attributes
- Conserve computer memory
- Ease input work
- Ease verification of data entered
- Simplify subsequent searches for data in databases

BUILDING REGISTER	
Identity:	• Building number
Location:	• Address • Representative coordinates
Description:	• Builder/owner • Status (under construction, in use, demolished) • Type (detached house, semi-detached house, farmhouse, garage) • Function (residence, farm, shop) • Structure (concrete, frame, brick) • Water supply (public, cistern, stream) • Sewage (public, subsoil disposal, closed system) • Heating (electric, oil, wood) • Available area
Dating:	• Year built
References:	• Reference to property register • Reference to address register

Figure 4.34
Typical contents of a building register.

The coding of geographical data is not new. Systems have been established in many fields for coding pipes, manholes, streets, properties, buildings, the names of towns, and so on. Indeed, codes have been used for many reasons, not least as file access keys or to conserve the space used on file cards.

Coding of attribute data often includes data structuring. Codes are often assigned according to a hierarchical classification system devised to ease such data operations as searching and sorting. Examples include the official codes widely used for addresses, names of towns, highways, and so on. For example, a land-use application might involve a hierarchy such as that illustrated in Figure 4.35.

Attribute data may also be tabulated, as illustrated in Figure 4.36. Tabulation and coding ease sorting, such as finding all wooded objects or all transportation objects by using their respective land use category codes. The column headings in the table are the names of attribute types. The type of data may be specified for each field, such as integer (land-use code), decimal (area), and text (name). Code tables may be compiled and used with the main table to produce more meaningful printouts from the system.

Attribute data may be entered relatively easily in most GISs, either manually via a keyboard or by importing data from an existing register; see Chapters 9 and 10 for further details. ID codes are usually entered together with the attribute data. They may also be registered or edited into compilations of attribute data which initially have no codes.

Level 1 Code	Attribute	Level 2 Code	Attribute	Level 3 Code	Attribute
100	Built-up	110	Industry	111	Light
				112	Heavy
				113	Other
		120	Transportation	121	Roads
				122	Railways
				123	Airports
				124	Parking
				125	Terminal
				126	Other
		130	....	131	....
200	Wooded	210	Coniferous	211	Fir
				212	Pine
				213	Ohter
		220	Deciduous	221	Oak
				222	Beech
				223	Other
		230	....	231	....

Figure 4.35
Land-use code hierarchy. Attributes are often coded in accordance with a hierarchical classification system.

4.6.2 Storing attribute data

Attribute data are usually most easily and expediently stored in tabular form. Each line in a table represents an object, each column an attribute. Attribute data are therefore often called *tabular data* and are normally stored in a relational database; see Chapter 8, especially Section 8.8.3.

Data on different types of objects are usually stored in separate tables, each dedicated to a single object type. In each table, line formats and lengths are identical throughout. The number of columns may be extended by combining several tables, either by using a common access key or by entering new attributes manually.

In principle, table design is independent of whether the geometrical data to which attributes refer are in the form of vector data or raster data. However, table content must be relevant to the objects, so each

ID code	Land-use code	Area (ha)	Township code
1	222	22.67	0914
2	211	1.45	0916
3	121	46.80	0923

Figure 4.36
Tabulated attribute data. Attribute data can be sorted and arranged in various tabular forms.

object or line must have a stable identity or access key. Data available in existing computerized registers are not always in convenient tabular form. As a result, conversions and roundabout methods must often be used to access such data for GIS uses; see Chapters 9 and 10.

4.7 Linking Digital Map and Register Information

Common identifiers in map data and attribute data permit moving from map data to attribute data, and vice versa. Attribute data which basically lack georeferencing may be linked to geography. As illustrated in Figure 4.37, this is possible if the attribute data that lack georeferencing have access keys in common with attribute data that have other access keys in common with map data. The connection is then from attribute data to other attribute data to map data.

This illustrates one of the distinctive capabilities of GIS. Data that initially contain no geographical information or referencing may be given geographical dimensions and may therefore be used to enhance and present data in new ways, in maps or on screen. It is not always necessary to link geometry with attributes. In some instances, the geometry can be stored directly together, with the attribute data linked to each register object. This can occur in the case of, for example, a building register with coordinates representing each building or

DIGITAL MAP DATABASE

Point no.	Theme code	Building	X coordinate	Y coordinate
1567	10	589	10,112,00	6,771,00

BUILDING REGISTER

Building no.	Lot no.	Owner	Year built	Type of building
589	44/113	Peder Ås	1952	House

LAND REGISTER

Lot no.	Landowner	Date of birth	Area (m^2)	Property address
44/113	Nils Nilsen	200519 291 89	800	Toppen 1, 1099 Fjel

PERSONAL REGISTER

Date of birth	Name	Sex	Address
200519 291 89	Nils Nilsen	male	Svingen 3, 4999 Vika

Figure 4.37
How elements of data in one register can be used as an access key to another register, thereby acting as a link between dissimilar data. A precondition is that identical information must be available in a minimum of two registers.

a register of measurement points for use in registering pollution levels. We have also mentioned that identification codes and theme codes are, in principle, attribute data though stored together with geometry.

Geographic attribute linkage

Map data may be used not just to link maps and attributes, but also geographically to link dissimilar attributes. Superimposing dissimilar data, such as geological data and vegetation data, is often hampered by a lack of commonality between the observations made in the field. That is, the observation areas listed in the respective attribute tables cannot be listed together because they refer to different sets of locations. In GIS this problem can be solved by using cartographic integration, in which overlay techniques are used to combine geometry from two dissimilar thematic maps into a single synthesized map. The synthesized map contains numerous new objects and areas, all of which are related to the two original thematic maps. Hence the objects in the synthesized map comprise the least common units between the original maps and are therefore called *integrated terrain units* (ITUs; Figure 4.38). An attribute table is associated with the ITUs. In it, the ITUs are

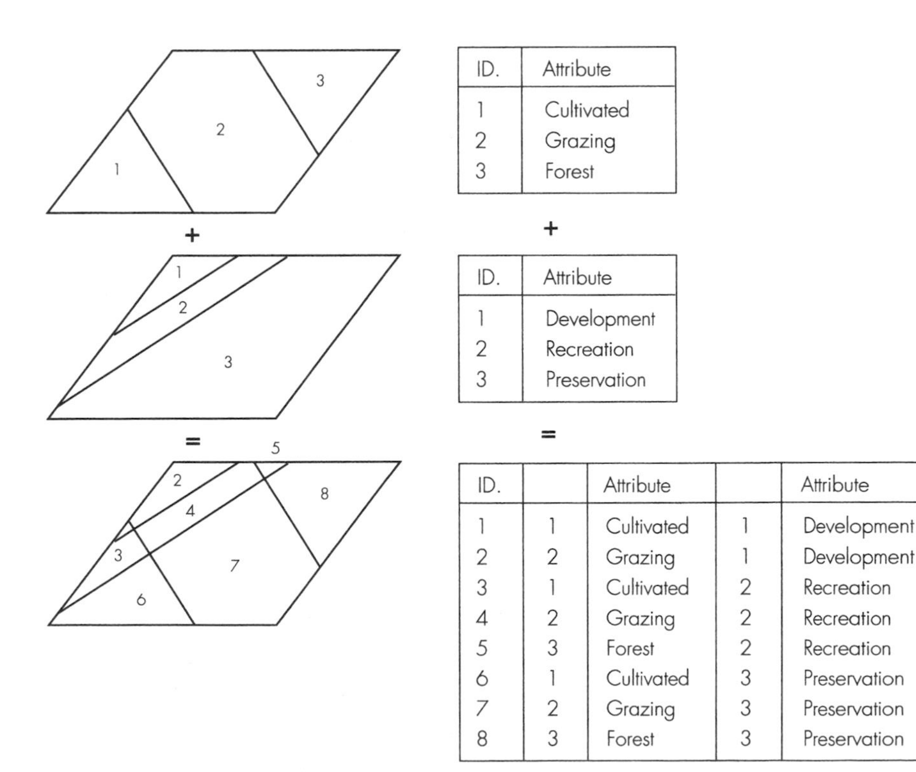

ID.	Attribute
1	Cultivated
2	Grazing
3	Forest

+

ID.	Attribute
1	Development
2	Recreation
3	Preservation

=

ID.		Attribute		Attribute
1	1	Cultivated	1	Development
2	2	Grazing	1	Development
3	1	Cultivated	2	Recreation
4	2	Grazing	2	Recreation
5	3	Forest	2	Recreation
6	1	Cultivated	3	Preservation
7	2	Grazing	3	Preservation
8	3	Forest	3	Preservation

Figure 4.38
Attribute data can be rendered comparable by superimposing geometry from dissimilar geographical units over each other to find the integrated terrain unit.

listed on the lines and the elements of the original thematic maps are in the columns. This table contains all the relevant attributes and therefore may be used in further analyses of the data.

Cartographic integration is straightforward when the areas of the original map data contain more or less homogeneous data, such as property data, land use, vegetation, and geology. Complexities arise when properties are not evenly distributed over an area. Consider, for example, a typical township with an unevenly distributed population that averages 100 persons/km^2. An ITU might locate in an uninhabited area of the township and hence misrepresent the facts of its population. In all such cases, rules must be contrived to designate how attributes shall be divided among ITUs. The ITU is also called a *basic spatial unit* (BSU) and defined as a fundamental area unit which has homogeneous properties in the context of a particular subject.

CHAPTER 5

Advanced Data Models

5.1 Surface Representation

The models discussed in Chapter 4 describe limited parts of two-dimensional real world. Several other data models used in GIS can extend the real world to include the terrain surface, the time factor, and movable objects. The digital representation of a terrain surface is called either a *digital terrain model* (DTM) or a *digital elevation model* (DEM). In GIS disciplines, the term DTM is often used not just for the model itself, but also for the software used to manipulate the relevant data. Only the model is discussed in this chapter; data manipulation is discussed in detail in Chapter 15.

The terrain surface can be described as comprising two basically different elements. The random (stochastic) elements are the continuous surfaces with continuously varying relief. It would take an endless number of points to describe exactly the random terrain shapes, but these can be described in practice with a network of points. It is usual to use a network that creates sloping triangles or regular quadrants.

The systematic part of the terrain surface is characterized either by sharp cracks in the terrain, such as the top or bottom of a road cut, or by characteristic points such as spot depression and spot height. The systematic part is thus best represented by lines and typical single points. Prominent terrain features can be verbally described using many terms, such as smooth slope, cliff, saddle, and so on. Geometry, however, has only three terms: point, line, and area. One cannot describe continuously varying terrain using only three discrete variables, so all descriptions are necessarily approximations of reality.

Essentially, DTMs comprise various arrangements of individual points in $x-y-z$ coordinates (Figure 5.1). Often, their purpose is to compute new spot heights from the originals. A terrain model can be realized by linking height as an attribute to each point (x,y). This type of elevation model can only describe a surface and cannot handle more z values to the same point. Therefore, the term *2.5-dimensional* is often used to describe the DTM dimension. This model is most suited to visualization. In a three-dimensional elevation model, elevation is an integral part of position (x,y,z) and the model can handle several z

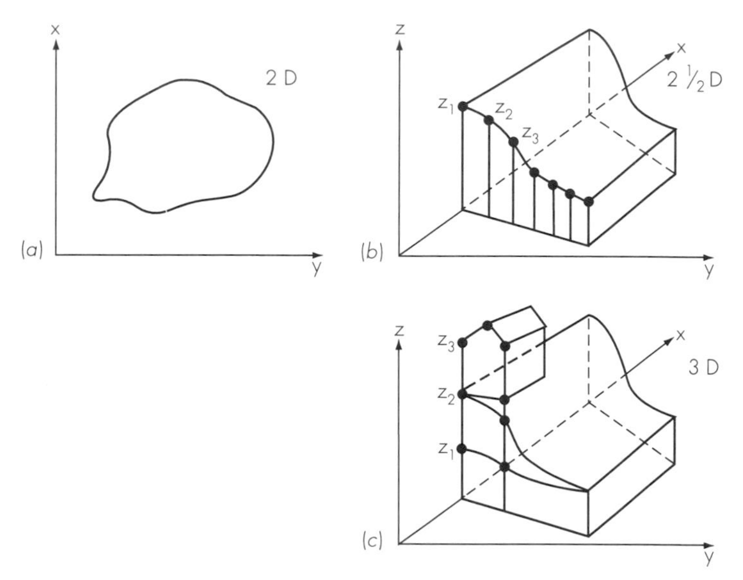

Figure 5.1
With 2.5-dimensional models (*b*), it is only possible to store one height value for each point (*x*, *y*): that is, it is only the terrain surface itself that can be described. In real three-dimensional models (c) several height values can be stored on the same point (*x*,*y*) and the relationship above and below the terrain surface can be described in the same model. Compare with the two-dimensional model (*a*).

values for the same *x*,*y* pair. That is, it can handle different geological layers, roof heights on buildings, roads that cross each other, together with the terrain surface. A three-dimensional model is also suited to volume calculations.

The *z* value of a new point is calculated by interpolation from the *z* value to the closest existing points. If the points are stored in an unstructured way, *all* registered points will have to be searched to be able to calculate the *z* values to a new point. This can be very time consuming even for a powerful computer. It is therefore usual to use data structures which also describe the contiguity between the points. This is achieved by using data structures based on single points in a raster (grid) or triangles covering a surface.

Grid model

A systematic grid, or raster, of spot heights at fixed mutual spaces is often used to describe terrain (Figure 5.2). Elevation is assumed constant within each cell of the grid, so small cells detail terrain more accurately than large cells. The size of cells is constant in a model, so areas with a greater variation of terrain may be described less accurately than those with less variation. The grid model is most suitable for describing random variations in the terrain, while the systematic linear structures can easily disappear or be deformed. One possible solution

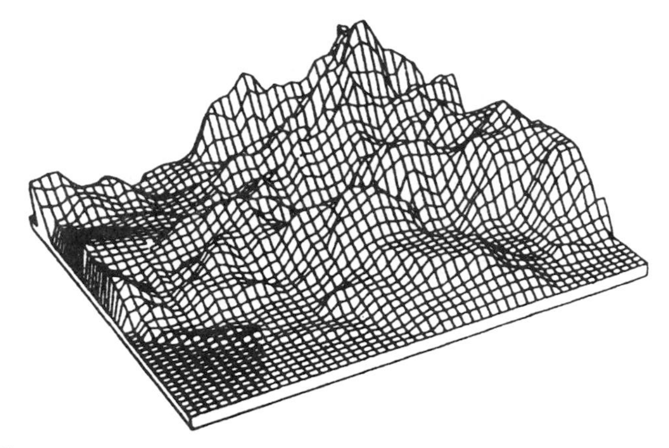

Figure 5.2
Digital terrain models describe the terrain numerically in the form of *x*, *y*, and *z* coordinates. Graphic presentation can be either in the form of a grid (as in the figure) or as profiles.

can be to store the data as individual points and generate grids of varying density as required. It is debatable whether the grid model represents samples on a grid and can therefore be called a point model, or represents an average over raster cells. In the United States the former seems to be the most usual.

Elevation values are stored in a matrix and the contiguity between points is thus expressed through the column and line number. When the data points are dispersed, the averages of the elevations of those closest to grid points, within a given circle or square, are assigned to the grid points with inverse weighting in proportion to the intervening distances involved. When the data relate to profiles or contours, grid point elevations are interpolated from the elevations at the intersections of the original data lines and the lines of the grid.

Terrain may also be described in terms of chosen or arbitrarily selected individual points (i.e., a *point cloud;* Figure 5.3). In principle, the

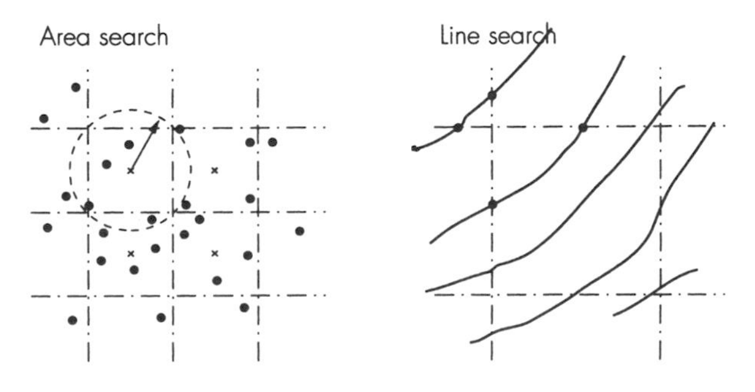

Figure 5.3
The DTM-based grid model is created either from an area search in a point cloud or from intersections with contour lines.

characteristics of the terrain between points are unknown, so it follows that point densities should be greatest in areas where terrain features vary the most. Only elevations are stored for the points of a regular grid, but both point coordinates and elevations must be stored for point clouds. So for given terrain coverage, the amount of memory storage required for the two point arrangements differs. For describing abrupt terrain variations, such as the top and bottom of a road cut, point models are inferior.

TIN model

An area model is an array of triangular areas with their corners stationed at selected points of most importance, for which the elevations are known. The inclination of the terrain is assumed to be constant within each triangle. The area of the triangles may vary, with the smallest representing those areas in which the terrain varies the most. The resulting model is called the *triangulated irregular network* (TIN; Figure 5.4).

Insofar as possible, small equilateral triangles are preferable. To construct a TIN, all measured points are built and the model thus represents lines of fracture, single points, and random variations in the terrain. The points are established by triangulation and in such a way that no other points are located within each triangle's converted circle. In the TIN model, the $x-y-z$ coordinates of all points, as well as the triangle attributes of inclination and direction, are stored. The triangles are stored in a topological data storage structure comprising polygons and nodes, thereby preserving the triangle's contiguity.

Various algorithms are available for selecting representative points from the basis data (grid, contours, point clouds) and for creating appropriate triangles. Should the basis data be available in grid form, it is possible to move a window (one point and its eight contiguous

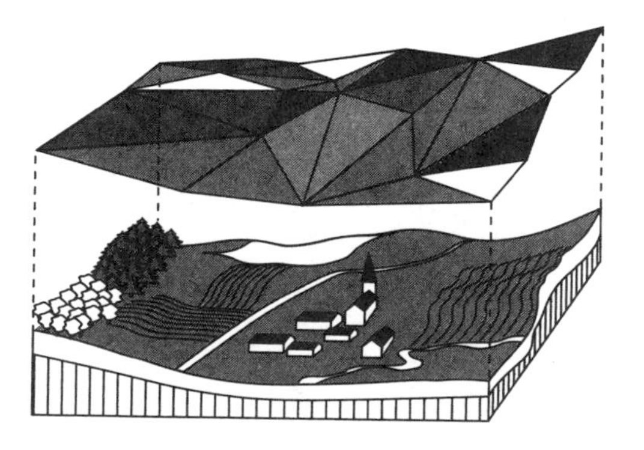

Figure 5.4
The surface of the terrain can be described as inclined triangular areas, starting from points selected for their importance. This method of describing terrain is called the *triangular irregular network* (TIN).

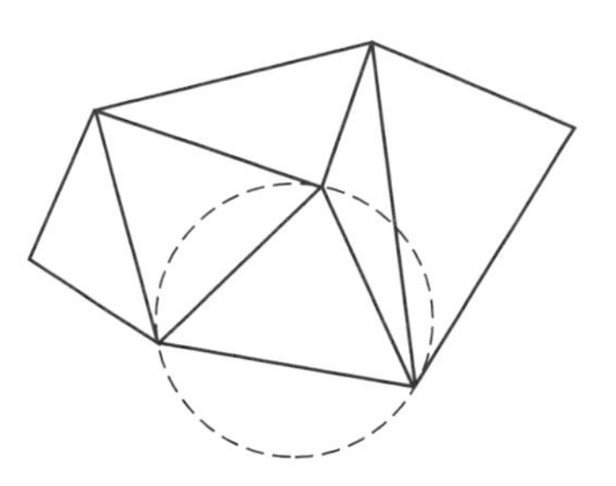

Figure 5.5
Delaunay triangulation is a method used to fit triangles in a point cloud. The circle described ensures that the triangles have good geometry with least possible variation in the page lengths.

points) step by step over the data and remove the points that are least characteristic in relation to their contiguous points. Triangles can be formed by laying circles through three points and testing whether there are other points within the circle. If other points are not available, a new triangle will be formed. This method produces triangles with a low variance in length; it is known as the *Delaunay triangle* (Figure 5.5).

Compared to the grid model, the TIN model (Figure 5.6) is cumbersome to establish but more efficient to store because areas of terrain with little detail are described with fewer data than similar areas with greater variation. However, the TIN model normally requires considerably larger storage capacity than the grid model. TIN models are good for describing terrain because the sharp breaks of slope between uniform-slope facets fit certain types of terrain well.

Other models

Isolines—continuous lines connecting points of the same elevation—may represent terrain in much the same way as contour lines depict terrain on conventional maps (Figure 5.7). The point densities should be greatest in those areas in which the terrain varies the most. As the intervening terrain between successive isolines is unknown, smaller elevation increments between isolines result in greater accuracy of description.

Although an isoline model may be compiled readily, amending its data is involved. In practice, the methods used are determined by the data compilations. Parallel profile lines connecting points of varying elevation may be used to describe terrain. The density of points along profile lines should be increased in areas where there are major variations in the terrain. In principle, the terrain between successive profile lines is unknown, so the closer the lines, the greater the accuracy of description.

A combination of isolines and individual points may also be used to describe terrain, especially when specifying such point features as

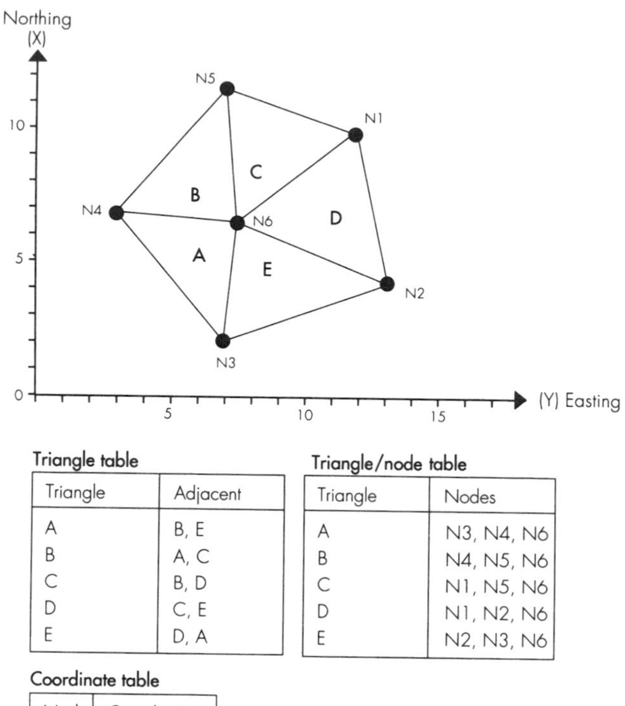

Triangle table

Triangle	Adjacent
A	B, E
B	A, C
C	B, D
D	C, E
E	D, A

Triangle/node table

Triangle	Nodes
A	N3, N4, N6
B	N4, N5, N6
C	N1, N5, N6
D	N1, N2, N6
E	N2, N3, N6

Coordinate table

Node	Coordinates
N1	X_1, Y_1, Z_1
N2	X_2, Y_2, Z_2
N3	X_3, Y_3, Z_3
N4	X_4, Y_4, Z_4
N5	X_5, Y_5, Z_5
N6	X_6, Y_6, Z_6

Figure 5.6
In the TIN model, the triangles are stored in a topological structure.

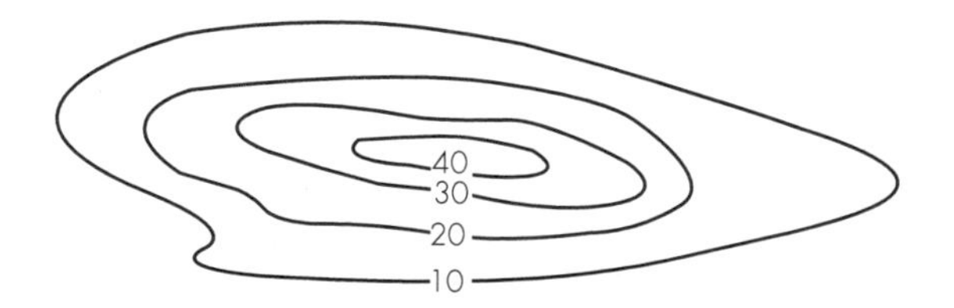

Figure 5.7
Lines that link points with the same terrain height can be used to represent the terrain surface. This corresponds to traditional elevation contours. This structure is poorly suited to the calculation of elevation values for new points.

peaks and valley floors, or vital terrain lines, such as the top and bottom of a fill. As mentioned previously, the grid and TIN models are best suited for calculation of the z value of new points.

Practical observations

Grid models and TIN models are always less accurate than the original data from which they are derived. In some GIS/DTM, therefore, original data are stored in point clouds. Models with accuracy suiting specific tasks are compiled from these as required. For example, a grid model for estimating road construction excavations might be more accurate than one intended to detail the general vegetation of the region.

GIS based on vector models can easily manipulate elevations stored as spaghetti data, but can handle elevation grid data less easily. Only a GIS based on a topological model can manipulate TIN data. Terrain data are usually compiled from survey point elevations, from isolines digitized from existing maps, or from photogrammetric point and/or line (contour or profile) registration. Normally, photogrammetric point registration is 30% more accurate than line registration.

Various interpolation programs compute new z coordinates for new $x-y$ coordinates, thus facilitating specific computations such as estimating cut-and-fill volumes in road planning, or assessing reservoir volumes for hydroelectric plants. Various GIS may also be implemented with functions for calculating slopes, drawing in perspective in order to visualize the impact of works, computing the runoff, or perhaps, draping in colors to enhance visualization; see Chapter 15 for a more detailed discussion.

The ways in which data are represented and stored are decisive in determining the type and efficiency of the computations. For example, digital isolines are ill suited to calculating slopes and relief shadowing; draping and runoff calculations are most expediently performed using a TIN model; the TIN model is ill suited to visualization without draping; and so on.

The methods used to describe terrain surfaces may also be used to describe other continuously varying phenomena. Thus population density, prevailing temperatures, or biomass production can be described quite simply by assigning the parameter involved to the z axis of all the observation points located in $x-y$ coordinates (Figure 5.8).

Accuracy

High accuracy is required in all terrain models to be used for engineering purposes. The accuracy of terrain descriptions is determined primarily by random variations in the terrain, spreading of measured points, distance between measured points, and accuracy of points by the method of generating the model grid and triangular surfaces; and by the method used to interpolate between points in the model. In DTM, errors in the x and y coordinates result in errors in the elevation.

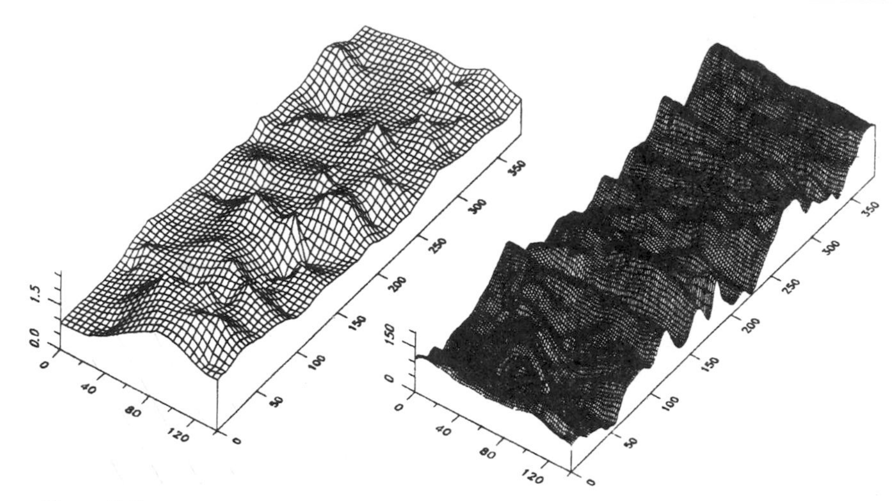

Figure 5.8
By placing various themes on the *z* axis, the distribution of varying amounts over the terrain can be illustrated, such as (*a*) differences in elevation in an area; (*b*) biomass production in the same area.

For a grid model, the following degrees of accuracy are typical:

Source	Accuracy in elevation and ground plan
Surveying	± 5 cm
Photogrammetric data from 1:6000 images	± 20 to 30 cm
Digitized 1:1000 maps	± 50 cm

In those models in which cells and profiles are recreated from a point cloud with each computation, accuracy depends on the cell or profile density. Profiles must be closer to each other to represent more rapid terrain variations, but greater profile densities naturally call for the processing of greater amounts of data.

5.2 Three-dimensional Objects

All physical phenomena are located in space; thus the world can be described as three-dimensional (Figure 5.9). A complete data model should be based on these three dimensions: ground, position, and elevation. This applies not only to terrain surfaces but also to buildings, borders, addresses, accidents, and all manner of data; a complete data model should manipulate georeferenced data in three dimensions.

The realization of three-dimensional objects in GIS still has theoretical and practical limitations (Figure 5.10). Topological data are needed for such procedures as color filling (and photo texture) of vertical ar-

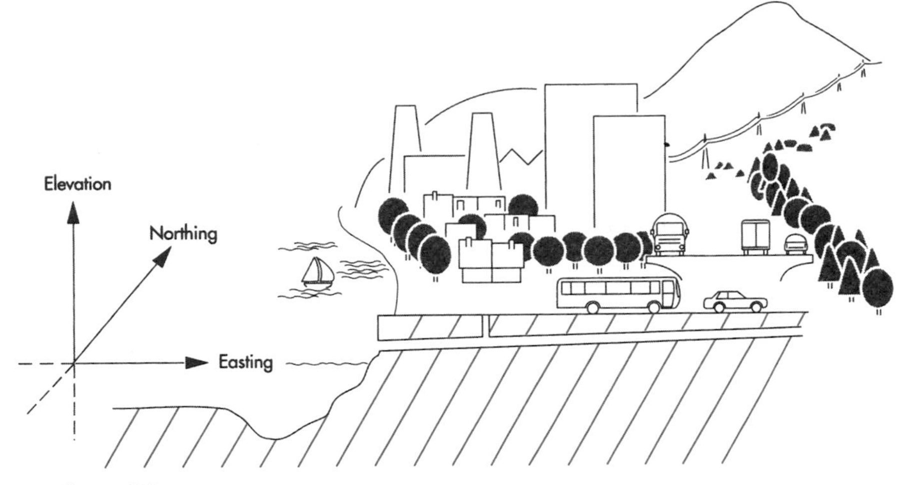

Figure 5.9
The world is three-dimensional, with phenomena having a location and a surface area in both elevation and ground plane.

eas and for data search. It is a theoretical and mathematical problem to establish topology for three-dimensional objects. The topology will be very complex and present opportunities to establish objects which are illegal (objects that cannot be oriented) (Balstrøm et al. 1994). It is also difficult to establish satisfactory routines for checking whether declared data exist in three-dimensional topology.

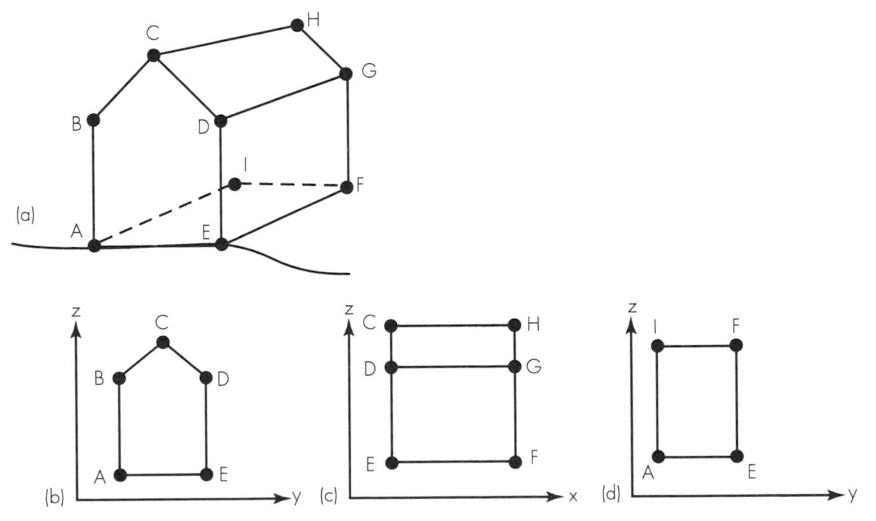

Figure 5.10
Establishing three-dimensional topology is demanding. Large volumes of data are required, which can create theoretical problems (nonorientable surfaces) that are difficult to solve. Traditional GIS operates only with two-dimensional topology, whereas in three dimensions you have to relate to three levels (*y,z*; *x,z*; and *x,y*).

Specification of all objects in three dimensions can easily increase the amount of data collected beyond that which is needed; the amount of data will in any case be considerably larger than with the use of two and 2.5 dimensions. It may also influence the techniques used to collect data. The collection of photogrammetric data provides free elevation data in addition to the northing and easting data of the ground plane. When existing maps are digitized, however, elevation data must be entered manually (and sometimes inaccurately, as exact elevations may not be available for all objects). Today, we see that users have an increasing need for digital three-dimensional map data. This applies in particular for applications connected to urban areas. At present, commercial GIS is still only capable of handling two-dimensional topology. Even though relational databases support binary large objects (BLOBs) for storage of texture (building facades or similar), this type of data cannot be searched for as with other data. Therefore, new database techniques have to be developed in addition to relational databases and the special databases which are generally used today. Models can be constructed, but should in this case be carried out in systems for computer-aided design (CAD).

5.3 Representation of Time

In the real world, time is a factor that concerns us deeply. It was studied by ancient Greek philosophers such as Plato and Aristotle, and also plays an important role in Einstein's theory of relativity. Most things change with time. The same applies to geographical data. Take, for example, property data. New properties are continuously under development because of the division of existing properties; therefore, new geometry occurs. However, most changes are related to title ownership, with the resulting changes in attribute values. One example of an area in which extreme changes occur is that of the transport sector with vehicles continuously changing position and where frequent on- and off-loading of goods change the attribute values that are linked to the vehicle.

In addition to the fact that both the geometry and attributes of objects change over time, the reference system can also be changed. In Chapter 6 it is discussed how this occurs when roads are changed in a distance-based reference system, but it also happens when the geometry of administrative reference units is changed (e.g., with changes in municipal borders and postal zones). In addition, topological changes often occur as a consequence of geometrical changes.

In practice, it is difficult to create a data model that is capable of incorporating all imaginable changes. The time factor is relatively often neglected in GIS, probably because we are more concerned with documenting our current situation than we are with historical changes. Databases are updated continuously, so unless special measures are

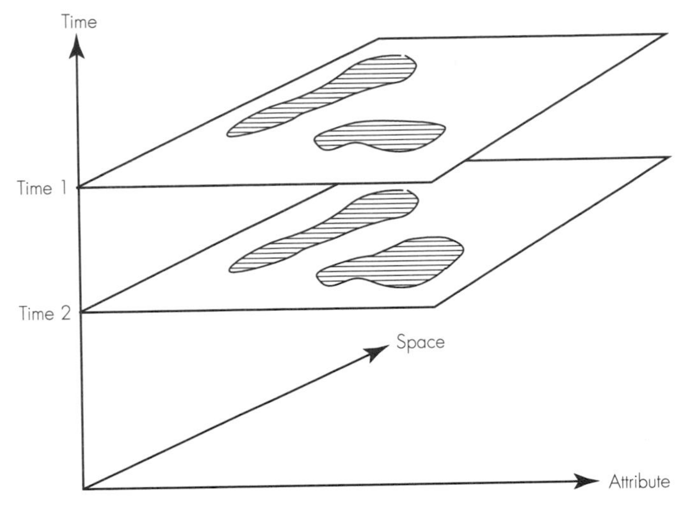

Figure 5.11
In GIS it is normal to realize time in a model with time layers. Reality is thus simplified to allow changes to occur spasmodically and not continuously.

taken, the time picture will be fleeting. The use of analog technology can document time changes via different printed versions of maps. To the extent that maps have been archived, it is also possible to preserve special versions of history, although they might be unsystematic. If the time factor is not incorporated in the data model for GIS, we run the risk of losing important historical data.

The most usual way of handling the time factor in GIS is to look on time as an attribute to the objects in the same way as for other attributes (Figure 5.11). This view corresponds with the usual way of presenting geographical data (i.e., as two-dimensional time overlays) and can thus be realized for both vector and raster data. However, this simple approach will not necessarily create a logical connection between the various time layers. It can therefore be extremely difficult to assess what the situation might have been between two time layers.

Possible practical solutions will therefore be:

1. The attributes of the objects will be changed.
   a. Historical data are stored only in fixed or variable time intervals (e.g., every second year, every fifth year, etc.). The attribute values between these intervals may have to be interpolated.
   b. All changes are registered and stored for selected types of objects (i.e., historical data have to be preserved by date stamping).

The type of object will decide whether all changes should be registered or whether time intervals are sufficient. For example, time intervals would be most suited for registration of changes in population density, while changes in the form of new property title holders should have complete registration.

2. The geometry of objects is changed.
   a. Historical data are stored only in certain time intervals (e.g., every second or fifth year). The geometry between these intervals may have to be calculated. As in the case of attribute values, the object type will decide where it is possible to interpolate new geometry.
   b. All changes are registered and stored for selected object types (i.e., all historical data have to be preserved by date stamping).

Changes in geometry can lead to changes in the relationships between objects and the resulting changes in topology, which also have to be preserved.

Registration in time intervals is more of a practical solution, where the main aim is to maintain rapid access to data and limited data volumes rather than realization of a basic data model, of which time is an integral part (Figure 5.12). Time models specify how changes in terrain over time can be preserved. Updating comprises the routines to be followed for registration of changes and the speed at which changes can be loaded into the database(s) and, to a certain extent, can be viewed independent of the models.

Several prototype systems and even a few commercial systems are available which provide some temporal support (Skjellaug 1996), but it would seem that there is a fair amount of research and development work remaining before a complete data model for the time factor can be realized in GIS. Even though a data model can be created that can handle the time factor satisfactorily, we are, in practice, reliant on changes being registered and stored in the database within a reasonable time. The updating and organizational procedures around this are dealt with later in the book.

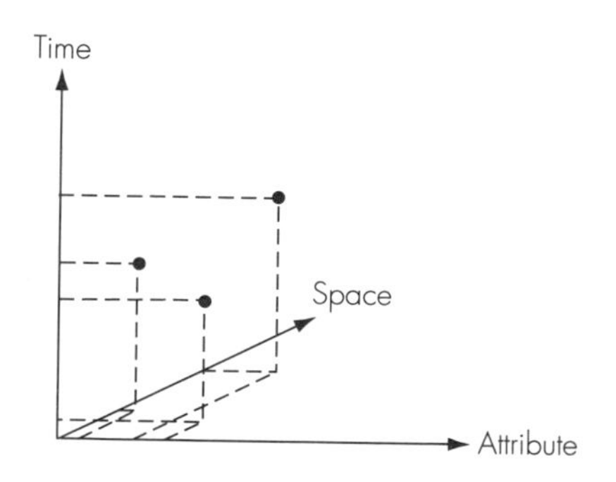

Figure 5.12
In reality the changes occur more or less by chance and are thus scattered randomly in time, attribute, and space. However, it is difficult to create suitable models that can be realized in GIS.

5.4 Models for Movable Objects

A considerable part of the real world consists of movable objects: vehicles on a road network that carry passengers or goods or water running over the terrain surface. One should also be able to realize this aspect of the real world in GIS. Special models have therefore been developed to handle these conditions.

5.4.1 Network model

The network model comprises road systems, power grids, water supply, sewerage systems, and the like, all of which transport movable resources. The most usual type of network for which GIS models are developed is road systems; the following description therefore refers to road systems. For most purposes, reality can be simplified to a model that can handle two different situations:

1. Displacement of resources or objects from one place to another
2. Allocation of resources or objects from or to a center.

As with other GIS data models, this model is based on geometry and attributes (Figure 5.13). The geometry in the network is represented by lines consisting of connected lines of vector data. The geometry of a road system will be represented by the center line of the road. This model assumes topological data built up of links and nodes. Every link and every node in a network must have a unique identity. The attributes are tied to links and nodes in a linear system through these IDs and are intended to describe the total accessibility of the system. We can thus state that the model is based on three basic relationships:

1. Continuous, connected networks
2. Rules for displacements in a network
3. The possibility of attribute value accumulations due to displacements

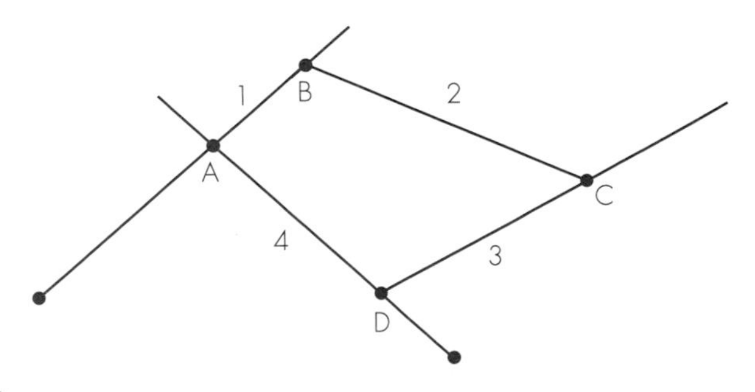

Figure 5.13
A road network can be split up into links and nodes, where a link is a line without logical intermediate intersection and a node is an intersection point where two or more links meet, or a start/end point.

LINK	DISTANCE	RESTRICTION
1	350 (m)	40 (km/h)
2	500	60
3	400	50
4	450	40

NODE	RESISTANCE
1	1.0 (min)
2	0.2
3	1.0
4	0.2

Figure 5.14
Attributes can be attached to links and nodes and the resulting data displayed in tables.

Attributes are connected to links and nodes and consist of two main categories. One sets conditions for transfer in the network, while the other specifies which resistance occurs at different locations in the network (Figure 5.14). Attributes that determine how objects can be moved in the network can have direction predetermined (one-way streets, closed roads, weight limits etc.). Attributes that specify resistance in the network can be speed limits, road works, peak traffic, traffic lights, bus stops, sharp curves, and so on. The accumulation of all resistance occurs along a route in the system and indicates the transfer speed from the start to the finishing point.

Once the model has been constructed, it is possible to simulate the quickest and/or the shortest route between points A and B based on the route with lowest accumulated resistance. For example, it is possible to define areas that are covered within a driving distance of 2 km and 5 km, respectively, from, say, a school and to simulate personnel transport by different means. Certain relationships need to be taken into consideration when establishing network models. Links are customarily selected to carry information, which may complicate the task because data volume and complexity increase in proportion to the number of links. The initial step, how a road is divided into links, determines the nature of the nodes. If all intersections, events, and features along the road result in nodes, the number of links may be enormous, resulting in the need for large storage capacity and slow data retrieval.

The network model represents a real-world model different from those described in Chapter 3 since it is based neither on entities/layers nor is it fields/object oriented. Information in the network model is based on links and nodes, which are not found in the entity model. Nor is it object oriented, since new nodes and links will not be established wherever attributes change; they are established only where it is practical to measure the flow of resources through the network.

5.4.2 Model for movement over surfaces

In the network model, movements are limited to the network. There are, however, some situations where access is otherwise in the terrain, such as water that flows on the surface (drainage) or the movement of

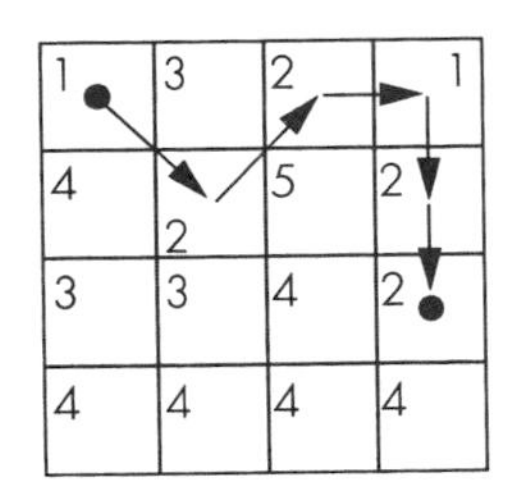

Figure 5.15
Optimize route location on raster data.

cross-country vehicles. The free flow of resources in the terrain can also be modeled by using geometry and attributes, but in this case, it is practical to use the full-coverage raster model instead of vector data. The geometry is thus represented with regular cells, and attributes are represented with coded values for each cell (Figure 5.15).

The cells are coded with attribute values which characterize the terrain in relation to the phenomenon to be studied. In the case of drainage, direction of slope can be one theme and angle of slope another. The accumulation of cell values based on certain rules will thus give the total drainage values for different parts of the terrain.

Connective models of raster data may also be used to determine travel distances, to identify areas of given shapes and sizes, and so on. The raster model for movable objects is in many ways not unlike standard raster models. It is a question of the suitable coding of cells relating to that part of the real world which the model is intended to reflect.

5.5 Combination of Models

No models, of any degree of complexity, are perfect in relation to the real world; they are only more or less successful approximations. However, they can be better in combination than singly. This can be utilized in the creation of hydrological data models by combining three dimensions, the network model, and the raster model (see Chapter 15 for a more detailed discussion).

The technique of multimedia integrates several types of models: vector, raster, 3D, time, and so on. The multimedia technique helps the user to develop complete mental models of spatial problems and gives the user the ability to navigate in a GIS-derived information space. However, there is still a good deal of research and development work to be carried out before the time factor and three dimensions are fully integrated in commercial GIS.

CHAPTER 6

Georeferences and Coordinate Systems

Before dissimilar geographical data may be used in GIS, they must be referenced in a common system. The problems associated with dissimilar geographical data are hardly new: there are numerous georeference systems that describe the real world in different ways and with varying precision. Georeferencing is defined as positioning objects in either two- or three-dimensional space. We have previously looked at several probable georeferencing systems. In this chapter we give a more systematic presentation of basic methods for georeferencing, map projections, coordinate systems, and elevation systems. There are two principal methods of georeferencing:

1. Continuous georeferencing systems
2. Discrete georeferencing systems

6.1 Continuous Georeferencing Systems

Continuous georeferencing implies continuous measurement of the position of phenomena in relation to a reference point with no abrupt changes or breaks. It involves resolution and precision. Resolution refers to the smallest increment that a digitizer can detect; it is similar to precision, which is a measure of the dispersion (in statistical terms, the standard deviation). Accuracy, however, is the extent to which an estimated value approaches a true value (i.e., the degree to which it is free from bias).

Theoretically, there is no limit to how precise continuous measurements may be. Precision depends entirely on the measuring method. Distances may, for instance, be measured to within 1 kilometer, 1 millimeter, or 1 nanometer. Many geographical phenomena, including property boundaries, manhole locations, building details, and many map details, are measured on a continuous basis.

Continuous systems include direct georeferencing, which involve:

- Coordinates on the curved surface of the Earth
- Geocentric coordinates
- Rectangular coordinates

For obvious reasons, the direct method is also known as the coordinate method. This includes three familiar terms: points, lines, and areas (polygons and volumes). The other continuous system is relative georeferencing, which includes:

- Polar coordinates
- Offset distance
- Measurement along (road) networks

6.1.1 Direct georeferencing

Datum

As we know, the earth is not a perfect sphere, but more like an ellipsoid with flattening (ellipticity) at the poles (Figure 6.1). The shape of the Earth is therefore expressed by the shape of the ellipsoid. Various sizes of ellipsoid have been used depending on how accurately it has been possible to measure the shape of the Earth. In 1924, an international standard size was established for the ellipsoid, but it has been proved recently that the Earth's radius at the equator is 251 m less than was believed in 1924.

A datum (reference level) is a model (ellipsoid) of the Earth used for geodetic calculations. The common datum for a country or area requires that there are specific coordinates for the datum origin (starting point), while this point has to have a height in relation to a given sea level. Today, many countries have adopted WGS84 (World Geodetic System 1984) as a datum; this is based on the center of the Earth's mass.

Map projection

Georeferenced data may be drawn on maps only when referenced to a plane surface, not to the curved surface of the Earth. Various projections are used to represent the curved surface of the Earth on the plane surface of a map. They are classified in three groups according to the underlying geometrical transformations involved: cylindrical (Mercator, UTM, etc.), conical, and azimuthal (Figure 6.2). It should be re-

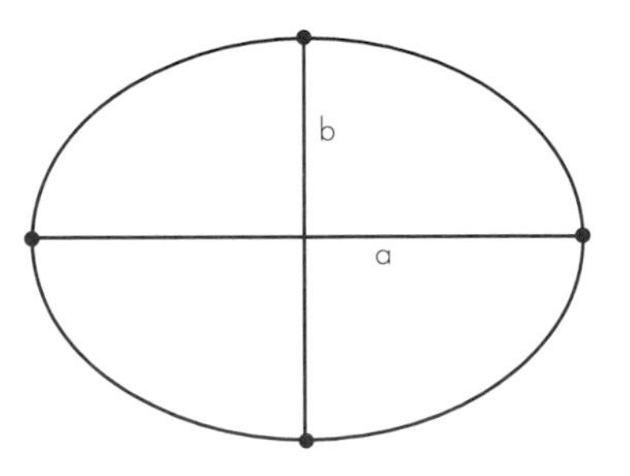

Figure 6.1
The Earth is often described using a rotation ellipsoid. The length of the long *(b)* and short *(a)* half-axes has gradually changed through history as more exact measurements of the Earth's shape have become available.

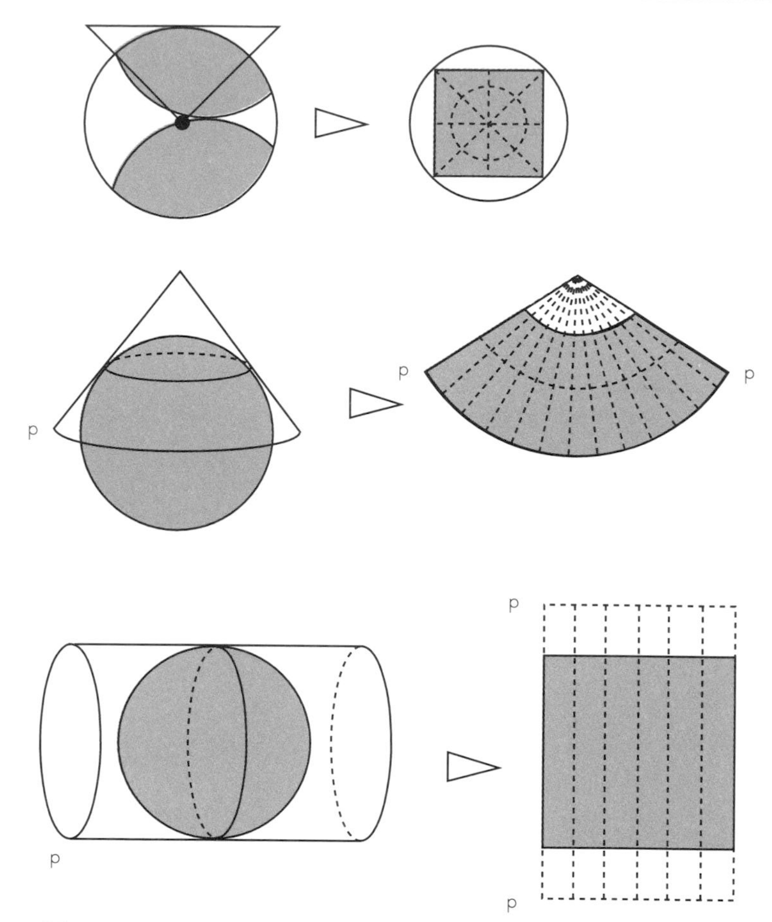

Figure 6.2
Map projections: *(a)* azimuthal projection; *(b)* conical projection; *(c)* cylindrical projection.

membered that all projections incur errors which, using the projection method, will affect distance, area, direction, or shape, and that these errors multiply with the increasing size of the area represented.

Coordinate systems

The geographical coordinates on the surface of the Earth are *latitude,* measured in degrees north or south of the equator, and *longitude,* measured in degrees east or west of Greenwich (Figure 6.3). Positions in latitude and longitude are only relative; distances and areas must be calculated using spherical geometry and the Earth's radii to the points in question. In applications, latitude and longitude are usually used in describing major land areas.

Many countries have national, even local, georeference systems of rectangular cartesian coordinates which permit locations to be given in units of length relative to a selected origin. Most systems comprise x and

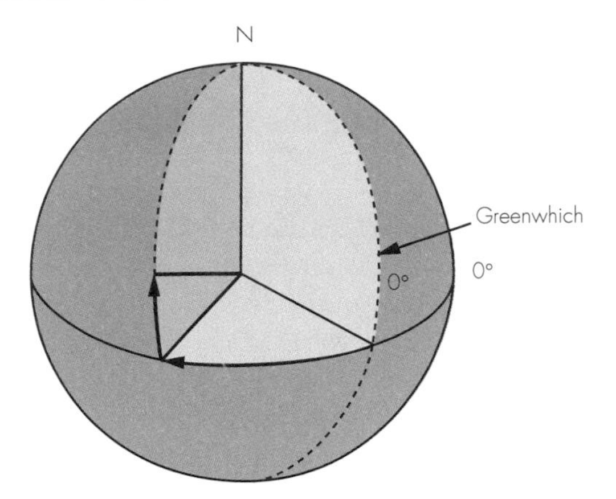

Figure 6.3
The geographical coordinate system originates from the Earth's center and defines a location in degrees in relation to the equator and the zero meridian at Greenwich.

y axes and coordinates. Coordinate system's orientations may differ, so coordinates should always be identified unambiguously, for example, in terms of compass directions (northing, easting) from the origin.

Ordinary geometrical computations may be made in rectangular coordinates, but errors caused by the map projection will always arise. This difficulty is most often overcome by constraining axis systems to relatively small areas. Larger areas are then represented by several axis systems, displaced in relation to each other. In areas where neighboring axis systems meet, data must be transformed to a common axis system to avoid working in two systems. Many GISs have facilities for transforming data from one coordinate system to another, based on common points in the two systems. When the common point is unknown, the parameters for the datum, project method, and coordinate system should be ascertained.

The best-known coordinate system is the UTM Grid (Universal Transverse Mercator Grid; Figure 6.4). UTM covers the entire surface

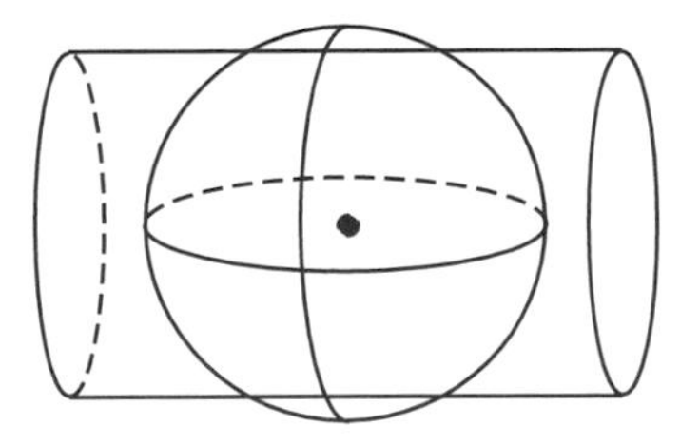

Figure 6.4
UTM is based on a cylindrical projection cutting through the globe. The zero point for the *x* axis is located on the equator, and that for the *y* axis, to the meridian through Greenwich.

of the Earth with the help of 60 zones or axes, each with a width of 6°. The mid-meridian in each zone represents the zone's north–south axis, and the zone's origo lies at the meeting point of the axes and the equator. Thus the equator represents the zone's east–west axis. To avoid negative eastern values, the origo for each zone has been given an addition of 500,000 m in eastern direction. To avoid negative northern values, the origo has been given an addition of 10,000,000 m for areas south of equator. Since there are in total 60 axis systems or zones with the same origo value, it is necessary to specify in which zone we are operating. The meridian through Greenwich is used as the basis for zone numbering in running numbers eastward from 1 to 60. In addition, for rough georeferencing the Earth's surface is divided into 20 belts starting from the equator, lettered from C to X from south to north. A grid is thus created of zones and belts (Figure 6.5).

Geocentric coordinates are based on a rectangular coordinate system with an origin at the center of the Earth. The z axis is coaxial with the axis of the Earth's rotation and is positive in the direction of the North Pole. The x axis goes through the zero meridian in Greenwich; the y axis is orthogonal to the left of the positive x axis, as shown in Figure 6.6.

The advantage of the geocentric coordinate system is that it covers the entire Earth, which is why it is used for GPS georeferences (see Chapter 10). Geocentric coordinates can be transformed to Cartesian coordinates and be used together with them. The weakness of the geocentric coordinate system is that the elevation reference system is not the same as for normal national elevation systems. This is described in more detail in Section 6.1.2.

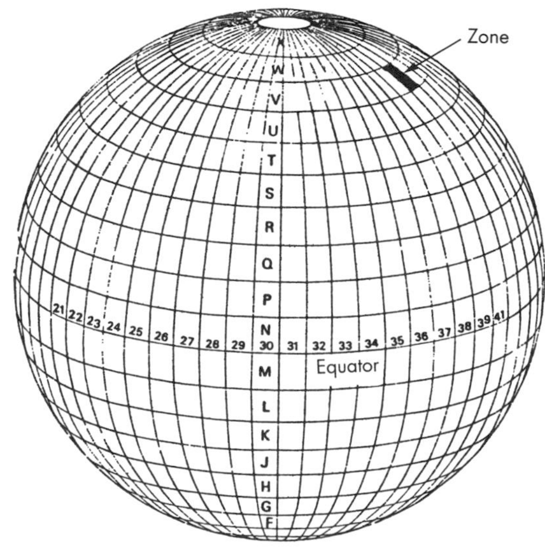

Figure 6.5
In the UTM system the Earth is also divided up into reference zones and sections.

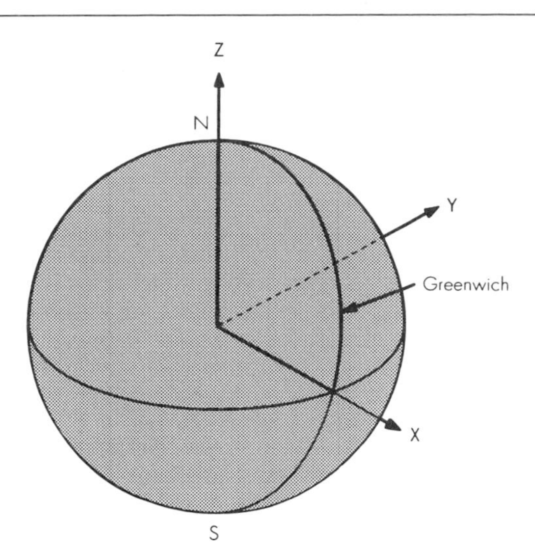

Figure 6.6
The geocentric coordinate system is rectangular, with an origin at the center of the Earth and axes as shown.

6.1.2 Elevation referencing

We established previously that reality is three-dimensional. The example above shows that this must be allowed for in practice. Most countries have established a national vertical reference system, called a *datum*. Normally, the zero point for an altitude system is based on mean sea level (Figure 6.7). This area, which goes through all points at zero altitude, is called the *geoid* (Figure 6.8). The geoid is affected by the mass of the Earth and therefore follows the Earth's contours: upward at mountains where there are large amounts of land mass and downward where there is less. Geoid can be defined as the hypothetical surface of the Earth formed from mean sea level and its continuation through the continents at the same level of gravitational potential.

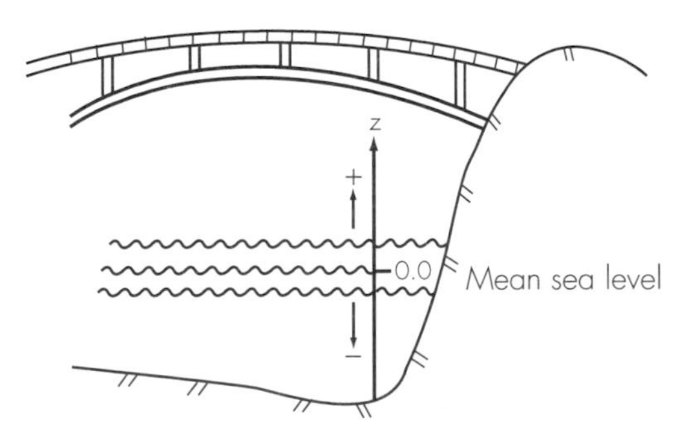

Figure 6.7
In many countries the zero point for elevation references is based on measurement of the variations in the ocean's surface.

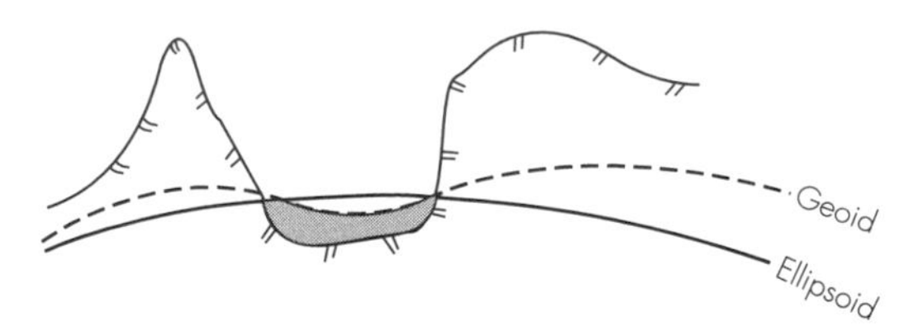

Figure 6.8
Because of variations in the Earth's mass, the geoid will not coincide with the rotation ellipsoid. National elevation references are normally linked to the geoid.

The fact that a geoid does not have an even surface is not a problem as long as normal theodolites are used which are adjusted with the help of levels affected by the same forces of gravity. However, problems arise as soon as GPS is used where heights are measured in relation to a mathematical point in the center of the Earth. Heights on the surface of the Earth, based on geocentric coordinates, can only be calculated in relation to a mathematical surface—the rotation (reference) ellipsoid—which only approximates the surface of the Earth. This ellipsoid differs from the geoid. The transformation of ellipsoidal height to geoid undulations is therefore impossible without knowing the gravity at either the actual point or the transformation parameters. Since measurements of gravity are seldom carried out, the elevation of GPS points is most frequently measured by traditional surveying instruments or by finding the transformation parameters from measuring the ellipsoid height (using GPS) at points with a known geoid height.

Few GISs can handle three-dimensional data satisfactorily. Most systems register height only as an attribute of the objects. Good topological models for three-dimensional objects are still to be developed.

6.1.3 Relative georeferencing

Relative georeferencing includes:

- Polar coordinates
- Offset distance
- Measurement along (road) networks

Polar georeferencing is based on the measurement of a distance in relation to a reference point and a direction in relation to an axis, usually the north axis (Figure 6.9). This is known as an *indirect* georeferencing method. However, if the measurements are carried out in a Cartesian coordinate system, a simple calculation can convert data from polar to Cartesian coordinates.

Another indirect method is to specify location in relation to details in the terrain. Using the offset distance method, location will be specified using either direction and distance or only distance(s) from speci-

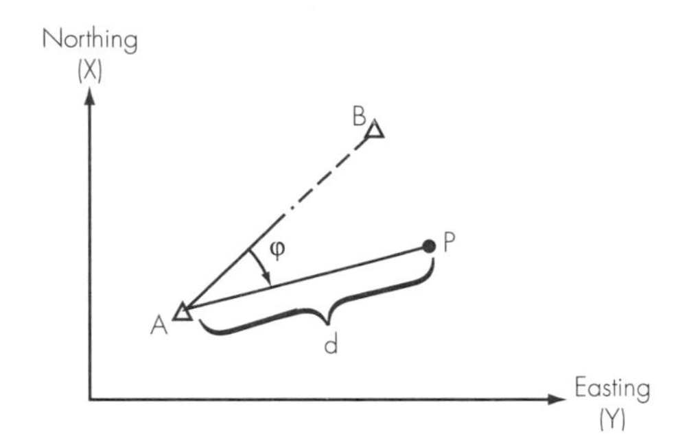

Figure 6.9
The polar location of a point is expressed as the direction and distance from other fixed points in the terrain.

fied objects in the terrain (Figure 6.10). For example, the locations of buried pipes are often given relative to terrain features such as road edges, buildings, or poles. Where several pipelines have been laid in the same trench, their location can be specified in relation to the center line of the trench. Such measurements can be converted to existing rectangular coordinate systems where the coordinates are known for the details from which they are measured. For example, these coordinates could be taken from detail maps.

Most people will have noticed the milestones that stand along main roads in many countries. The function of these milestones is to indicate the distance (in miles) from a given zero point and act as a visible point of reference for measurements in the road network (Figure 6.11). A network can have many zero points, often defined from different road junctions. Where the total layout of the road system is known, measurements (taken by measuring tape or odometer) along roads give good but indirect georeferencing. A road network system, and other types of continuous networks, such as railway and tram, in many ways provides a schematic or topological georeferencing.

Customarily, links are selected to carry information. Since data volume and complexity increase in proportion to the number of links, how a road is divided into these links determines the nature of the

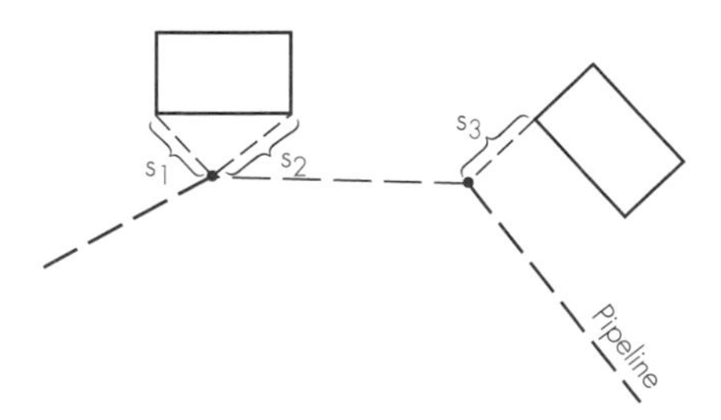

Figure 6.10
The location of a point can be expressed as the distance from physical objects in the terrain, known as the *offset distance*.

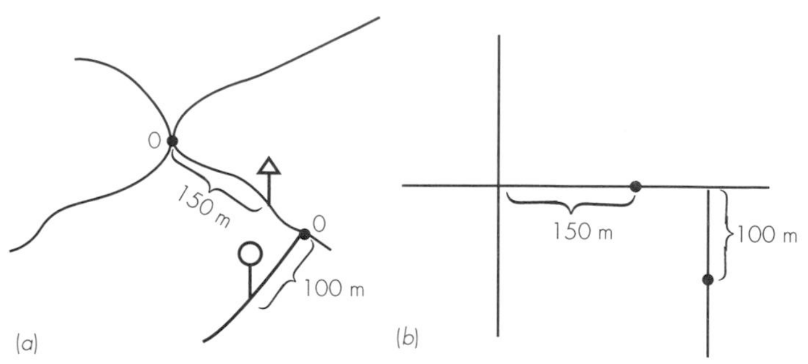

Figure 6.11
a) In a road network the reference system is often based on measurement of distances from a given intersection. b) This will only give in principle the topology of the network; the road network can therefore be presented only schematically.

nodes. If all intersections, events, and features along the road result in nodes, the number of links may be very great.

The transformation of measured distances in the road network to a Cartesian coordinate system can be beset with problems. First, there must be sufficient coordinate determined points which describe the road. Second, a distance of, for example, 899 m along an inclining road will not correspond exactly to 899 m on level ground. It is not possible to give an exact transformation without knowing a road's vertical and horizontal Cartesian coordinates.

Systems using relative georeferencing are unstable and vulnerable since the references are linked to physical objects in the terrain. Once the referencing objects are moved or removed, the reference system is also changed. For example, should a road be changed by rerouting, or straightening a curve, the distance value of the reference points along the road will also be changed.

6.2 Discrete Georeferencing Systems

In discrete georeference systems the positions of phenomena are measured relative to fixed, limited units of the surface of the Earth. With this type of system, we know that each object is located within the specified reference unit, but the location within the reference unit is unknown. Objects are linked to geography using the reference units as tags; this is known as the *tagging method.* Typical reference units include:

- Address and street codes
- Postal codes and area name
- Administrative zones and statistical units
- Grids and map sheets

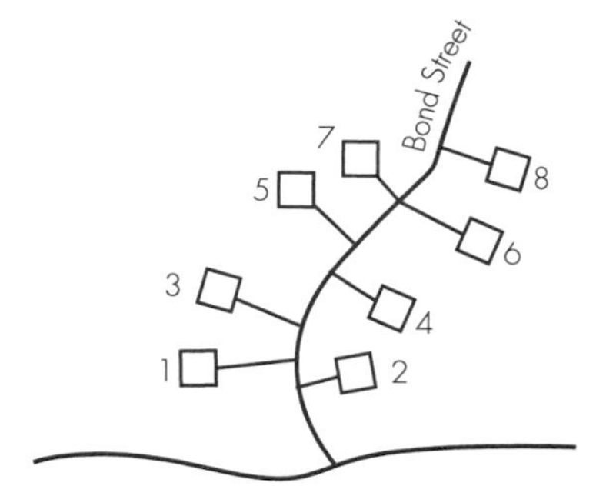

Figure 6.12
Street addresses are often used as simplified georeferencing.

The unit size determines the accuracy of registration: the smaller the units, the greater the accuracy. It is assumed that, in principle, the units are given a standard specification. In practice, however, the division between units can be relatively diffuse (e.g., it can be difficult to define exactly which geographical area is covered by an area name).

Many geographical phenomena are registered in discrete systems. Address and street codes, for example, may be used to indicate locations along roads (Figure 6.12). Reference to a whole street gives a clear discrete georeference. Roads and streets are often divided up into selected registration units, such as city blocks, or fixed or variable distances. In addition, each house has its own number. As the road is gradually split up, georeferencing becomes almost continuous, although there will still be clear geographical gaps between units.

Postal codes, statistical and administrative units, locate phenomena within geographical districts of variable shapes and sizes. This type of division into areas can be described as irregular division, as opposed to a grid that is regular division (Figure 6.13). The relevant irregular geographical districts are often arranged in a hierarchy (e.g., county, municipality, and statistical census areas).

Grids give positions in the same way as grids used in GIS. Typical examples include the indices common on city maps or on national sheet map survey series (Figure 6.14). Most countries have an official system of division and numbering of map sheets in different scales. Reference to a map sheet number will thus provide an approximate discrete georeference.

Discrete reference systems are often easy to use and are therefore expedient when accuracy is not a major concern. For example, finding the address of a new house is easier than locating the exact coordinates of its four corners. However, problems can arise whenever boundaries are moved, unit shapes and sizes are changed, or name/title changed. This means that should an administrative boundary, or a road name, be changed, all references to the old unit will be of limited

Figure 6.13
Discrete location registration may be based on a reference system of administrative or statistical geographical units.

value vis-à-vis the new georeference. Since data sets are not linked to the same geographical unit, it is difficult to compare data from different periods of time.

Discrete reference systems are often based on code indices in a hierarchy that makes it easy to group objects linked to each unit and even

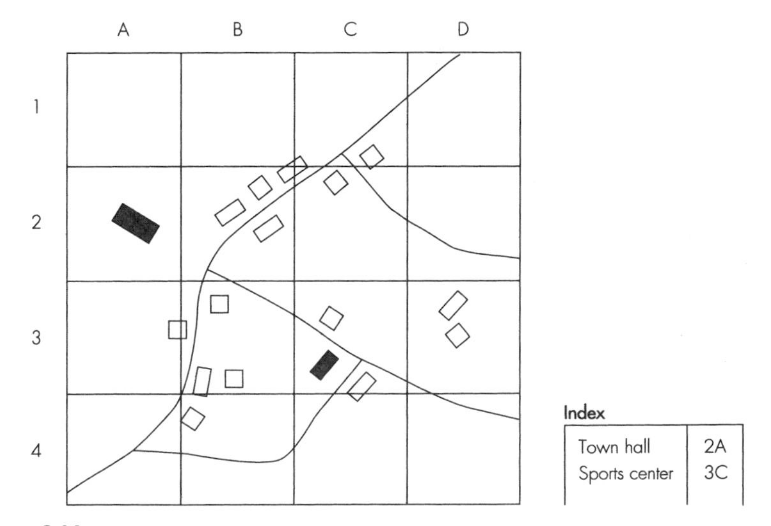

Figure 6.14
A typical example of a discrete reference system is the index system common on city maps.

combine them into larger units. Since georeferencing does not involve geometry, it is not possible to calculate directions, distances, and topology in a discrete reference system. So, save for regular grids, they are ill suited for GIS applications. Nevertheless, the data registered may be linked to rectangular coordinates through reference transformations or through cartographic fixing of reference unit area limitations (e.g., by digitalization).

CHAPTER 7

Hardware and Communications for GIS Applications

GIS tasks are many and varied, so no one system assembled from currently available computers, peripheral equipment, and software can be expected to meet all present and future needs. Moreover, the variety of computer hardware, software, quality, and prices on the market is broad and expanding. So matching a system to defined needs remains crucial in implementing GIS facilities (Figure 7.1). As described below, the main physical parts of a GIS system collect, store, manipulate, and present data.

7.1 Computers

The heart of a computer consists of one or more integrated-circuit microprocessors, or central processing units (CPUs), which contain the circuitry necessary to interpret and execute program instructions for processing data and controlling peripheral equipment. *Reduced instruction set computer architecture* (RISC) is a relatively new technological development based on high-level programming and offers a number of improvements compared to the traditional CPU. For example, it has a wider range of basic functions and can carry out more complex operations with greater power and speed. The choice between RISC and the traditional CPU is more a matter of division of labor between the processor and a low-level operating system. Users will notice no particular difference other than the speed.

Data are stored in the working memory together with the operating system and the application program, which may be entered from secondary storage as needed. The working memory is completely solid-state electronic, and therefore access to it, and retrieval from it are rapid.

Random access memory (RAM) is the most common type of working memory and is relatively expensive. Professional GIS users today normally use a PC with 64 to 128 MB of RAM. The number of operations

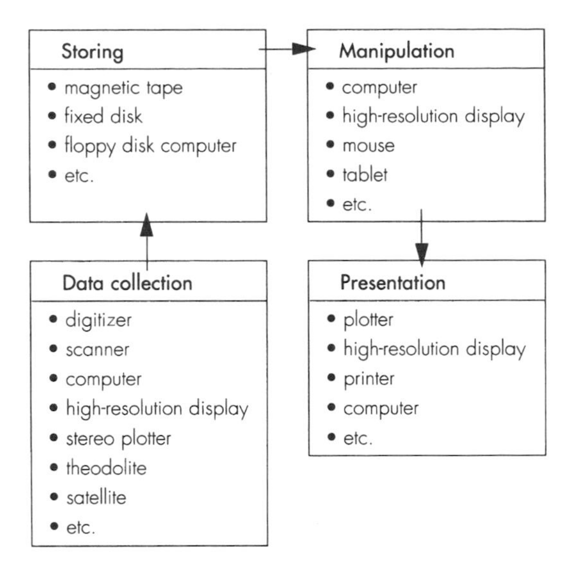

Figure 7.1
A geographical information system can comprise equipment such as that shown. It is intended to meet requirements for data collection, storage, manipulation, and presentation.

per second is expressed in megahertz. A typical operation might be to send 8, 16-, 32-, or 64-bit bytes through the processor, or from the processor to RAM. In 1997, PCs were available with a processor speed of 200 MHz. Performance has doubled every two years over the last decade and there is as yet no sign of a change in this trend. For GIS applications, RAM is normally more important than processor speed.

A byte comprises usually 8 adjacent bits, processed as a unit and capable of holding one character or symbol. *Mega-* is a prefix normally indicating a magnitude of 1 million in the decimal system. However, as shown in Figure 7.2, in the binary system used in computers it indicates a magnitude of 2 to the twentieth power.

Prefix	Meaning in decimal system *Example*: frequency (hertz)	Meaning in binary system *Example*: capacity (bytes)
kilo-	10^3 = 1,000 Hz kilohertz (kHz)	2^{10} = 1,024 bytes kilobyte (kB)
mega-	10^6 = 1,000,000 Hz megahertz (MHz)	2^{20} = 1,048,576 bytes megabyte (MB)
giga-	10^9 = 1,000,000,000 Hz gigahertz (GHz)	2^{30} = 1,073,741,824 bytes gigabyte (GB)

Figure 7.2
Common prefixes in decimal and binary systems.

Secondary storage is used to store permanently large quantities of data or program systems, such as GIS data that are not actively being processed. Secondary storage may consist of:

- Hard disks, with capacities from 240 MB (in PCs) up to several hundred gigabytes
- Floppy disks, with a capacity of 360 KB to 1.44 MB
- Magnetic tapes, with a capacity of from a few hundred megabytes to 20+ GB (tapes are usually described as tertiary storage)
- Optical disks, with a capacity of some hundred megabytes to 1 or 2 GB
- CD-ROM (compact disk—read-only memory) disks with a capacity of 650 MB

Solutions are being developed today which will quickly increase tenfold the storage capacity of CD-ROMs. In 1998, a typical PC for GIS work has a hard disk of 2 to 5 GB.

A secondary storage data search is mechanical and therefore slower than a working memory (RAM) search. Magnetic tape searches (and writing) are particularly slow, since the search has to be carried out from the beginning of the tape. They are therefore better suited to storage of backup data than for data that are in frequent use. Search time on disks is considerably shorter since data can be read from a given point on the disk. ROM is based on a special technology that allows the data to be written out only once onto the storage medium, thereby eliminating the possibility of making changes at a later stage. For this reason, CD-ROM is unsuitable for GIS data that will need to be changed frequently and is better suited for the distribution of large quantities of data. The storage capacity of magnetic storage media seems to more than double every three years. This is of great advantage for GIS applications that require the frequent handling of extremely large volumes of data.

Two types of computer are normally used in GIS: the workstation and the personal computer (PC). The PC was designed for individual use, while the workstation was intended for multiple users and to provide users with access to a larger computer. However, developments in the late 1980s and early 1990s diminished the differences between the two types; PCs are now used in networks, and workstations can be single units for individual use. The only remaining difference is that PCs use operating systems developed for PC use, while workstations use operating systems originally developed for larger computers. The new "power PCs" can run powerful operating systems such as Microsoft's Windows NT. Windows NT can run on both PCs and workstations and thus eliminates the traditional division between the two types of machine.

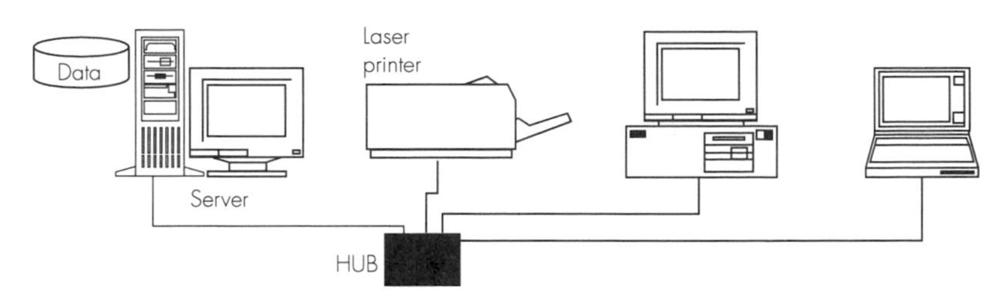

Figure 7.3
Typical local area network (LAN).

This trend is due to the explosive developments in both PC technology and markets during the late 1980s and 1990s. Most important in terms of processing speed and storage capacity (the two main measures of computer capability), the PCs of the late-1990s outperform the minicomputers of the late 1980s and the mainframes of the late 1970s, at a fraction of the cost. Second, the growing popularity of PCs has made hardware and software products for them readily available at highly competitive prices. Finally, many of the features formerly used only by minicomputers and mainframes are now available in PC systems. Among the most widespread PC systems are local area network (LAN) configurations (Figure 7.3), which facilitate multiuser access without restricting local computing ability. Some of the traditional limitations of PCs, such as the 640 KB working memory restriction imposed by the MS-DOS system, have been overcome by a variety of software artifices. Finally, computer interconnections are now common, so users need no longer be constrained by the choice of computers within one of the four major categories. Systems now available and operating may include several PCs and workstations or even a mainframe computer connected to a LAN with one or more servers.

7.2 Networks

Information is one of the most important strategic factors facilitating development. Modern political and economic systems cannot function without a continuous interchange of reliable information. Worldwide, the communication and distribution of data is a large and rapidly growing sector.

The term *network* is used here to mean data systems linked together by electronic lines of communication and software. There has been a considerable increase in the use of networks in recent years, due to the following factors:

- Exchange of data between systems (users)
- Sharing of expensive resources (equipment)

GRID

Global environmental problems are both large and complex. Until the mid-1980s, most environmental problems were addressed individually and locally. However, as the underlying problems are global, there is a need for the global approach that is now possible using computer technologies. In 1985 the United Nations Environmental Program (UNEP) established the Global Resource Information Database (GRID) as part of an overall effort to collect, manipulate, and promulgate global environmental and natural resource data. By using GIS, remote sensing, telecommunications, and other modern technologies, these data may be made readily available to professionals, government officials, decision makers, managers, and the general public. GRID's mandate is to acquire data and information relevant to prudent management of the global environment.

GRID finds it increasingly important to make maps, datasets, and other environmental information products available on the Internet/WWW. The State of the Environment of Norway is the first comprehensive national State of the Environment report made accessible on the WWW. The GRID-Arendal home page also gives hyperlinks to related institutions worldwide and will be an important tool for any Internet user searching for environmental information.

The basis of network technology is that individual units in the system comprise common resources (server) for the rest of the network (clients). The server will normally have a greater data storage capacity than the other computers in the system and is capable of handling complex program systems. Local networks are available that link up local data equipment—for example, in the same building, for sharing of data storage capacity and software, as well as printers, plotters, and so on.

HUBs are used as a node between computers enabling different machines to communicate with each other. Local networks are often called LANs (local area networks), and remote networks, often known as wide area networks (WANs), link together computers over greater distances (Figure 7.4). WANs are based on the services of national/international telecommunications companies.

Several standard local network systems are available, some of the commonest being Ethernet (based on bus topology) and token ring (based on ring topology) (Figure 7.5). A number of LANs can be linked together to form larger systems [metropolitan area networks (MANs)] with the help of *bridges.* A local network can be connected to the national/international network (a WAN) with the help of *gateways.* Satellite communication is also used to connect networks.

In a LAN the distance between units is normally limited to a few hundred meters. However, this can be increased to 3–4 km using amplifiers or bridges. Data transfer over greater distances has to be car-

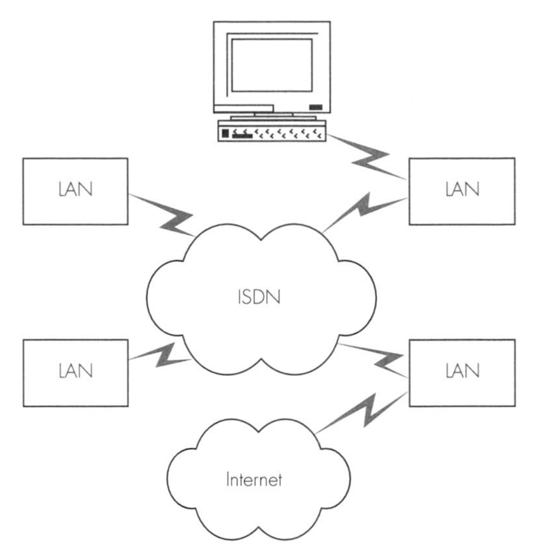

Figure 7.4
Typical wide area network (WAN).

ried out over the public telephone network system. In this way it is possible to transfer data throughout the world. A modem (modulator/demodulator) is used to connect digital data to and from an analog telephone network.

The bandwidth is the measurement of transfer capacity in a network and is measured in bits per second (bps). The transfer capacity in a local network depends on how many users are using the network at the same time. However, typical maximum values are 5 to 10 megabits per second (Mbps). With the use of an analog telephone network and a modem, the transfer speed of a WAN is often limited to 300 to 14,400 bps. With the use of digital data systems modems are not required and capacity increases immediately to 64 kilobits per second (kbps). For example, the speed on the Internet is normally from 64 kbps to

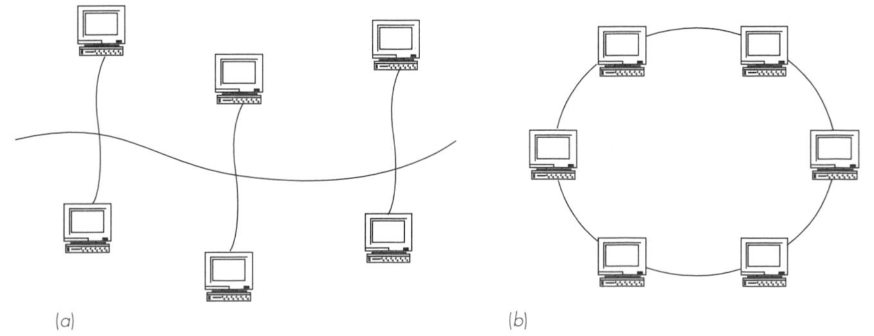

Figure 7.5
Ethernet (*a*) is based on bus topology; token ring (*b*) is based on ring topology.

1.5 Mbps. However, there are special solutions [asynchronous transfer mode (ATM)] with speeds of up to as much as 150 Mbps. Ethernet, with a capacity of as much as 100 Mbps is now being taken into use. Using fiber-optic cables, it is now possible to achieve a transfer speed of 150 Mbps, which makes multimedia—the simultaneous transfer of sound, image, and data—a reality.

A network requires control mechanisms in the form of software to establish the necessary connections, known as *access protocols*. Special network operating systems such as Novell Netware and Microsoft LAN Manager have been developed which can operate independently in relation to equipment and protocols. This type of operating system carries out configuration control, security control, error control, accounting functions, and so on.

The Integrated Service Digital Network (ISDN) has a larger bandwidth and has been developed specifically for the transfer of various digital data such as text, image, and sound over greater distances. ISDN, called a *point-to-point digital carrier,* represents a physical transfer technology. The system is based on the technology and software of national and international telecommunications companies. ISDN requires a fully digital solution within a digital telephone system.

Internet

The Internet comprises many networks (LAN, WAN, etc.) which are linked together by lines in the telecommunications network (Figure 7.6). Internet is a service based on established physical networks such as those of national and international telecommunications companies. The physical elements of the network can be a modem (for home users), ISDN, frame relay (fixed lines), Ethernet, ATM, and so on. The lowest level in the Internet is TCP/IP, a set of protocols (language) that enables machines to communicate with each other (Figure 7.7). All computers that are linked to the Internet have a unique address (IP number). The TCP protocol splits information into single packages of fewer than 1500 characters before they are sent. The IP protocol acts as an envelope with an address on each package and ensures that each package arrives at the right address. On arrival at their destination, the packages are regrouped into information with the help of the TCP protocol.

The Internet was started by the U.S. Defense Department toward the end of the 1960s; it is a service that is fully available to anyone. Users connected to the Internet can search for and retrieve information from databases throughout the world. They can also make their own data available to others and send electronic mail (E-mail) and send and receive normal data files, functional data programs, and multimedia files in the form of speech, music, images, animation, and videos. Many organizations have adopted Internet services for internal activities, known as *Intranet*. The Intranet is separated from the

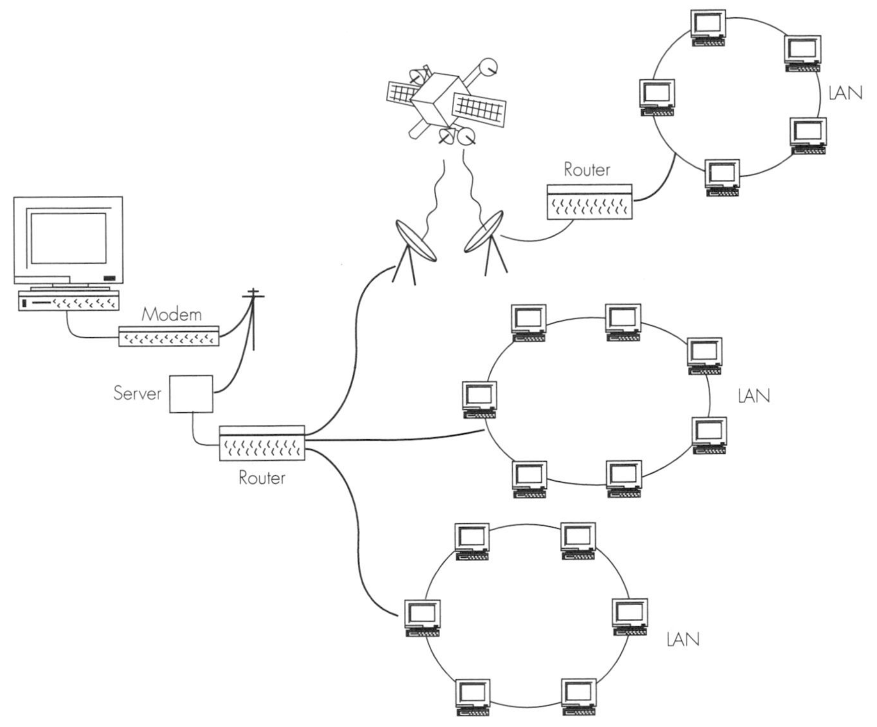

Figure 7.6
The Internet consists of many networks (LAN, WAN, etc.) linked together through *paths* in the commercial telephone network.

open Internet by a "fire wall," a combination of hardware and software aimed at securing the organization from external users.

Many Internet applications are useful in GIS. The Internet/World Wide Web (Chapter 8) can be used for distribution of information on geographical data in the form of *metadata bases*. User-friendly Internet

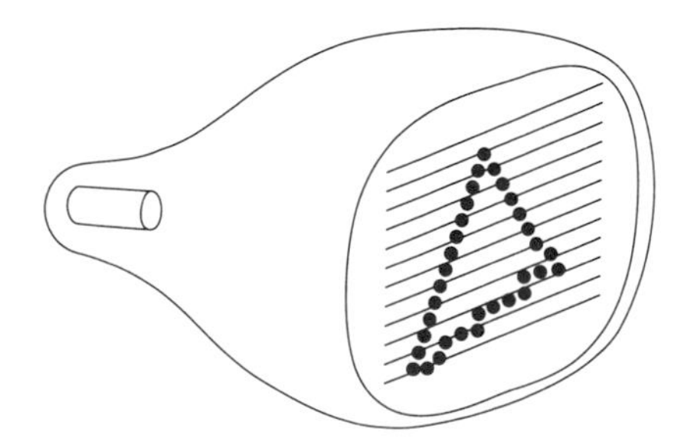

Figure 7.7
TCP divides the data into small packages for shipment. Each is then packed into IP *envelopes* and sent in different directions through the network. On arrival at the recipient, the packages are reassembled by TCP.

solutions for data search and transfer are currently under development. In the long term, the Internet may become a "GIS trading exchange," a marketplace for the purchase and sale of geographical data. In addition, simple GIS functions are currently accessible on the Internet, and it is possible to operate GIS software available on a server from terminals.

7.3 Displays

The display (monitor) of the computer is the user's primary visual communication medium in all computer work. A display consists of a screen and its associated electronic circuitry. The screen itself is either the face of a cathode ray tube (CRT), similar to those long used in television sets and radar, or one of the newer, flat arrays of semiconductor elements, such as the liquid crystal displays (LCDs) used in laptop and notebook PCs. At the time of writing, LCD technology is advancing, but the CRT remains the mainstay for almost all GIS applications because it produces more light, has higher resolution, and has a greater range of possible display colors than do LCDs. All CRTs are vacuum tubes and operate on the principle that a beam (ray) of electrons emitted by a cathode is focused and directed at an electroluminescent screen. The information displayed is refreshed periodically. Regeneration screen images are refreshed up to 60 times a second. As with TV images, the displays are based on rasters, and GIS vector map data are converted to raster data for screen presentation (Figure 7.8).

Color screens employ the three basic colors red, green, and blue (RGB) in varying intensities, which reproduce a wide range of hues. A high-resolution 19-inch (diagonal) screen designed for computer graphics may resolve 1280 points horizontally and 1024 points vertically. Screens are now available with processing functions that include vector-raster conversion, storage of vector data, and color look-up tables (LUTs) for rapid alteration of color patterns. High-definition color screens of 17 to 21 inches are used most frequently in GIS applications.

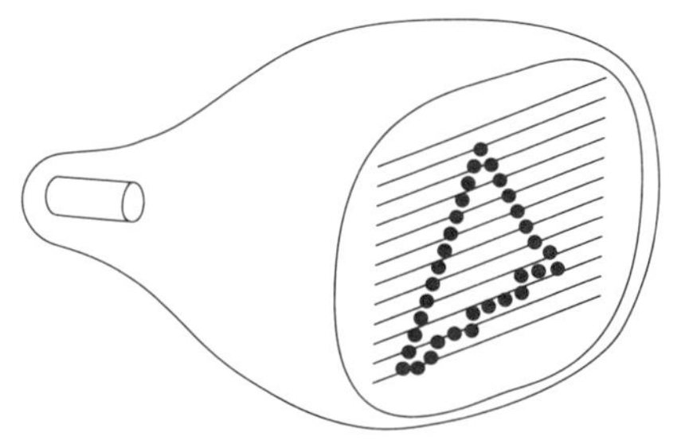

Figure 7.8
Raster regeneration screen.

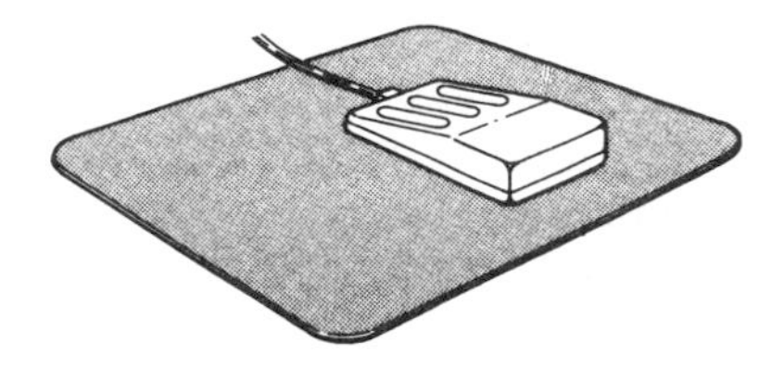

Figure 7.9
A mouse is used to communicate with application software. By moving the mouse, the cursor on the screen is moved. Program commands are activated by pointing and clicking the mouse buttons.

The on-screen position of the cursor, the movable, sometimes blinking, symbol that indicates the position of the next operator action, is controlled in many ways. On keyboards, this is usually accomplished by four arrow keys and various function keys. Provided that the computer involved is suitably equipped, the cursor position may also be controlled by either a joystick or a mouse, a palm-sized button-operated device which when moved across a flat surface produces a corresponding movement of the cursor on the screen (Figure 7.9).

Tablets are also used to control cursors and communicate with application software. A tablet is a small digitizer arranged to accept overlays containing command displays. By moving either a penlike device or a mouse, the cursor on the screen is moved and software instructions are initiated by pointing at fields in the menu.

Small penpad computers have been developed. These consist of a penlike unit, combined with a screen which is sensitive to touch and special character recognition software (Figure 7.10). These replace the

Figure 7.10
A penpad computer makes it possible to handle data in the field as they are produced.

keyboard as the input unit. The operator uses the pen to make selections from menus and dialogue boxes, and the software interprets handwritten input on the screen. The system can also receive data from GPS, total stations, and other field equipment. The penpad system can display an existing map and can be used in the field for inspection, data input, calculations, and simple editing.

Some computers support the use of a light pen, which when its tip is pointed at a feature in the on-screen display, registers its coordinates and permits selective functions to be performed, such as the retrieval of information on the objects pointed out.

7.4 Quantizers

Quantizers are devices that convert analog data, such as data on ordinary maps, into digital data, as used in GIS. The quantizers used most frequently include the:

- *digitizer:* a combination of manual positioning and electromagnetic sensing on a plane surface; historically, the first analog–digital converter for GIS applications
- *scanner:* an automatic electromechanical system that converts a picture into a raster of points (pixels)
- *video:* the electronic equivalent of motion-picture photography, often followed by rasterization similar to that employed in a scanner
- *automatic line tracer:* a function performed through software programs on raster data that have been scanned and displayed on screen

7.4.1 Digitizers

A digitizer consists of a flat nonconducting surface in which is embedded a grid of wires (Figure 7.11). Electric currents passing through the wires give rise to electromagnetic fields, which together form a pattern on the surface. A small magnifying glass fitted with crosshairs is used to locate objects on a map placed on the table. The glass is also fitted with a sensor coil, which detects the position of the crosshair center with respect to the underlying wire grid. Thus positions are entered continuously or at the command of a keystroke. A keypad with up to 25 keys connected to a microprocessor allows identifiers, thematic codes, and so on, for objects that have been located to be entered, and also facilitates interaction with the program involved.

Digitizers are available in various designs and sizes. Overall size ranges from the 27- × 27-cm tablet size up to 1.0 × 1.5 m. Map digitizing requires a relatively large surface capable of accepting the com-

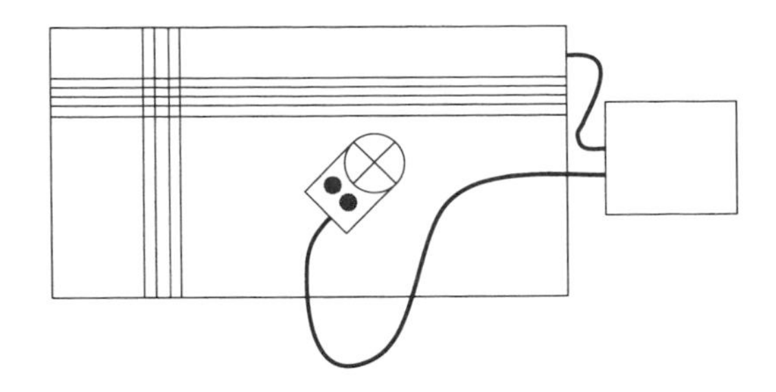

Figure 7.11
Most digitizers are basically a flat surface in which a grid of thin wires is embedded for conducting electric current. The sensor coil detects the electromagnetic field around the wires, thereby registering the position of the crosshair center.

mon 84.1- × 118.9-cm ISO A0 format (according to the international ISO 216-1975 standard). Those used for encoding map data must feature an accuracy of at least ± 0.5 mm. A common set of characteristics includes a resolution of 0.02 mm (the smallest unit of registration), an absolute accuracy of ± 0.15 mm, and a repetitive accuracy of ± 0.025 mm. Some digitizers are fitted with background illumination (as are ordinary light tables), an advantage in carrying out many digitizing tasks. A digitizing table is ideal for accurate digitization of all types of map information.

7.4.2 Scanners

Scanning is a means of encoding through conversion to a digital image comprising a regular grid of pixels rather like the array of dots that comprises a newspaper photograph. Scanners are classified into three main categories according to their basic mechanism and physical size: flat-bed, drum, and continuous-feed.

Flat-bed scanners resemble automatic drafting machines. Map image information is registered via a photosensitive pickup that rides on a track, which is mounted on a beam that itself moves in the transverse direction (Figure 7.12). Scanners can be extremely accurate and capable of high resolution. Resolution is normally measured in dots per inch (dpi). Flat-bed scanners usually have very good resolution, normally 500 to 1000 dpi. Pixel size is then as small as 25 μm (0.025 mm) and 50 μm (0.050 mm), which corresponds to a resolution of 2.5 or 5.0 cm in detailing terrain from a map with a scale of 1:1000. Accuracy is very good, ordinarily ± 0.10 mm.

Drum scanners employ the same process as flat-bed scanners except that the map to be encoded is mounted on a drum that rotates beneath a fixed track on which the pickup rides (Figure 7.13). Normally, the pickup digitizes one grid line for each drum rotation. Most drum scanners have a resolution of from 400 to 800 dpi, some even as much

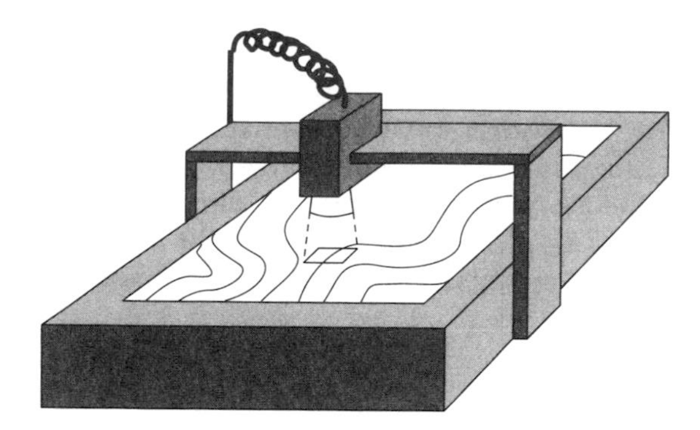

Figure 7.12
Scanners are operated by means of a photosensitive pickup being moved with great precision over the surface of a map. Photosensitive cells register the different gray tones on the map, thereby creating a raster image showing all the lines and white areas on the map.

as 1600 and 2000 dpi. Drum scanners are usually less accurate than flat-bed scanners but adequate for all normal GIS use. Degree of accuracy is normally expressed as a percentage, such as ±0.1%, which would correspond to ±0.6 mm for a map width of 60 cm. Drum scanners can scan an entire map in the course of 2 minutes.

With *continuous-feed scanners,* documents to be scanned are fed past a row of sensors as wide or wider than the maximum width of the documents scanned. The basic structure is simpler and less expensive than that of table or drum scanners, so accuracy and resolution are limited accordingly. Therefore, continuous-feed scanners are best suited to ordinary document scanning, which is why they often are called *document scanners.*

Scanners can register color, usually in a three-pass process involving transillumination of the scanned map and successive magenta, yellow, and cyan blue filtering. Some scanners convert raster data to vector data in the scanning process itself, simultaneously presenting the results on screen.

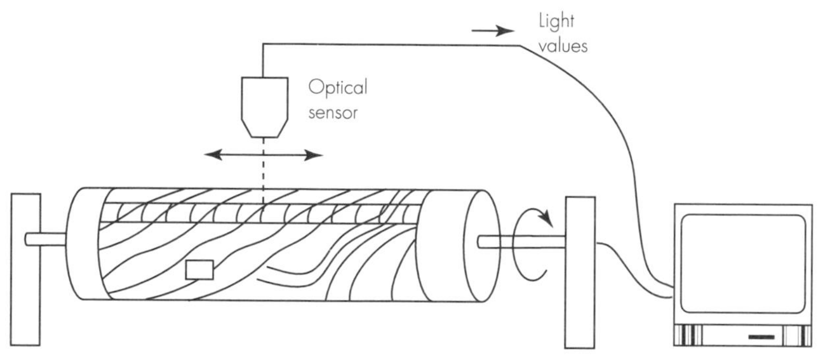

Figure 7.13
Drum scanner principle.

Scanners have the drawback that scanning senses everything on a map, including labels, coffee stains, wrinkles, and other unwanted information. Therefore, extensive editing is almost always necessary unless a map is drawn as needed, which is rarely the case. As editing is expensive, many organizations now routinely redraw maps to be scanned. Scanners are ideal for the quick and cheap production of digital background maps for later improvement by vectorization and pattern recognition. With more efficient software for pattern recognition, scanners will probably be able to handle many more demanding data conversion tasks of the future.

7.4.3 Automatic line tracing

An automatic line tracer may be used rather than manual tracing on a digitizer whenever the data call for individual lines to be scanned. The old solution involved an "intelligent" scanner, a narrow-beam laser that follows lines. A technique for replacing the physical line tracing with a software function is used today. Software programs can perform an automatic line tracing function on raster data that have been scanned and displayed on screen. To a certain extent, an operator has to control the process for following special lines (such as water pipes), for solving conflicts between crossing lines, and so on.

Raster data are transferred to vector data automatically by coordinates being computed continuously for all points on a line, as they are identified by the software. Automatic line tracing is slower than pure raster scanning but more rapid than manual tablet digitizing, and the results are lines in continuous form (vector data). By on-screen digitizing it is possible to benefit from the zoom functions on screen to inspect (view) details.

7.4.4 Video

Video cameras are also used to digitize maps, documents, and pictures. The resolution and accuracy attainable are relatively poor (typically, 512 × 512 pixels), although adequate for some applications. The use of video images in GIS is very limited at present.

7.5 Plotters and Other Output Devices

7.5.1 Pen plotters

Pen plotters are output devices that produce continuous lines. Sometimes called *vector plotters*, they represent the oldest form of plotting technology because they can only process and plot vector data. There are two types: flat-bed plotters and drum plotters. Larger plotters often contain built-in hardware and software, which simplify drawing and improve their graphic capabilities.

Flat-bed plotters are relatively accurate, typically with an absolute position accuracy of ± 0.075 mm and a repetition accuracy of ± 0.015 mm in both axial directions. The largest plotters can handle formats up to ISO A0 (84.1 × 118.9 cm). The accuracy of a typical drum plotter is between ± 0.075 and ± 0.25 mm.

Smaller plotters, for ISO A3 (29.7 × 42.0 cm) and A4 (21.0 × 29.7 cm) formats, are often hybrids of the flat-bed and drum designs, in which the medium is fixed by a drum to a bed. Accuracy is usually on the order of ± 0.25 mm. Pen plotters are generally poorly suited for surface filling and are today little used in GIS.

7.5.2 Ink-jet printers

Ink-jet printers form images by projecting droplets of various colored ink onto paper. The technique has long been used in the printing and labeling industries and was first used in computer printout devices in the mid-1970s. One variety, the pulsed ink-jet printer, in which one or more columns of ink nozzles are mounted on a head that traverses the medium, is capable of high printing speeds, up to 400 characters per second. From a cartographic viewpoint, most of the early ink-jet printers had relatively poor accuracy. They were therefore used mainly for such low-accuracy applications as replicating screen displays. However, in the early 1990s ink-jet printers with high resolution, up to 400 dpi, became available, both in letter sizes and in larger sizes suitable for use as a plotter for GIS applications. Today, the color ink-jet plotter is the most popular plotter for GIS applications, being relatively economical, fast, and having a satisfactory accuracy (± 0.25 to 0.5 mm) and cartographic quality (400 to 600 dpi). Ink-jet plotters can be used for both vector and raster data.

7.5.3 Electrographic plotters

Electrographic plotters include electrostatic, electrophotographic, and electrosensitive plotters. The electrostatic and electrophotographic processes, which are used most frequently in GIS applications, involve an intermediate electrostatic image. The electrosensitive process involves direct interaction between an electric current and the surface of a special paper and is used in various high-resolution applications such as photo-typesetting.

In an electrostatic plotter (Figure 7.14), the image is first written as a pattern of electrostatic charge, either directly onto specially treated paper or onto a drum coated with aluminum oxide. The pattern is then made visible by washing the charge with a colloidal suspension of particles of pigment carrying the opposite charge. In the drum design,

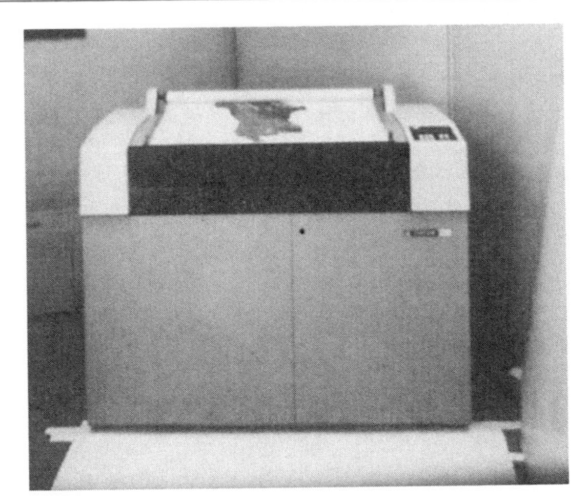

Figure 7.14
Electrostatic plotter: Calcomp Model 5835XP.

the visible image is then transferred to plain paper by the application of pressure and an electric field.

Electrostatic plotters, which are frequently used in GIS applications, have an advantage over pen plotters in that they can plot from both vector and raster data. The drawback, however, is that most models require special paper, which means that existing maps cannot be amended. Electrostatic plotters require climatized rooms, as the paper is affected by humidity. The resolution of a good electrostatic plotter is usually on the order of 200 to 400 dpi.

In an electrophotographic plotter, a beam of light writes the image onto the electrically charged surface of a drum. The light discharges the charged surface to form the image, which becomes visible once the drum is dusted with a dielectric powder (toner). The image is then transferred to paper by direct contact and the application of an electric field. The toner particles are finally affixed to the paper by heat and/or pressure. Electrophotographic plotters plot in large formats and normally have a resolution of 400 dpi.

The electrophotographic process, also known as xerography, is the same as that used in most office copying machines. It is also used in laser printers, in which lasers are the light sources. A wide range of laser printers is now available for use with text and graphics, including the printing of map data in monochrome and colors. Most handle paper formats up to standard letterhead: ISO A4 (21.0 × 29.7 cm) or the U.S. $8\frac{1}{2}$- × 11-inch standard, but printers for ISO A0 (84.1 × 118.9 cm) are also available. Laser printers have resolutions varying from 300 to 600 dpi.

The light-emitting diode (LED) is a new technology similar to a laser. The main difference is that instead of using a laser beam, an array of LEDs is used to transfer digital data to the paper through several processes.

All electrographic plotters can produce final images in black and white or color, depending on the pigment color (magenta, yellow, cyan blue, and black, in successive passes). Wherever particularly high resolution is needed for GIS applications, electrostatic plotters are the selection of choice.

7.5.4 Thermal plotters

Thermal plotters and thermal transfer plotters use heat to produce images. Thermal plotters use local heating to warm thermo-sensitive paper, which is coated with two separate, colorless components. Once heated, these combine to produce a wide range of colors of a reasonable quality. Thermal transfer plotters transfer colored wax to produce high-quality images. Both types are available to handle sheet sizes up to ISO A3 (29.7 × 42.0 cm). Thermal plotters have a limited color coverage and the colors soon fade. Direct thermal plotters are a new kind of thermal plotter which give very good cartographic products and which are used quite extensively in GIS.

7.5.5 Film printers

In a film printer, a laser or LED draws a raster image on a photosensitive medium. The image resolution is very fine, from 400 up to 2500 dpi, and the accuracy is good. The technique is normally used to produce film originals for printing, as used for printing maps in color and for production of photographic films from satellite data. Digital map data may be entered directly into devices that produce films for printing.

Comments

Output speed for all types of plotters is determined by the time needed to transfer data to the plotter and plotting speed. Plotters for GIS applications should have relatively large memory since it can often be necessary to transfer 20 to 30 Mb at a time for plotting. Plotter speed is regulated by the complexity of data and the plotting technique. The fastest plotting is achieved by electrophotographic, laser, and direct thermal plotters, whereas ink-jet and pen plotters in particular are slower. Black-and-white plotters are normally faster and cheaper than color plotters. (This does not apply to pen plotters.) In addition to machine costs, one should also consider operating costs, such as paper, toner, and maintenance. Experience shows that these can be quite high and quickly approach the purchase price of the equipment. For a single A0 plot, running costs today can vary between $1 for an ink-jet plotter and $0.5 for a laser device (Ireland 1996). Today, PostScript and HPGL are the normal format for color plotters.

CHAPTER 8

Software for GIS

GIS software is under continuous development, with respect not only to the specific GIS program functions, but also operating systems and other general program tools. For this reason it is not possible to give a comprehensive description of a GIS software program which is relevant in the long term. In the following section, therefore, we present just a few of the basic relationships. More detailed and up-to-date information can be obtained from suppliers' manuals and brochures. Individual GIS functions are explained in Chapters 13, 14, and 15.

8.1 Operating Systems

The operating system is a basic program that administers the internal data processing in a computer. Important elements of the operating system are the file manager and disk manager, which deal with the file structure and page structure of the computer, respectively (Figure 8.1). Different operating systems, such as MS-DOS, Windows 3.x, Windows 95, Windows NT, OS/2 for PCs and UNIX for workstations, are used in different types and makes of computer and for different GIS software. As application programs cannot be moved from one operating system to another without extensive modification, the choice of application program is often determined by the choice of computer.

It is no longer possible to buy professional software made for DOS. Windows NT is platform independent by taking advantage of the 32-bit addressing technique of the 386-, 486-, Pentium-, and RISC-based processors. There are no technical reasons today why Windows NT cannot be ported to run on any computer platform; the system is capable of carrying out more than one task at a time (multitasking). Windows NT has the same user interface as Windows 3.x and can run most MS-DOS and Windows-based GIS software.

Current trends are providing more options, and many suppliers of GIS programs now offer versions matched to different operating systems. Computer manufacturers are increasingly offering wider ranges of equipment, such as UNIX in workstations and Windows in PCs. There appears to be a clear trend toward increased adoption of Windows NT as the standard operative system for GIS software.

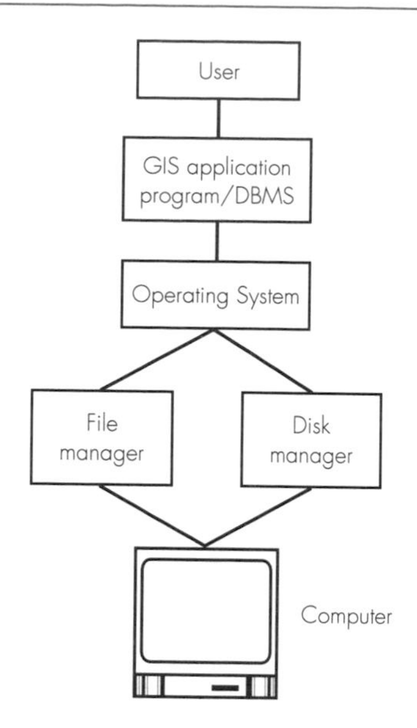

Figure 8.1
A DBMS runs on a computer just like an application program. The operating system, with file manager and disk manager, lies between the DBMS and the physical computer.

8.2 Communications Between Users and Computers

Users and computers communicate at the human–machine interface (MMI). The features of MMI are vital in GIS applications, where users need to control communications with their computers by deciding what is to be done at what time and in what order—always with the option of canceling an action midway through a communication.

In particular, MMI communications require input/output devices, such as ordinary keyboards, special function keys, mice, joysticks, and so on, along with their supporting software. More recently, tablets have been used in interactive human–machine communications. A tablet is a graphical input device in which positions can be selected by using crosshairs instead of pointing on screen.

The dialogue between a user and a computer is *interactive* when user commands result in immediate system responses in the form of messages, error messages, new choices, and so on. On-screen dialogue may be controlled by:

- User commands to the system
- User responses to system messages or queries
- Menus of options that elicit user input
- Icons (pictorial symbols used in menus) to which the user points the cursor

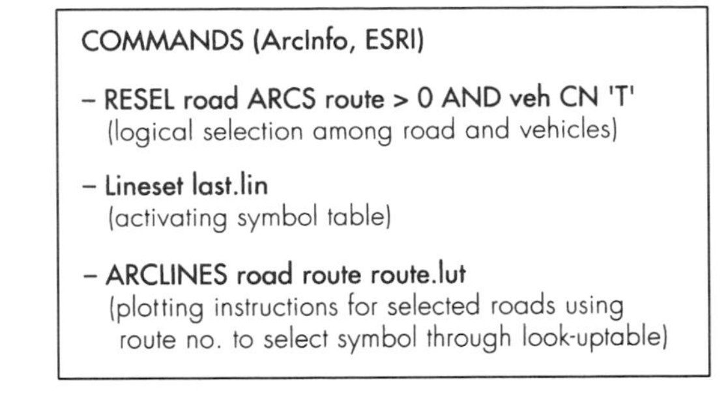

Figure 8.2
System commands can be given in the form of long strings of command instructions.

Commands

The first software systems were controlled by relatively cumbersome sets of commands and had no interactive features. Many GISs contain remnants of these earlier complexities, although choices between commands and menus are often available. To use commands, users must enter relatively long instructions, which in turn require a knowledge of numerous basic commands (Figure 8.2).

Modern systems are often structured so that users may compose and store sequences of composite commands compiled from large registers of basic commands. The programming involves system macro instructions and makes it possible to develop customized applications. Composite commands are best suited to repetitive operations for a specific product or application. Although software systems are now menu driven, many experienced operators prefer commands to menus because commands permit greater flexibility and often achieve final results more rapidly.

Queries and messages

The dialogue may involve the user either answering system queries or following instructions given in system messages; examples are given in Figures 8.3 and 8.4. Dialogues of the type illustrated in Figure 8.4 are well suited to the needs of many end users, such as those who use PCs or workstations while serving the public. Standardized query

PROMTS

Give number for selected route of roads : < >
Select roads maintained by lorry (L) or truck (T) : < >
Plot the roads selected, Y / N : < >

Figure 8.3
Command activation can be in the form of questions to the operator. These questions can often be adapted by the user for special functions.

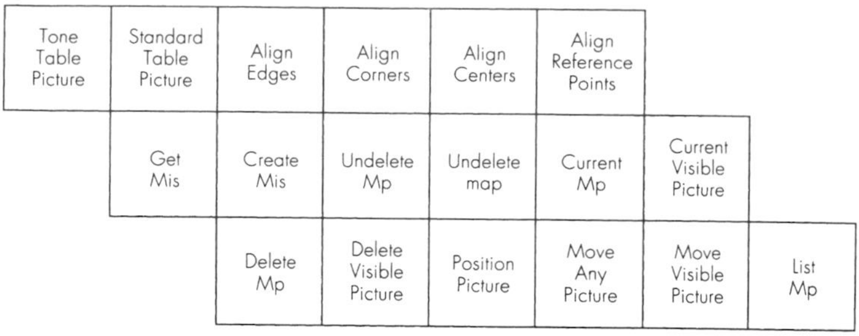

Tone Table Picture	Standard Table Picture	Align Edges	Align Corners	Align Centers	Align Reference Points		
	Get Mis	Create Mis	Undelete Mp	Undelete map	Current Mp	Current Visible Picture	
		Delete Mp	Delete Visible Picture	Position Picture	Move Any Picture	Move Visible Picture	List Mp

Figure 8.4
Plotter menu from the Swedish Teragon image-treatment system. Commands are activated by positioning the menu table on the digitizer and pointing and clicking the mouse in the appropriate fields.

routines, known as *Structured Query Language* (SQL), have been developed for relational databases (Chapter 12).

Menus

Communication between users and computers via menus is now commonplace. A menu is a program that elicits user input through an on-screen display of options, which may be commands or keys to submenus (Figure 8.5). The software system is then said to be menu driven. An option may be selected from a menu by moving the cursor to it, using the arrow keys on the keyboard, and striking a single key; or as is now more common in GIS applications, a mouse or other pointing device may be used to position the cursor on an option, which is then selected by clicking the mouse button.

On-screen menus are commonly organized with a row of the titles of submenus across the top of the screen. The submenus are *pull-down*

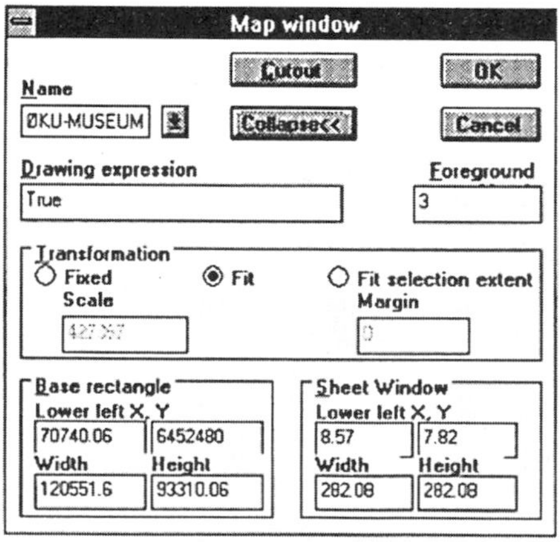

Map window
Cutout
OK
Name
ØKU-MUSEUM
Collapse<<
Cancel
Drawing expression
True
Foreground
3
Transformation
Fixed
Scale
Fit
Fit selection extent
Margin
Base rectangle
Lower left X, Y
70740.06
6452480
Width
Height
120551.6
93310.06
Sheet Window
Lower left X, Y
8.57
7.82
Width
Height
282.08
282.08

Figure 8.5
In modern GISs, dialogue with the user is generally based on screen menus where one can point and click in the command fields.

Figure 8.6
Command fields in screen menus can be presented in the form of easily comprehensible symbols called *icons.*

menus; because when selected they appear to be pulled down from the top row, as a window blind is pulled down from the top of a window frame. Options in the pull-down menus may, in turn, trigger more pop-up menus. Functions such as these are available in a variety of "windows" programs (e.g., MS-Windows, X-Windows, Motif). Many GISs allow the user to create his or her own menus, composed of a series of commands. Thus the user interface can be tailored to fit an individual application. A special macroprogramming language is used for this purpose.

Icons

Icons are selected by pointing and clicking. Each pictorial icon on screen represents a command. The icons may be simple, such as a set of boxes for color selection; or they may be more complex, such as a rubbish bin to represent discarding (Figure 8.6).

Windows

Windows techniques permit simultaneous operation of different programs by displaying information on them in separate rectangular areas on-screen (Figure 8.7). Windows, menus, and dialogue boxes may be moved around on screen as needed, for example, when they obscure part of an active application on screen.

Ancillary features

Many systems contain help functions that users may activate for further information on commands. Once a command is activated, an explanation of it can be requested using a contextual search. Users may often select various function parameters. If a parameter is not speci-

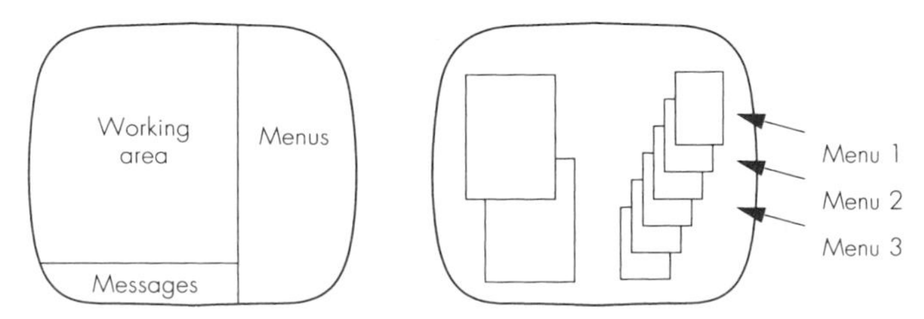

Figure 8.7
The screen can be split into several fields by means of the windows technique. This allows several different commands to be activated simultaneously.

fied by the user, the system will revert to its own standard (default) value.

Response time is crucial in all interactive work, since response times that are longer than $\frac{1}{10}$ second may lead the user to doubt that the system has understood the last command entered. However, batch processing may be advantageous in some cases, as with final drawing. In batch processing, commands are collected in a file and executed when computer resources become available.

GIS software

Geographical information systems often comprise complex program systems for various applications. This is because many systems were originally devised for special, limited tasks, such as mapping the network of a particular energy supplier. However, the current trend is toward more general systems (tool kits) from which users may select components suited to specific applications. The tasks involved can be demanding, so general systems often have relatively high entry levels. Some GIS software suppliers can supply their own development tools for programming of special applications.

Some software systems are limited to use with specific computers and operating systems. There are also constraints on the types of graphic displays, plotters, digitizers, and other peripheral devices that may be used. Moving a software system from one computer to another can be extremely difficult because of the difference in the respective operating systems. However, the process is becoming easier now that standards have become established in the market.

Modern software systems are flexible and to a large extent based on official or de facto standards. One such standard is the graphics kernel system (GKS). This set of computer graphics routines is intended for use by application programmers and is standardized by the International Organization for Standardization (in the ISO 7942 and ISO 8651 standards). GKS ensures consistency; a screen can be identified as a screen regardless of its maker. De facto industry standards have also been established. An example is HPGL, a plotter language developed by Hewlett-Packard, and PostScript, a page description language developed by Adobe Systems for word processors and desktop publishing packages.

Many software systems boast libraries containing a wide range of types and models of digitizers, screens, plotters, and other peripherals, which the software supports through having the relevant drivers. A driver is a software routine that handles the specific details and characteristics of a single peripheral device so that it can work correctly with the system. The user may select peripherals from the library listings in order to customize the software to a particular set of peripheral devices.

8.3 Computer-Aided Design

Computer-aided design (CAD) has many features in common with GIS, but they are intended to solve completely different tasks. CAD contains data for construction of a future reality or construction of technical devices and is therefore well suited to graphical construction of artificial objects. GIS normally contains data observed in the field for existing or previous reality, suited to the analysis and simulation of reality. Such analysis functions (e.g., overlay) which are based, among other things, on topological data structure, are not available in CAD.

Many of the artificial objects in CAD have identical geometry and the same nongeometrical values. General descriptions common to many objects are therefore often stored only once. One of the differences between CAD and GIS is that the presentation parameters in CAD, such as type of symbol, color, and scale, are stored together with the objects, whereas in GIS these parameters are normally defined in connection with the presentation.

We have already seen that GIS is based on a two-dimensional model of reality and, in addition, 2.5 dimensions in digital terrain models. CAD allows access to a three-dimensional tool (i.e., the same x,y pair may have several z values). CAD may therefore be better suited to handle DTM than GIS. With CAD it is possible, as in GIS, to produce maps and all kinds of terrain perspectives, but CAD can also be used to control machines for computer-aided manufacturing (CAM).

8.4 Multimedia

Multimedia has a user interface comprising text, sound, and image and is making inroads in the GIS world (Figure 8.8). Multimedia program systems and equipment comprise:

- Processing, storage, and exchange of data
- Processing "all" types of data: numbers, text, graphics, images, animation, sound, and video
- Integration of all kinds of data
- Interactive use of data under machine control

In a GIS connection multimedia includes such elements as hypermaps, cartographic visualization and animation, digital images and satellite data, sound, hypertext, virtual reality, and hypermedia spatial databases. Multimedia techniques enable the establishment of information from images, maps, and so on, by direct manipulation. The use of animation techniques allows multimedia to show spatial changes over time. The most extreme virtual reality experience is obtained through

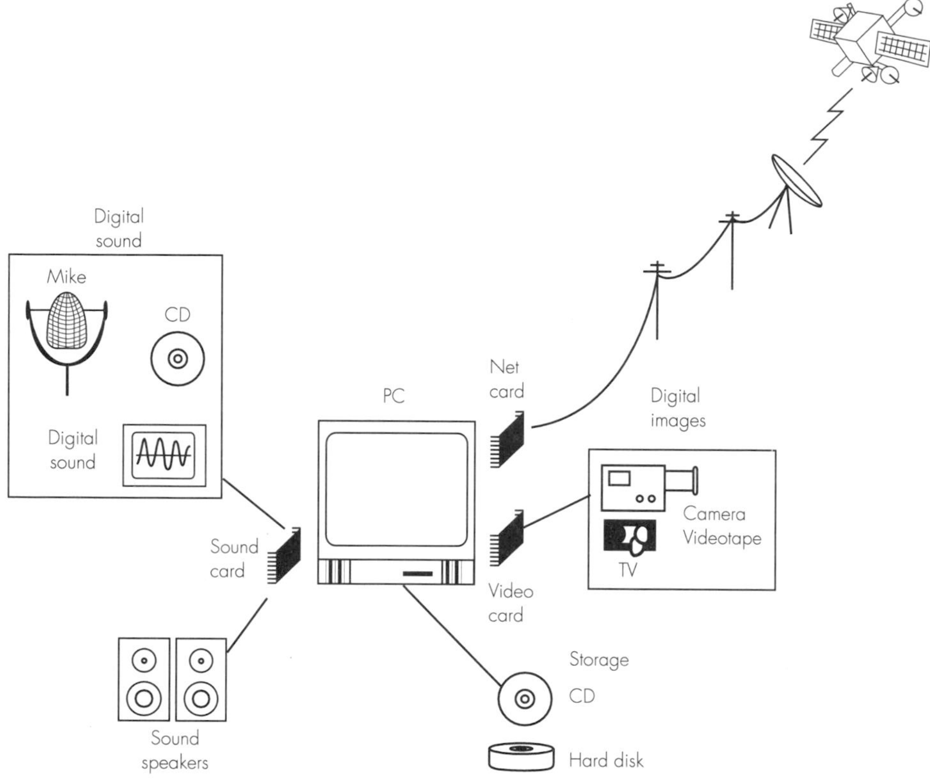

Figure 8.8
Multimedia comprise software and hardware which include integrated data in the form of sound, images, numbers, text, and video that can be treated interactively.

turning on a computer and submerging oneself in its reality. Equipped with two screens for eyes, three-dimensional sound, and data gloves, one can be transported into a world where almost all exterior sensations are artificial.

Sound, image, and video are a long stream of binary data that can be of random length. In relation to databases, therefore, we can talk of binary large objects (BLOB). Database systems must be capable of handling BLOBs of random length and of deciding how they will be stored, depending on size, access time, and transfer speed. Many GISs are already able to store and display a BLOB quite easily.

8.5 World Wide Web

World Wide Web (WWW) is software developed for use on the Internet. WWW (originally developed by CERN in Switzerland) is a freely available program system where users can design screen images on the Internet to display their own data in the form of text, pictures, video,

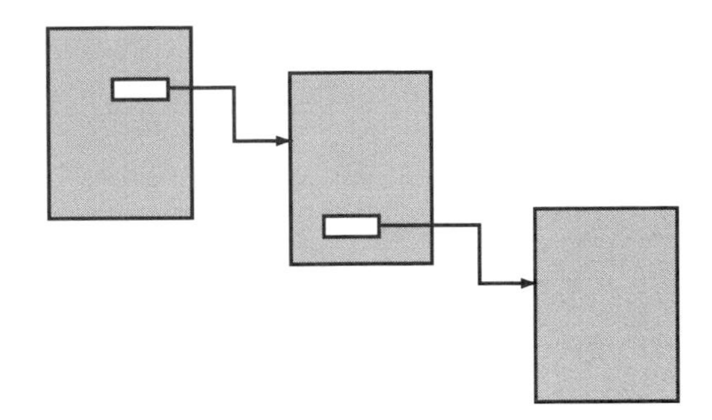

Figure 8.9
Using the hypertext technique, it is possible to point and click on selected text segments (or other objects) and obtain more information via the Internet than was clicked on.

and sound. WWW is a set of standards for the treatment of communication protocols and for structural hypertext. Marked text and objects interconnect the client machine and data lying on the server(s) at other locations on the network. A special program called Hypertext Transfer Protocol (HTTP) is used for linking and transfer of data (Figure 8.9).

Web home pages are created by using a marking language known as HTML (Hypertext Mark-up Language). With the help of Hypertext, home pages can be linked to other home pages and to other Internet resources. Hypertext is a nonsequential text where special references are created between different parts of the text (links), which can be traced via a computer. The hyperlink technique is also used to create a multimedia system based on special references between different segments of information (e.g., from picture to text and from text to sound. WWW has a simple technology using a fixed user interface and has therefore become very popular with public and private institutions/users alike. The Internet can be compared to a national highway system, whereas WWW is a delivery service for packages and letters to customers/subscribers who use the telecommunication network as a transport system.

8.6 User Requirements

GIS users may be divided into groups according to their functions and needs. These groups might comprise:

- End users, who seek problem solutions and see final products only in the form of maps and reports
- GIS specialists, who advise and solve problems for end users, GIS operators, and data compilers
- GIS operators, who understand the functions of a specific system so as to manipulate the data
- Data compilers, who understand the data but not the system

The value of GIS depends on how well users state the problems and how well GIS specialists provide solutions. End users frequently need only a limited knowledge of GIS. However, GIS specialists must be familiar with the general principles of GIS, data and data compilation, cartographic facilities, and not least, with the potentials and limitations of the system employed. In other words, a specialist must be able to sketch alternative production routes.

A GIS specialist has to be expert in several fields, although not everyone in a GIS organization will need to have the same capabilities. Outside consultants may also be engaged. Normally, it will take about a year for an operator to become thoroughly familiar with a larger system. However, he or she can be productive long before that. In many cases, it is possible to complete a task without specific knowledge of all the available functions. The initial training of operators preparing to work on larger systems usually lasts 14 days, and is followed by additional training for specific parts of the system. Furthermore, whenever new computers are installed, training courses are held to familiarize staff with the operating system, backup routines, and other facets of their operation.

The need for computer knowledge depends on the size of the system and the task confronting the operator. A single-user PC is easier to operate than a multiuser workstation system. Professional staff members need only moderate computer knowledge to learn how a system works. That said, extensive computer knowledge is a prerequisite for staff who will be responsible for operating the system and will require an in-depth understanding of the geographical information system, the operative system, and the peripheral equipment.

Most organizations that have implemented GIS have been obliged to conduct some development for matching older data systems (such as those containing older registers), improving existing programs, interfacing new peripheral equipment, and so on. Developments such as these require computer expertise, which may either be acquired in-house or contracted externally. Digitizing personnel and other data compilers should be familiar with the attributes with which they are dealing, but need only a limited knowledge of the GIS involved.

8.7 Working Environment

Work at a GIS station can be demanding; therefore, the working environment involved in any GIS implementation should always be taken into consideration. Digitizing maps can be physically tiring and mentally tedious. Clearly, such work cannot be performed continuously for an entire working day and should be split up into shifts through rotation or with longer breaks. Spring-loaded armrests are available to ease arm loading in digitizing. The same precautions apply to the prolonged use of a mouse for pointing and clicking. A good working position and sufficient breaks are also necessary for good work at a screen.

The working environment should be relatively compact, so that manuals, maps, manuscripts, notes, and the like are all within reach. A U- or L-shaped layout is often best. Screen height above the floor, distance from the user, and keyboard and tablet positions should all be freely adjustable. Lighting should be indirect to minimize screen reflections.

As concentration is vital in GIS work, workplaces should be shielded from disturbance. Printers, plotters, and other noisy equipment should be located away from the operator. In some cases where the equipment produces heat, air conditioning or fans may be required. Staff training is essential to the success of the system, and system handbooks and other documentation should be in the operators' native language, if at all possible.

8.8 Database Management Systems

8.8.1 Traditional libraries and information banks

Map data have traditionally been recorded in the form of lines and symbols on foil and paper. Similarly, descriptive data have been recorded in written form on index cards, on lists and various documents, and in various types of photographs. These traditional data repositories are organized in varying systems of filing cabinets and drawers, loose-leaf binders, folders, and other document files. Each data repository may be regarded as a *library* or *bank* from which users may retrieve information.

Conceptually, traditional information libraries and banks render data readily accessible because recorded data can be physically handled and viewed. The database management system used for filing and retrieval of information is therefore carried out manually. However, there are many drawbacks. For GIS purposes, the major disadvantages are:

- *Data are dispersed.* Data may be located in the files of several agencies or organizations, with no ready means for transferring the data from one storage facility to another.
- *Structure and storage methods are dissimilar.* There are so many different ways of compiling and storing data that data from disparate sources can seldom be used together without extensive manual translation from one form to another.
- *Verification is uncertain.* There are no ready means of recording or authenticating data verification.
- *Retrieval is slow.* Data have to be retrieved physically: the physical

search and the transport of sheets and documents are time consuming.
- *Data are normally available to a few users only.*
- *Data are restricted largely to the uses for which they were originally compiled.* Using such data for other purposes is difficult, if not impossible.

Data libraries and banks

Computerized data libraries and data banks differ from their traditional counterparts in a variety of ways, most noticeably in that all data are recorded in digital form and stored physically on magnetic or optical media. Initially, these electronic means of recording and storage may seem less accessible than their traditional counterparts because the recorded data cannot be handled physically and viewed.

The conventional repository for data is a *data bank,* a computerized system for the "deposit" and "withdrawal" of data. A data bank may either be available to a wide range of users or restricted to a few authorized users only. The data deposited (entered) may be in the form of one or more *files* or in the form of a *database.* The difference between a file and a database is semantic and varies somewhat from one discipline to another. For GIS purposes, a *file* is regarded as a single collection of information that can be stored, while a collection of files is regarded as a *library.* A *database* comprises one or more files that are structured in a particular way by a database management system (DBMS) and accessed through it (Figure 8.10). DBMS can therefore be defined as a software package for storage, manipulation, and retrieval of data from a database (Figure 8.11). Thus several databases can be created from a single file. Apart from such temporary working files as word processing text files, which may be entered in GIS, almost all computerized GIS information is stored in databases.

As discussed in preceding chapters, real-world phenomena are structured (i.e., grouped and described) in order that they may be understood more easily. Structured geographical phenomena are de-

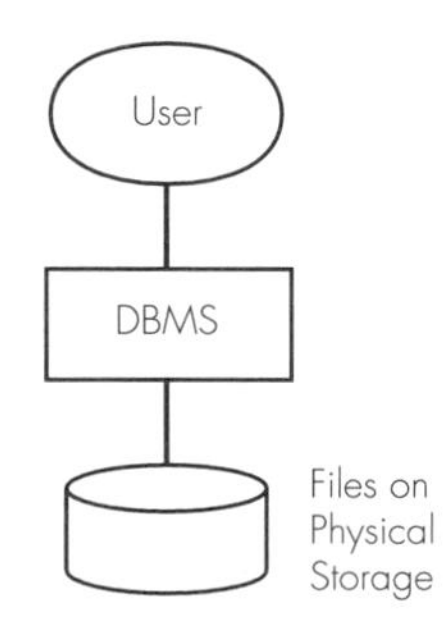

Figure 8.10
A DBMS is located between the physical storage of data and the user.

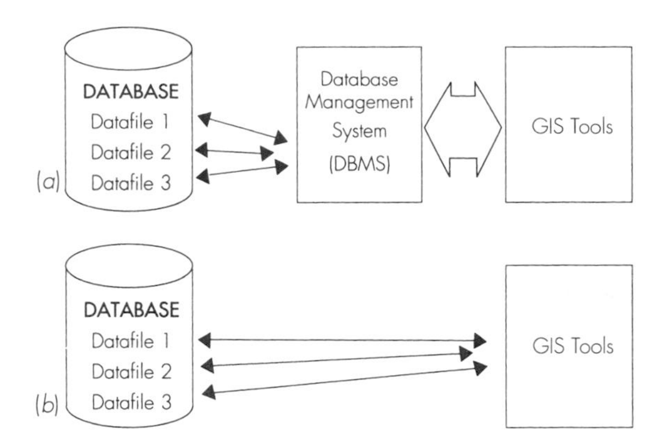

Figure 8.11
A DBMS (*a*) ensures easier communication between a database and GIS tools than through direct manipulation of files (*b*).

scribed in terms of geometry, attributes, and relationships in a data model (Figure 8.12). A data model may be delineated digitally using coordinates for points, lines, and areas, to which codes may be assigned.

The *advantages* of databases and DBMSs compared to traditional libraries and information banks are that:

- Data are stored in one place.
- Data are structured and standardized.
- Data from dissimilar sources may be interconnected and used jointly.
- Data are amenable to verification.
- Data may be accessed rapidly.
- Data are available to many users.
- Data may be used directly in many different applications programs, including programs whose purposes differ from those for which the original data were compiled.
- In principle, the data storage method can be considered to be independent of the program (DBMS) that accesses the data.
- Standardized data access through DBMS ensures integrity and consistency between data files.
- It is possible to introduce security measures in regard to data access.
- It is possible to realize multiple use of data.

Figure 8.12
The data fed into the database have to be structured so as to produce useful applications and products.

The disadvantages of database storage compared to traditional storage are:

- Use of databases requires expertise.
- The products used are relatively expensive.
- Users may have to adapt to a data flow that differs from the traditional.
- Users may have to adapt to a data organization that differs from the traditional.
- Data may easily be misused.
- Data may be easily lost, necessitating special security measures.

In other words, despite many advantages, databases are no easy option. They can be both misused and overused. The homeowner who uses a database on a PC to keep an updated record of the contents of the home freezer is solving the daily "what's for dinner?" problem with technological overkill; a pencil and paper next to the freezer is more efficient.

Similarly, the choice of a DBMS ill suited to an application may escalate both work and costs. Database structure is perhaps most important because it affects the response time, storage capacity, flexibility, and other parameters which together determine the user friendliness of a database.

Structures differ and no single structure is best for all applications. The various methods of data storage differ primarily according to the facilities or operations supported by the DBMS:

- *Files* (no database management): tabulations of data in files. Databases called *flat-file databases* may comprise files only, with no supporting hierarchy or other interrelationships.
- *Database systems:* distinguished according to the type of database management involved:
  - Relational database systems
  - Hierarchical database systems
  - Network database systems

8.8.2 Files

Files comprise *records,* each of which contains *fields* (Figure 8.13). Each record contains data concerning a single topic or affiliation; each field contains an item of data consisting of one or more characters, words, or codes that are processed together. *Keys,* which are codes used to access information, help to retrieve records from files. Keys are associated with one or more fields of a record.

Data in a file may be stored sequentially, as in a line (Figure 8.14). New records are simply added at the end of the file. Records may be of variable length, in which case the beginning of each record must

	Field 1	Field 2	Field 3
Record 1	110	Oslo	10.61
Record 2	115	Stockholm	9.15
Record 3	116	London	18.33
–			
–			
Record n			

Figure 8.13
A file is divided up into records and fields.

contain information about its length, although such information is not necessary when the length of the record is fixed.

Searching for data in sequential files is time consuming. For example, if the search is for points stored in a random sequential file, all records in the file must be queried to ascertain whether the desired points have been located. Therefore, storage in unsorted sequences is recommended only for relatively small quantities of data.

The search time may be reduced if the data in a file are structured: for example, when they are sorted by ascending coordinate magnitudes. The search may then be limited by the largest magnitudes entered. A search could also start in the middle of a file, thereby allowing the search to progress successively, by dividing intervals into two, deciding each time which half contains the point sought, and carrying on until the point is located. This process, known as *binary search,* is considerably faster than sequential searching.

The address, or position of an item in a file, may be stored in another file. Known as an *index file* or *pointer file,* it contains a set of links

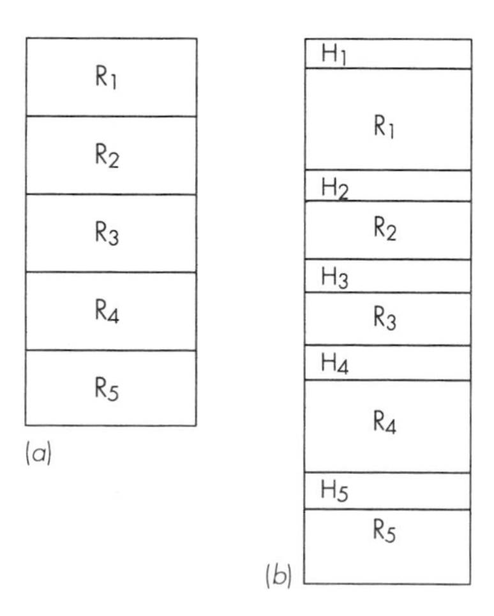

Figure 8.14
There are two categories of sequential files: (*a*) fixed record length (R_1, R_2) and (*b*) variable record length. The headers (H_1, H_2, etc.) specify the length of the record.

that can be used to locate records in the data file. Searching for items in the data file may be compared to searching for entries in a well-indexed book, by first finding page numbers in its index. The largest practical disadvantage with files is that each program application has to be specially tailored to the structure in the files. When changes in the file structure are made, application software also has to be changed. When files have to be shared simultaneously among several users, a central system has to be established to control who has access to files and which operations can be carried out by each user. Despite this, the file-based systems have great significance within GIS. Many of the leading systems on the market use file systems and tailormade management systems for storage and management of the geometric data. However, special index routines have been developed to increase efficiency in pulling out data (Chapter 12). Attribute data are normally stored and handled by a standard DBMS.

8.8.3 Database systems

Usually, unstructured files are not suitable for handling large quantities of data or data that change continually. Therefore, DBMSs have been developed to handle the complex tasks of multiple database storage and data amendment. Database systems consist of:

- Data
- Hardware
- Software

The data in a database system are stored on disks with associated input and output devices; thus a database is a physical collection of files structured by DBMS software. We shall now look at the software aspects.

All database systems aim for ease of search and linking of tabular data. A DBMS must be able to manipulate several types of objects and various relations between objects. In GIS the data in a database system can be structured according to different models. Traditional models are: relation structure, hierarchical structure, network structure.

Relational database systems

A relational database system is one in which the DBMS supports a relational model of the data stored. That is, it is more of a guideline *to* representing data than it is a structure *for* data. In many ways it is comparable with a table, which is why *tabular database* is an alternative term for *relational database* (Figure 8.15).

In the relational data structure, there is no hierarchy and no specific keys point to the various records. All fields in a line may be regarded as being connected to each other permanently, yet individual tables contain no information on other tables. Nevertheless, any field in a

Building ID	Property	Owner	Year	Type
589	44 110			
610	44 50			
955	44 99			

Property ID	Owner	AREA	ADDRESS
44 99			
44 50			
44 110	Nils Nilsen	6,51	9999 Toppen

Figure 8.15
Relational database. Each field in a table can be the key to locating data in another table.

table may be a key for accessing data from another table; the system permits all objects and attributes to be related to each other. Thus a collection of data on many objects with complex relationships may result in a large number of tables—a weakness of the relational approach.

Relational databases have three basic attributes or functions: primary key, relational join, and normal forms. Values in a single column, or a combination of values in multiple columns can be used to define a *primary key* for the table and act as the addressing mechanism. Data that lie in different tables are linked by using a function called a *relational join* (Figure 8.15). *Normal form* is the identification of interdependence between the data. There are three normal forms that should be met in relational databases. The first criterion is that there should be no recurring fields, so that the same data or attributes cannot appear under different names. Breach of the second normal form occurs when the function of an attribute is dependent on only part of a compound prime key. The third normal form is achieved when a relation fulfils both the first and second normal forms and in addition has no attribute dependent on another attribute, except when the second attribute is a prime key.

There are no pointers in the data tables. Internal index tables are used to direct intertable communications, which means that the system must open at least one index table for each connection between data tables. This can result in relatively large databases and slow access. Moreover, as there are no pointers, searches for records or fields must be sequential through the tables. The result is that relational database systems are inherently slower than hierarchical or network database systems. However, a field may be located quickly in a given table, and the data accessed may be manipulated.

Relational DBMS are the type used most frequently for attribute data in GIS applications, primarily because of their simple, flexible structures. Another reason is that they support the complex relationships common among real-world geographical objects. Multiple entries are less frequent than in hierarchical or network database sys-

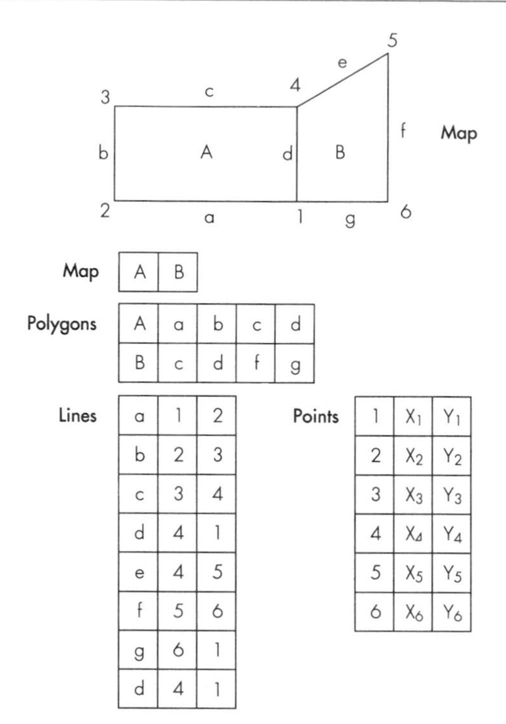

Map

A	B

Polygons

A	a	b	c	d
B	c	d	f	g

Lines

a	1	2
b	2	3
c	3	4
d	4	1
e	4	5
f	5	6
g	6	1
d	4	1

Points

1	X_1	Y_1
2	X_2	Y_2
3	X_3	Y_3
4	X_4	Y_4
5	X_5	Y_5
6	X_6	Y_6

Figure 8.16
Map data stored in a relational database.

tems, although search times tend to be longer. This is particularly noticeable in specific operations on digital map data: for example, when forming polygons and with automatic text placement.

Standard relational databases require that only single values be stored in each cell. They are therefore normally unsuitable for vector data, which often consist of lines with many points and would result in a lengthy data search time. The standardized question routines, structured query language (SQL), are not suited to geometric data but are quite effective for attribute data. The good qualities of relational DBMSs have led to the development of GISs that are based on custom-built (proprietary) relational DBMSs that are capable of handling both attributes and geometry (see also Section 12.5).

Commercial RDBMS have been developed specifically for handling geographical data (and binary large objects for handling of multimedia data). These are new solutions which so far have had very little application. The most usual database solution for GIS is a combination of file-based systems for geometrical data and standard relational DBMS for attribute data.

Hierarchical database systems

A hierarchical database system is one in which the DBMS supports a hierarchical structure of records organized in files at differing logical

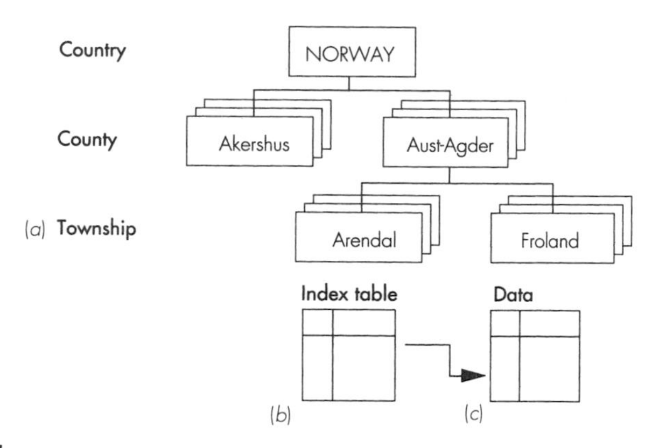

Figure 8.17
The storage structure (*a*) in a hierarchical database system. The pointer table (*b*) indicates where the data (*c*) are stored.

levels with connections between the levels (Figure 8.17). A record at a particular level contains data common to a set of records at the next-lower level. There are no connections between records at the same level. Each record contains a field, defined as the *key field,* which organizes the hierarchy.

Records of the same type are collected in files known as *elements.* Several types of elements may reside at the same logical level. Starting from the highest-order element, the hierarchy permits a set of elements to be accessed at the next-lower level. Any element may retain connections to only one set of lower-level elements and itself be a member of only one such set.

The construction of a hierarchical structure begins with a main object at its top (Figure 8.18). The main object has a range of characteristics that can be collected at the various levels of the hierarchy. Geographical data may be stored in a hierarchical structure in a manner that reflects the real world; for example, the levels of a hierarchical model may correspond to the levels of real-world administration at the country, county, township, or town/city level.

Hierarchical data structures are easily expanded and updated. However, they require large index files, require frequent maintenance, and are susceptible to multiple entries. Searches are rapid, but search routines are fixed and constrained by the structures. Thus records that lie at the same level cannot be searched, nor can new links or new search routines be defined. The elements or the structure are related only through one-to-many connections. This constraint imposes the presupposition that all queries can be known in advance and accounted for in structuring and entering data, something that is not necessarily either natural or suitable in GIS applications. As a result, hierarchical database systems are usually restricted to storing digital map data in GIS, a very effective solution where the data structure is naturally hierarchical.

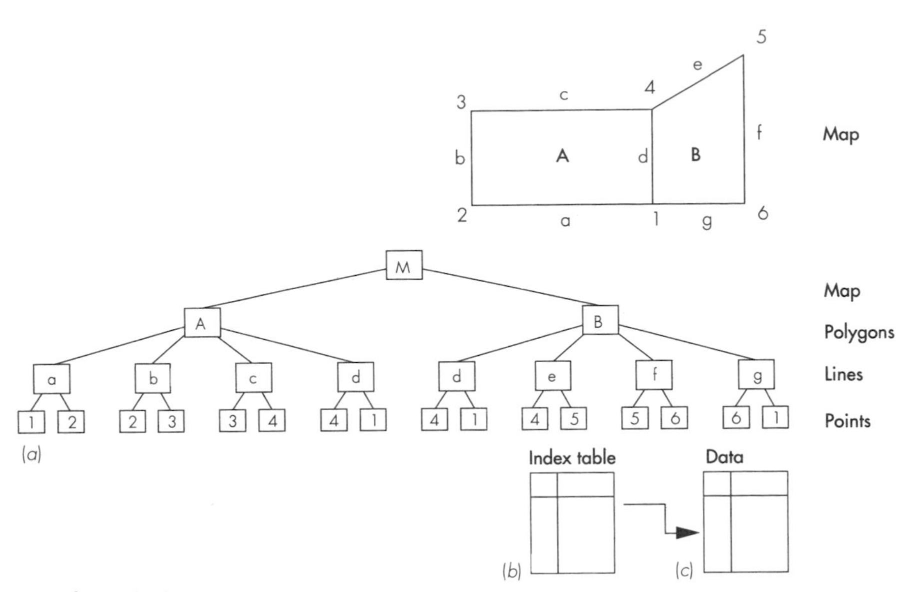

Figure 8.18
Map data stored in a hierarchical database system. Note the double storage of lines and points.

Network data structure

A network data structure is one in which the DBMS supports a network type of organization. Each element, or collection of like records, has connections to several different-level elements (Figure 8.19). Interconnections are made in the hierarchical organization, and a characteristic may be associated with two main objects. The resulting network structure closely represents the complex interrelationships that often exist between real-world geographical objects. The elements of the structure may be related through one–many, many–one, and many–many connections. A network data structure can be considered as an extended form of hierarchical data structure.

The purpose of the network structure is to improve the flexibility and reduce the multiple entries of the hierarchical structure. Points and lines are entered once only. Searches need not pass through all levels but can take shortcuts. However, the volume of index data is greater than that of the hierarchical structure. A network structure permits rapid connection between data that are stored in different disk sectors. Maintaining data stored in a network structure is complex, so although it is better suited to geographical data than is a hierarchical structure, it is used infrequently in GIS applications. Network structures are frequently used in administrative data systems.

Object-oriented database systems

Hierarchical, network, and relational database systems are intended for administrative tasks and are not particularly suitable for represent-

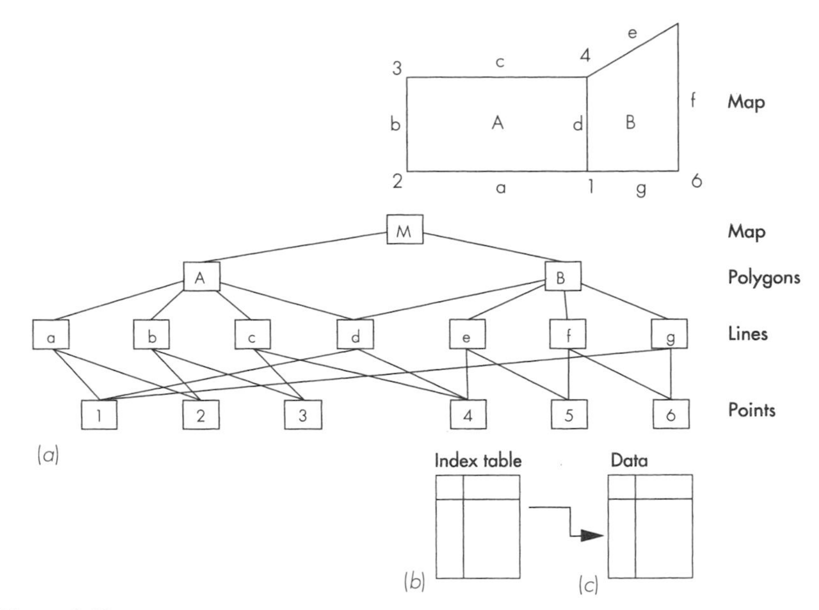

Figure 8.19
The storage structure in a network-based database system: (*a*) a map section; (*b*) polygons stored in the network system; (*c*) a pointer table that specifies where the map data (*d*) are stored.

ing a conceptual data model of geographical reality. All these systems, however, are record oriented; that is, the data they contain are filed record by record, rather like cards in a conventional card index file. For example, in a relational database, all the elements that together comprise a particular map object may reside in several records in various tables. Furthermore, geometrical data and attribute data are often separated, each residing in a separate database.

Object-oriented database systems attempt to overcome these difficulties by more faithfully representing the real world, which comprises complex homogeneous objects with varying internal and external relations (Figure 8.20). In object-oriented systems, data modeling and data manipulation are in many ways synonymous. Objects are manipulated as independent homogeneous entities, yet may comprise several other objects in a hierarchy. Flexibility in the object-oriented system also makes it easier to integrate the time factor in the data model.

Object-oriented database systems employ simple, albeit "intelligent" databases. The requirements for user expertise are, however, correspondingly higher because users must be able to define all relationships between objects in the database system. The chief advantage of object-oriented database systems is that they are easier to update on a consistent basis. For example, a line (which is an object) may be a property border, a road edge, or a zoning boundary. In an object-

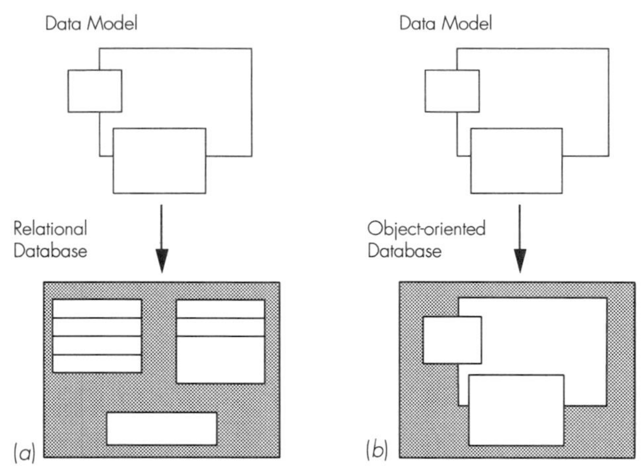

Figure 8.20
Object-oriented database (*b*) compared to a relational database (*a*).

oriented database, it needs to be updated only once, while in other database systems the data pertaining to it may be in various locations, requiring multiple updating operations.

At present, object-oriented database systems are not used widely in GISs; a few GISs do use them in manipulating data, because of their ability to handle lengthy transactions in a multiuser distributed environment. The implementation of database for GIS use is discussed at greater length in Chapter 12.

CHAPTER 9

Data Collection I

9.1 Introduction

At the outset, sources of data used in GISs may be (Figure 9.1):

- In various digital forms: vector, raster, databases, spreadsheet tables, satellite data, and so on
- Nondigital graphics, such as conventional maps, photographs, sketches, schematic diagrams, and the like
- Conventional documents in registers and files
- Compilations in scientific reports
- Collections of survey measurements expressed in coordinates or other units

The data sources for a comprehensive GIS are probably more numerous and of greater variety than in most other information systems. GIS is a true mixed-data system. Most often, the data mix makes for a lively system. Although most of the problems associated with collecting data for GIS uses had been identified by the mid-1970s, data collection remains the most expensive and time-consuming aspect of setting up a major GIS facility. Experience indicates that data collection accounts for 60 to 80% of the total cost (time and money) of a fully operational GIS, while the purchase of equipment accounts for not more than 10 to 30%. Other costs include training, development, and administration. As shown in Figure 9.2, the costs of data collection are dominant even when spread over an implementation period of several years.

There are five methods of producing digital map data:

1. Digitizing existing maps with a digitizer
2. Scanning existing maps
3. Digital photogrammetric map production
4. Entry of computed coordinates from field measurements
5. Transfer from existing digital sources

The first three methods, which produce digital data from ordinary maps, may all be divided into three phases:

1. Preparation
2. Digitization/scanning
3. Editing and improvement of data quality

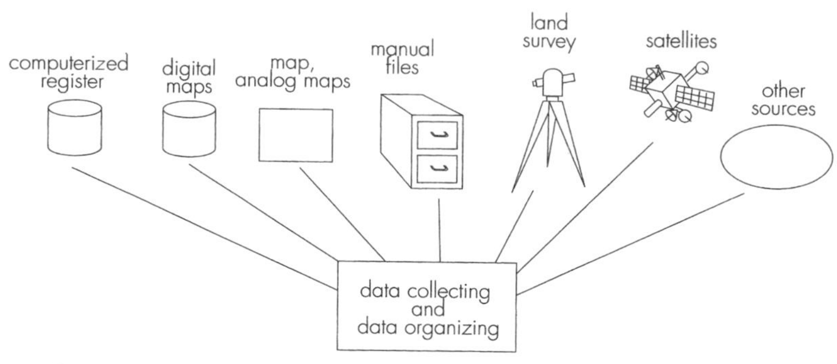

Figure 9.1
Sources of data can be both numerous and varied.

9.2 Digitizing Maps

Preparation

Data must be prepared at frequent intervals and carefully organized to facilitate the work of digitizing. Manuscript maps (Figure 9.3) may be used to group and highlight information relevant to the various themes involved. For example, roads shown on a base map may simply need highlighting in various colors, according to the type of road. Or the operation may be more comprehensive, as when collecting and grouping network information from various maps and sketches onto a single common map. In practice, these operations invariably demand some redrawing of maps prior to digitization. As a rule, changes and amendments to graphics are achieved more easily on paper maps prior to digitization than they are on digital data. Experience has shown (Aronoff 1989) that even with extensive redrawing, manuscript maps save as much as 50% of digitization costs. They are therefore

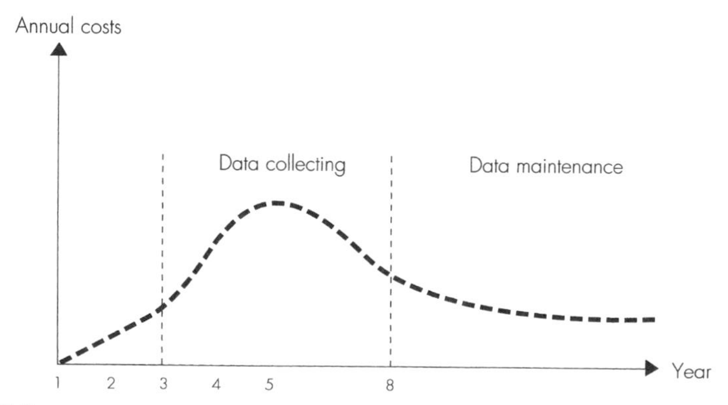

Figure 9.2
Data collection costs versus implementation time: a typical cost curve resulting from the introduction of GIS into a larger organization.

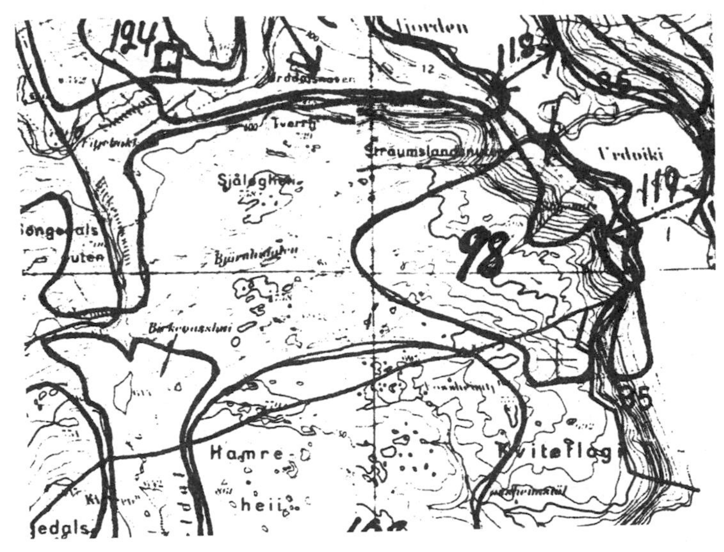

Figure 9.3
Typical manuscript map showing game registration in Norway.

considered vital in digitization. The identifier data necessary for relating digital map data to computerized attribute data are usually organized on manuscript maps.

Digitizing

As illustrated in Figure 9.4, digitizers are used in manual digitizing. Digitizers have their own internal coordinate systems, which may be related to terrain coordinates by cross-registering three to six or more coordinates with known terrain coordinates such as grid intersections, polygon points, or border markers. These cross-registrations permit the computation of transformation parameters, which may then be

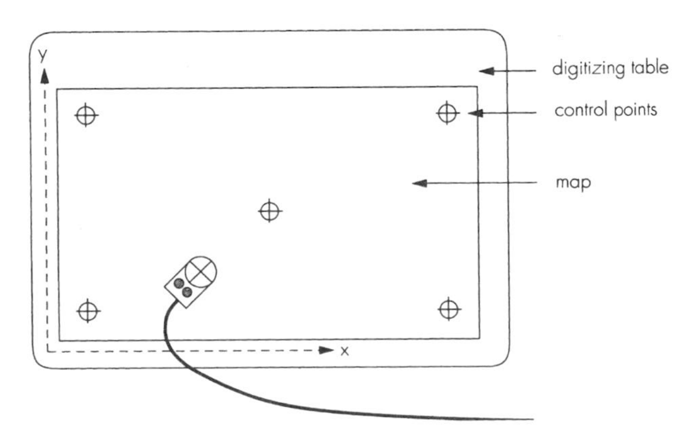

Figure 9.4
Map digitizing: a digitizing table complete with map and with the internal coordinate system defined.

used to compute terrain coordinates from digitizer coordinates. Some programs convert coordinates continuously during digitization, while others convert in the editing phase.

Conformal transformations involve transferring digitizer coordinates to terrain coordinates through parallel translation, rotation, and scale alteration. All objects then retain their original shapes. Affine transformations alter the shapes of objects through unequal x and y scale alterations and unequal rotations of the x and y axes. This method compensates for variable shrinkage in the map. The mathematical relationships are:

Conformal transformation:

$$X = c_1 + ax - by \quad (9.1a)$$

$$Y = c_2 + ay + bx \quad (9.1b)$$

Affine transformation:

$$X = c_1 + a_1x + b_1y \quad (9.2a)$$

$$Y = c_2 + a_2x + b_2y \quad (9.2b)$$

where X and Y are terrain coordinates and x and y are digitizer coordinates; a, b, c are transformation parameters.

Once the known points are registered, digitizing may begin. Two methods are used. In *point digitizing,* a crosshair is placed over the position to be recorded and a key is pressed to enter the datum. In *stream digitizing,* a crosshair is drawn along a linear symbol and data on positions are automatically entered at preset intervals of either time (such as five per second) or distance (such as five per centimeter). Data entry may also be curve fitted automatically, with more entries as the curvature radius decreases (Figure 9.5).

Point digitizing requires that the operator continuously select points and key in data entries. It is most suitable, therefore, for individual points and for lines comprising straight segments, such as property

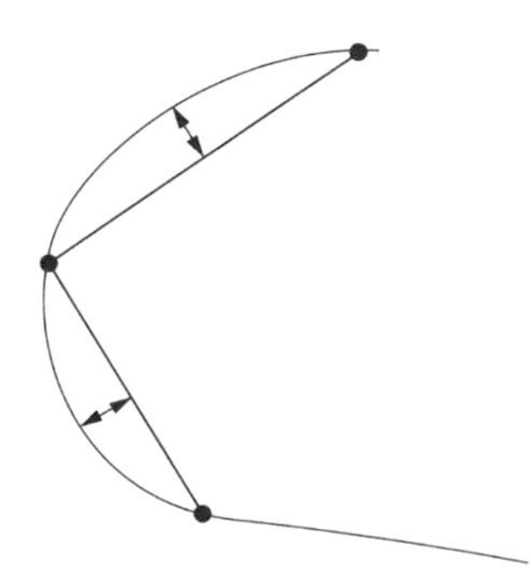

Figure 9.5
Principle of stream digitizing. Various methods can be applied in digitizing to select a point on the line based on, among other factors, a given arrow height value.

boundaries and building outlines, which need not be followed between successive data point entries.

Stream digitizing requires more sophisticated computer hardware and software (partly because data-entry speeds may be high) and greater operator skill. Once these requirements are met, stream digitizing is suitable for digitizing sinuous lines, such as contours and the borders between different soil types. The drawback is that unless its parameters are selected carefully, stream digitizing may generate unnecessarily large quantities of data. Point digitizing is preferable to stream digitizing for high-accuracy entry. Lines of fixed curvature, such as property borders on curving roads, may be described by radii of curvature rather than by their constituent points. Digitizing includes the entry of thematic codes for object types and ID codes that link object types to attribute data. For example, digitizing a building includes entry of the thematic code for buildings and an ID number for the specific type of building. A new ID number is entered for the next building, and so on. Codes are keyed in or entered via a menu.

In some systems the ID codes of map objects may be entered directly, provided that stored attribute data are available interactively on screen.

Digitizing accuracy and progress should be verified as work proceeds. Verification may be either manual or semiautomatic. The former requires objects on a manuscript map to be crossed off as they are digitized and for partially completed maps to be printed out as the work progresses. In some systems, results can be displayed continuously on screen, in effect a semiautomatic, human–machine variety of echo check (a comparison of received/entered data with original data). The screen display enables the operator to verify the work performed and to avoid double entries.

Digitizing often introduces error. Typical errors include contour lines that cross each other, gaps between lines that should be contiguous but dangle or fail to meet, intersecting lines that should not overlay yet do, lines meeting at the wrong point or intersecting each other, objects with improper or missing codes, and missing details (Figure 9.6).

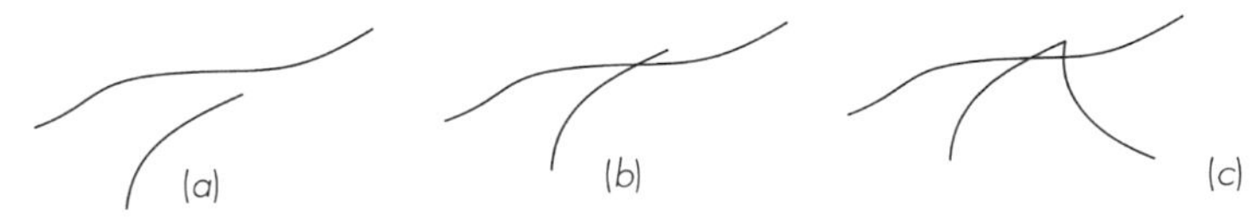

Figure 9.6
Typical errors that can occur during digitizing. Three types of error often occur, and these may require extensive correction work. The type of error illustrated in part (*b*), where dangles occur, is easy to correct automatically. It is therefore generally recommended that during digitizing, situation (*b*) be preferred to (*a*). The error shown in part (*c*) creates a new, meaningless polygon.

Some errors are corrected more easily than others. For example, it is better that the digitization of two lines contiguous at a point result in their crossing rather than in their not meeting at all. Crossed lines are more easily processed, particularly as the original contiguous lines (and their incorrect crossings) are normally parts of the border of a closed polygon, and a polygon of incorrect geometry is more easily corrected than a polygon that remains open. Some systems support "snapping" on a program for recognizing and connecting to nearby points or lines, so that nodes, closed polygons, and contiguous lines may be created quickly. Skilled operators can normally achieve crosshair placement accuracy of ± 0.05 to ± 0.06 mm.

The time required for digitizing depends on the complexity of the map in question and the extent of coding. A rule of thumb holds that digitizing may progress at a speed of one polygon per minute. In most cases, digitizing may be performed using a program run on a PC. This is both convenient and quick, first, because there is no need for other computing facilities, and second, because PCs now have both the capacity and the speed to suit the task. Error correction, editing, topology structuring, and other tasks may be performed using suitable programs.

Operator skill is decisive, both in terms of the time required for digitizing and for its quality. Most GIS facilities have found that operators experienced in cartography are preferable to operators experienced in computers or other disciplines. Operators must be trained to understand the logical connections between property borders, pipe networks, borders between soil types, and other themes that are digitized, and to master the software tools they use to the extent that they can correct their own digitizing errors.

Data editing and quality improvement

Normally, the product of manual digitization has to be edited and the quality improved. Tasks include:

- Error correction
- Entering missing data
- Forming topology

Data are verified visually on screen and/or on test printouts. Geometry, ID codes, and any thematic codes may be plotted and verified. A test printout may be laid on top of an original on a light table and compared to it, piece by piece. Where several themes are digitized, this comparison is made easier if each theme is printed out and compared to the original separately. Errors may be corrected and missing data may be entered either in the digitizing program or in the GIS programs. However, editing data is often time consuming, so that when the original data are poor, editing may take more time than the original digitizing. A program can, for instance, identify errors of the type illustrated in Figure 9.6 and will automatically delete unnecessary line segments, connect lines, and close polygons. However, automatic programs are no panacea. In practice, they often correct dangles and gaps that should be left alone.

Forming topology also reveals many errors. Typically, the same polygon number will be assigned to both sides of an unconnected line in a polygon structure. This is illogical, as all lines in a polygon structure must mark the borders between adjoining polygons. Several other control functions are also incorporated in the formation of topology.

9.3 Scanning

Maps are scanned to:

- Use digital image data as bases for other (vector) map information
- Convert scanned data to vector data for use in vector GIS

Preparation

Scanning requires that a map scanned:

- Be of high cartographic quality, with clearly defined lines, text, and symbols
- Be clean without extraneous stains
- Have lines of width 0.1 mm and greater

In scanning tasks, work is often moved from the preparatory phase to the editing phase. If, for example, the data encoded are to be used directly as digital image data, the quality requirements for the scanned map may be relaxed and preparation may be negligible.

Scanning

Scanning consists of two operations, scanning and binary encoding. Scanning a map produces a regular grid (pixels) of gray-sale levels, which range numerically from 0 (complete black) to 255 (pure white). Normally, map scanners use a pixel size of 25, 50, or 100 μm (1 μm = 0.001 mm). As a black-and-white line map comprises only black lines against a white background, scanning must clearly differentiate black from white. Accordingly, the scanner must be precisely adjusted for a gray-scale level corresponding to the map demarcation between black and white. If the level is set too high, vital information represented by thinner lines may be lost; if too low, extraneous information, such as dirt, stains, and dust, may be included (Figure 9.7).

Finding the most suitable gray-scale level can be time consuming, but once it is found, a map may be scanned in minutes. With a threshold set to differentiate black from white, a gray-level array may be binary encoded. As shown in Figure 9.8, each pixel is then binary coded as either

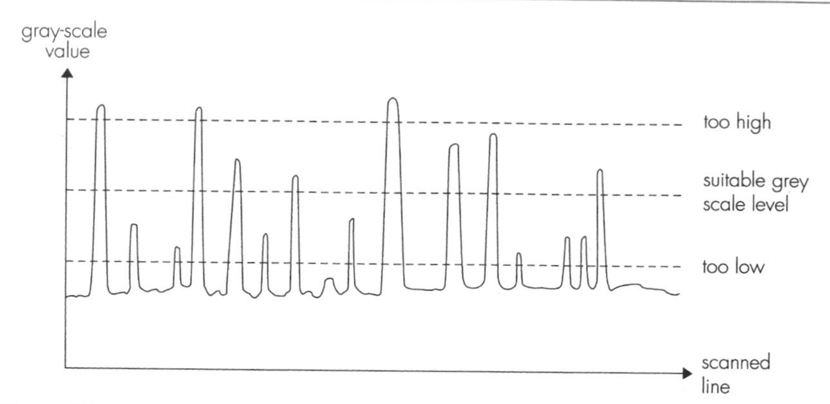

Figure 9.7
Gray-scale values along a scanned line. The peaks appear as lines on the map, whereas the bases are shown as white areas. Tests are carried out to establish the correct gray-scale value for each map being scanned.

one or zero, depending on whether the gray level is over or under the threshold set. Scanning produces voluminous data. With a pixel size of 50 μm, there are 400 pixels per square millimeter and six pixels in a line 0.3 mm wide. A 60- $\times$ 80-cm map encodes to about 200 million pixels.

Pixels are stored in rows and columns, and many adjacent pixels have identical values. Therefore, programs are available to compact the aggregate amount of data stored. Binary data are particularly suited for compression: run-length encoding compresses data by about 90%, and some advanced compression techniques can attain 98%.

Usually, a collection of data has a prefix header, which contains such details as the transformation parameters, terrain coordinates of the lower left and upper right corners, the pixel size, and so on. Each element of an

150	67	155	152	149	147	
139	59	158	160	162	154	
144	66	151	161	155	160	
162	158	72	162	153	154	
159	158	157	70	167	167	
153	145	156	75	166	165	
146	156	155	68	151	152	
155	157	156	161	65	160	
171	160	161	162	150	61	

(*a*)

0	0	1	0	0	0	0
0	0	1	0	0	0	0
0	0	1	0	0	0	0
0	0	0	1	0	0	0
0	0	0	0	1	0	0
0	0	0	0	1	0	0
0	0	0	0	1	0	0
0	0	0	0	0	1	0
0	0	0	0	0	0	1

(*b*)

Figure 9.8
Digital raster image: gray level array/binary (black, <80; white, >80). On the left is a raster image with gray-scale values. The corresponding binary-encoded raster image is shown on the right. All gray-scale values equal to 80 or more are lines on the map and have been given binary representation 1; the remaining gray-scale values are white areas on the map and have been given binary representation 0.

image contains only one gray-scale level or possibly one binary value but no information as to what this value represents or whether it has any logical connection with the adjoining element. A scanned map image thus contains no information on inner linear or area structures. Therefore, it is not possible to link attributes to structures in order to define whether they are roads or rivers, properties, or buildings. From a visual standpoint, however, it is possible to see structures in the data, but the data themselves supply no information on cohesive lines or areas. In popular terms, a scanned image can be described as "unintelligent."

Data editing and quality improvement

Where raster data are intended to be used only as a "dead" background image, the data will require minimal or no further treatment and can be used either as gray-scale images or binary data. There are several ways of carrying out quality enhancement of scanned raster data. These can comprise:

- Upgrading raster data through division into three categories: shapes, potential symbols, and noise
- Noise removal and pattern recognition of symbol candidates in raster data
- Thinning and vectorization
- Separating vectorized data into shapes, potential symbols, and noise
- Vector data pattern recognition of symbol candidates
- Error correction
- Coding
- Supplementing missing data
- Forming topology

The last four points are identical to those for quality improvement of data from a digitizing table. *Noise,* a term borrowed from acoustics, designates such information as stains and extraneous lines that are registered but not required. Data may be improved by structuring binary information in three categories: shapes, potential symbols, and noise. The noise category can then be filtered out, together with the potential symbols. If the data are to be used solely to compile a digital map image, only the noise need be filtered out.

Raster data contain no codes for map objects, but whenever map information is divided among various overlays (topography, inventories of facilities, land use, etc.), each overlay may be scanned individually to produce data relevant to its single theme. Each thematic overlay can then be converted to vector data with a common thematic code.

Raster to vector: vectorization

Some scanners vectorize as they scan. Otherwise, vectorization is usually part of the subsequent editing process. The vectorization of structures (lines, symbols, etc.) may be summarized in six steps:

1. A number of pixels forming a structure, such as a line, are registered.
2. All pixels transverse to the line, save those in its center, are stripped off (skeleton plotting).
3. Starting at one end, the pixels are connected one by one along the line (linearization).
4. Line curvatures are checked against set maxima which, if exceeded, indicate that the line is no longer straight, so linearization terminates.
5. Coordinates are determined for the start and end points of the straight-line segment, and a vector along the segment is formed accordingly.
6. Little by little, lines and structures are assigned coordinates and vectorization continues.

The result of vectorization is a continuous structure delineated by a sequence of coordinates designating the midline of the structure. A magnified illustration of a line of pixels is shown in Figure 9.9.

Vectorization may be subject to errors and defects, including:

- Deformations or interruptions of lines intersecting at nodes
- Vectorization of extraneous stains and particles on the original map
- Vectorization of alphanumeric information and text
- Unintentional line breaks, resulting in divided vectors
- Dotted-line symbols (trails, soil-type boundaries, etc.), resulting in many small vectors
- Smooth curves that become jagged (i.e., introduction of unwanted inflection points)

Many systems include error correction routines to circumvent these problems and assure scanning quality.

Some systems have a tracer too which, when started in a line structure of raster data, follows it more or less automatically on screen. The

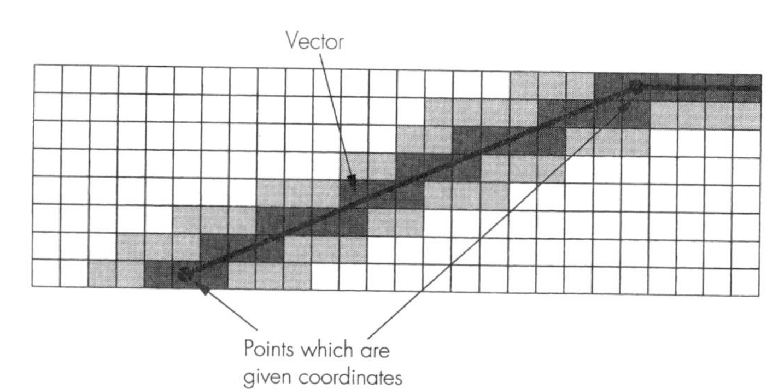

Figure 9.9
Vectorization of a scanned line on a map. The darkest pixels are the skeleton that is used under vectorization; the light rastered pixels have been stripped off; the pure white pixels appear as a white area.

process includes continuous vectorization of raster data. All operations are on screen, so the operator need intervene only in conflicts. The operator can steer the system into special line structures, such as tracing water pipelines on a scanned utility network map, thus allowing the data content to be separated into themes. No other map information, other than that which is being traced, will be vectorized.

Pattern recognition

In scanned data, numbers, letters, symbols, and other map information cannot be differentiated from one another directly; all information comprises pixels of gray levels. However, *pattern recognition programs* are available to analyze raster and vector data and to elicit features of a known pattern. The recognition of shapes and symbols is best performed on raster data, as vectorization invariably involves a loss of information. Nonetheless, raster data are usually vectorized prior to recognition.

To ease recognition, vectorized data are usually divided into data concerning shapes and relevant to potential symbols (after noise data have been filtered out). Recognition of letters and numbers is particularly useful, as is that of lines and specific map symbols (Figure 9.10).

Pattern recognition usually involves a system that is "trained" to detect specified patterns by assigning probabilities to their possible presence. Patterns are usually specified analytically or with the use of templates (model patterns) for comparison. For example, pattern recognition is used in the scanning of nautical charts in which the original registrations often comprise a multiplicity of depth numbers and letters. Recognitions as high as 95 to 98% have been attained. However, results of pattern recognition are highly dependent on the quality of the original.

Coding and other editing of vectorized raster data

Scanning and vectorization invariably introduce errors and obliterate some data. Consequently, data must be verified and corrected. Vectorized data have no thematic codes, so the various map themes cannot be distinguished (unless pattern recognition has been employed) and map objects have no ID codes that link them to attributes. The data

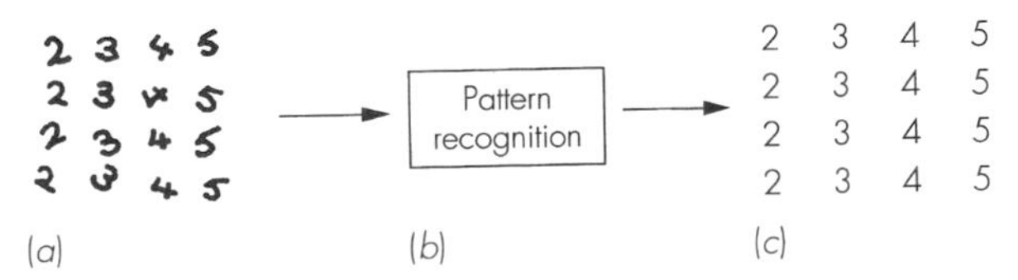

Figure 9.10
Pattern recognition. The figures in the binary raster image (*a*) are thinned, vectorized, and pattern recognized (*b*) and then translated to digital numerical values (*c*).

must therefore be coded by assigning codes to individual objects; this is usually achieved manually on screen. Once vectorized data have been coded, the subsequent manipulations are the same as for vector data. Where a map is scanned from several thematic layers or where line tracing is carried out on one or more themes, each group of vectorized data can be assigned a theme code.

Integrating image data in GISs

Some GISs can manipulate and display both vector data and raster (image) data on screen. In principle, raster data may be:

- Scanned maps
- Scanned documents, photographs, etc.
- Satellite data
- Video images of maps, photographs, aerial photographs, etc.

Vector data and raster data may be displayed simultaneously on a screen supported by storage split between raster and vector data and accompanied by an editing program for both types of data. Raster data are normally represented as gray-tone images. To be meaningful, the two sets of data entered must employ the same coordinate system for the images displayed. The relative positions of pixels, stated in row and column numbers, are transformed to terrain coordinates using selected points whose coordinates are known (e.g., the corners of houses, grid intersections, and road intersections). The transformations may be conformal, affine, or of another type, depending on the systematic errors in the scanned data (Figure 9.11).

Packed and stored raster data must be unpacked prior to being displayed on screen. Unpacking may be time consuming, but once it is

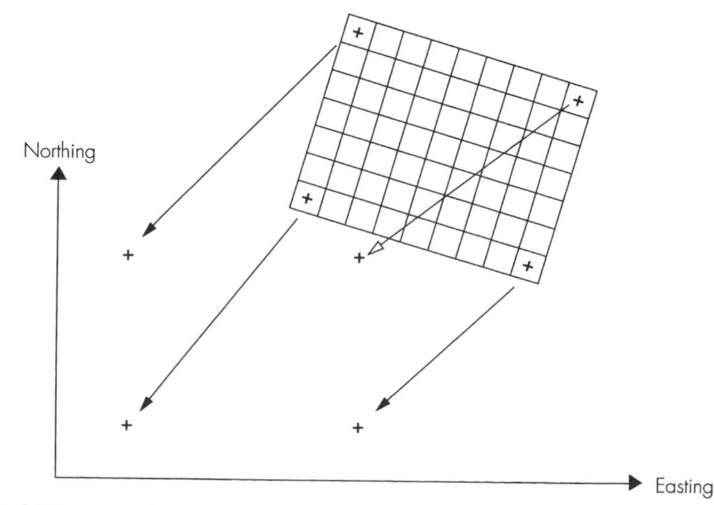

Figure 9.11
Transformation of raster registrations to a coordinate system.

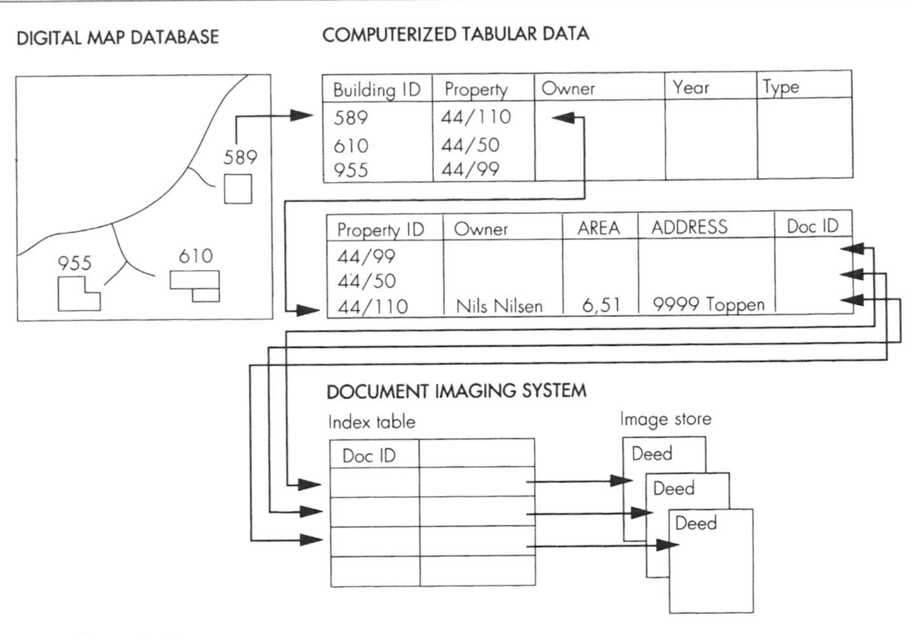

Figure 9.12
The integration of document and scanned images within GIS makes it possible to point at a building and bring up an image of the corresponding title deed.

completed and the unpacked data are stored in the working memory, the manipulation response time is satisfactory. Scanned map data may be used to compile a base image whenever the map generated is for orientation only. An example is the pipe data of a network to be documented in raster form, but for which the pipe routes are in vector form. Other examples include satellite data and image integrated in vector GIS. Applications of raster images include forming the bases for interactive correction and the amendment of vector data.

Scanned documents, video images, and other inputs may be integrated in GIS. Raster data may be stored on a video disk or a hard disk. A raster image and the GIS into which it is entered are linked via a table of identifiers of the map objects and pointers that indicate their relevant storage locations (Figure 9.12). Video images from a video disk can usually be screened more rapidly than scanned map data from a hard disk, which must be unpacked from storage.

9.4 Manual Digitizing or Scanning

The advantages of manual digitizing are that it:

- Can be performed on inexpensive equipment
- Requires little training
- Does not need particularly high map quality

The disadvantages are that it:

- Is tedious
- Is time consuming

The advantages of scanning are that it:

- Is easily performed
- Provides rapid compilation of digital map images

The disadvantages are that:

- Raster data files are large and consume memory space.
- Attribute data cannot be linked directly to line and polygon structures in raster data.
- Raster data lack "intelligence" and hence can be used only as bases for drawings.
- Improvement of data quality is labor intensive.
- The lines on enlarged raster maps may be annoyingly thick.
- Clean maps with well-defined lines are required.
- Relatively expensive equipment is required.
- Expert personnel are required.

Although scanning produces large quantities of data, compressed raster data need not require more memory space than vector data of customary format (i.e., ASCII). One reason is that raster data lack the voluminous drawing instructions that are essential in vector data. During the 1990s, manual digitizing has been the technique used most frequently, although scanning may have the greater potential, since in many applications speed is crucial. Further developments in pattern recognition will undoubtedly contribute to more widespread use of scanning.

9.5 Aerial Photographs and Photo Interpretation

The concept of aerial photography was patented in 1855 by French writer and photographer Gaspard-Felix Tournachon (1820–1910), who wrote under the pen name of Nadar. In 1856, Tournachon daringly ascended in a balloon from a village near Paris and took the world's first aerial photograph. In 1909, aeroplanes were first used. Since then, aerial photography has become a discipline with its own theory and practice. Because aerial photographs are taken from above, all objects in them appear as viewed from various vertical angles. So aerial photographs almost always require interpretation in order to provide information for mapping or depicting details as seen from the ground. Of course, almost all photographs used for their information content are interpreted in some way, but the interpretation of aerial photographs for mapping purposes is most common.

The combination of aerial photography and photographic interpretation provides information on relatively large areas without necessitating survey inspection on the ground. Roads, lakes and rivers, buildings, farmland, and forests are clearly visible in aerial photographs (Figure 9.13). Other characteristics such as various types of vegetation, soil and geological formations are more difficult to interpret. Accurate interpretation depends both on experience and verification by ground control. Consequently, photographic interpretation has become a skilled profession. Interpretation is based on extracting information from the shape of objects as well as size, pattern, shadows, gray tone or color, and texture (by comparison with larger contiguous areas). Indirect indicators, such as surface shape, runoff patterns, locations, and seasons may also be used.

Aerial photographs may be black and white on film sensitive to wavelengths of 0.4 to 0.7 μm; color on film with layers sensitive to the 0.4 to 0.5 μm (blue), 0.5 to 0.6 μm (green), and 0.6 and 0.7 μm (red) ranges; or infrared on film sensitive to the range 0.7 to 1.1 μm. Black-and-white film is most commonly used. Infrared film, however, may assist interpretation, particularly of vegetation because plants emit varying radiation patterns throughout the infrared spectrum. Most

Figure 9.13
Aerial photographs have a resolution of 0.02 to 0.03 mm and therefore contain a considerable volume of information. Panchromatic black-and-white photos are often used, as shown. Information from the photo can be interpreted by means of gray tones, shadows, patterns, textures, and so on.

aerial photographs have a resolution, or smallest discernible detail, of about 0.025 mm.

Experienced photographic interpreters can map the details of vegetation, geology, and other terrain features from aerial photographs and a few ground controls. Almost all natural resource mapping is based on aerial photographs and photographic interpretation, as well as topographic maps. Borders and characteristics of the various objects depicted in a photograph are drawn both in the field and in the survey office.

Like most photographs, aerial photographs are taken through a lens that views a relatively large area and replicates it, inverted, in a smaller area on a negative. The geometry involved in this process is termed *central projection* (Figure 9.14). A point on one plane (the ground surface photographed) is projected onto a second plane (the negative), so that the original point and its image on the second plane lie on a straight line through a fixed point (focal point of the lens) not on either plane. Aerial photographs always involve some distortion, because their vantage point is not exactly vertical and because terrain is seldom either completely horizontal or completely flat. Conse-

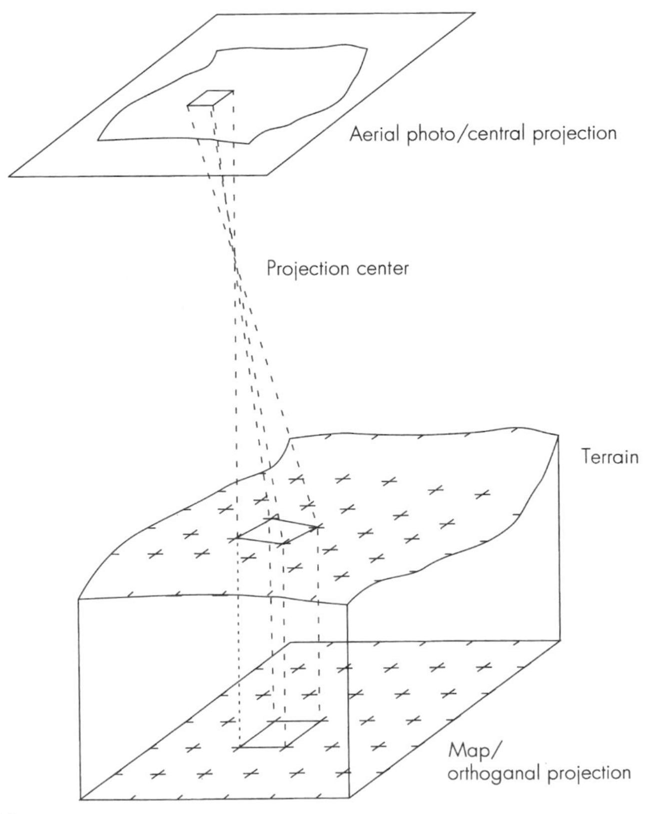

Figure 9.14
Central projection and orthogonal projection of terrain.

quently, the information in aerial photographs must be transferred to a ground coordinate system for computation of lengths and areas or before they can be entered in GIS with other map data. Where elevations in the terrain are known, aerial photographs can be transformed to true map images (orthophotos). Photographic data are used to create and revise maps using a variety of photogrammetric instruments (described further in Chapter 10).

9.6 Remote Sensing

Satellite remote sensing of the Earth began in the 1960s, when the technical capabilities of satellites operating in Earth orbit converged with the increasing ability of computers to manipulate large quantities of data. Two observation modalities evolved: *passive optical,* which deals with reflected sunlight and reemitted thermal radiation, and *microwave,* which, like radar, deals with the transmission and reflection of energy in the microwave portion of the radio-frequency spectrum. Apart from the nature of the phenomena observed, the difference between the two modalities is technical. The optical spectrum is considered to extend from the very short wavelength in the ultraviolet region to a wavelength of 1000 μm in the longer infrared region. This is the short-wavelength limit to radiation which can be generated and detected conveniently by microwave electronic devices. In mid-1991, the European Space Agency (ESA) launched the ERS-1 Earth Resources Satellite with a payload containing radar imaging instruments.

The two major earth resource scanning satellite systems in operation, which are relevant to GIS, are of the passive optical type. Landsat is the generic name for a series of five satellites launched between 1972 and 1984 by the United States; SPOT, an acronym for *Systeme Probatoire d'Observation de la Terre,* uses satellites launched by France between 1986 and 1997 (Figure 9.15). Both systems are used primarily for mapping vegetation, geological features, and soil types, and to some extent for water parameters and linear structures such as roads and rivers. SPOT can register three-dimensional images for processing to yield gross elevation data.

Every second a satellite carrying sensors that perceive reflected solar energy views a "scene" on Earth. In principle, the sensors might perceive energy at wavelengths anywhere in the optical region of the electromagnetic spectrum, from 0.4 to 1000 μm. Traditionally, the optical region is divided into subregions, each of which is termed a *spectrum* (Figure 9.16). Sensors are available for all spectra, from ultraviolet to far infrared, but not all may be used in optical remote sensing. This is because not all portions of the total optical spectrum are transmitted equally through the atmosphere; many are absorbed, some

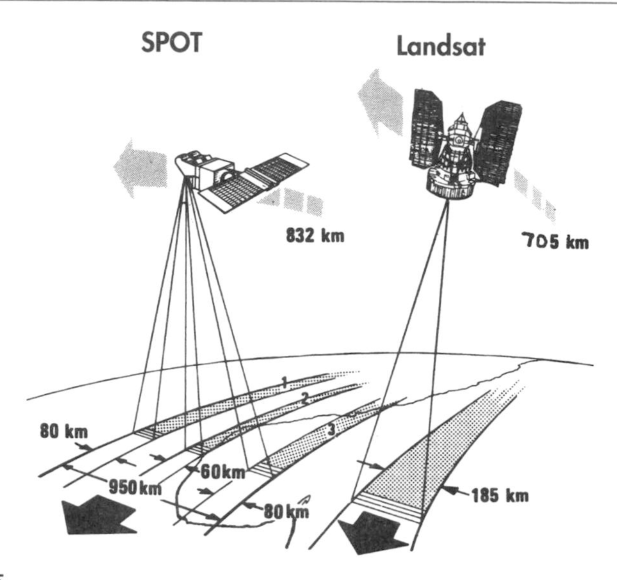

Figure 9.15
Satellite in orbit. Mapping satellites trace a continuous track over the surface while circling the Earth. The reflected electromagnetic radiation from each swath is registered as digital values from which images can be constructed. *(a)* SPOT; *(b)* Landsat (NOU 1983:24 Satellitfjernmåling).

almost completely (Figure 9.17). So the variations of atmospheric transmission, and absorption with wavelength, limit the optical spectrum that is useful in remote sensing to transmission bands, or *windows,* in the range 0.4 to 15 μm.

Sensors on satellites operate in a similar way to those in camera light meters, but instead of measuring energy only in the visual spectrum as a light meter does, they may be arranged in groups to measure energies in several spectra, from the visible through to the mid infrared (Figure 9.17). *Multispectral scanning,* as it is termed, results in Earth image data in each of the windows scanned. The spectral windows are chosen to monitor the wavelengths of greatest interest. Water, woods, cultivated fields, and other areas are distinguished by characteristic reflected and reemitted energy spectra, which are recognized in the satellite images (Figure 9.18).

Spectrum name	Wavelength (μm) range
ultraviolet	shorter than 0.4
visible	0.4–0.7
near infrared	0.7–1.0
solar reflected infrared	1.0–3.0
mid infrared	3.0–15.0
far infrared	longer than 15.0

Figure 9.16
Spectrum classes.

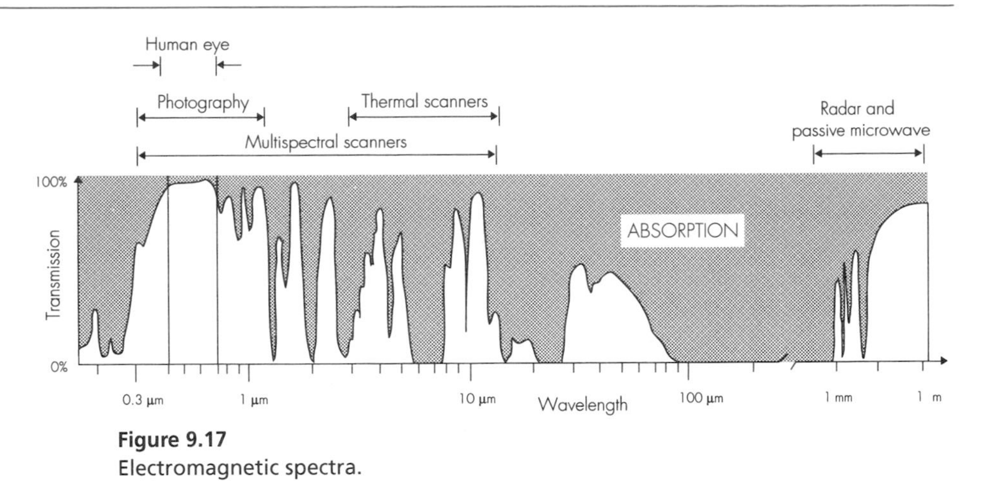

Figure 9.17
Electromagnetic spectra.

The images are compiled successively along a *track* (path) of satellite motion as projected on the surface of the Earth. Each sensor in a satellite comprises numerous detectors, each of which registers energy reflected from a small square on the surface of the Earth. Data from the squares are stored as image elements (pixels) in the satellite, with digital values corresponding to radiation intensity. Together, the sensors perceive a cross-track swath, with the motion of the satellite along the track generating the raster images of the surface scanned. Data are transmitted via microwave link to Earth stations.

The SPOT satellite orbit is polar (passing nearly over the poles) at an altitude of 832 km. The satellite is solar synchronous; that is, once every 26 days it passes over the same point on Earth at the same time

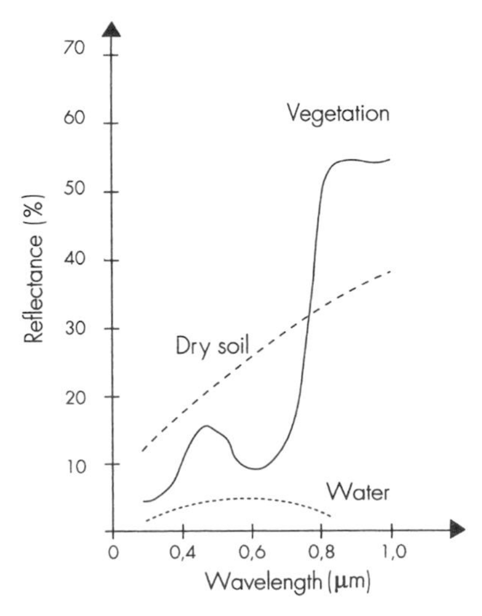

Figure 9.18
Typical spectral reflectance curves for vegetation, soil, and water.

of day. In other words, it scans the entire Earth every 26 days. SPOT has a two-sensor system. One sensor registers the visual spectrum with a resolution, called its *instantaneous field of view* (IFOV), of 10 × 10 m. The other is a multispectral sensor that registers three channels (green, red, and infrared) with a resolution of 20 × 20 m. A single satellite image, or scene, covers an area of 60 × 60 km. The SPOT sensors can be arranged to take three-dimensional images. SPOT can be programmed from the ground to change the recording angle so that the same geographical area can be covered from different positions in space. This allows for a much more frequent coverage than every twenty-sixth day.

The orbits of the five Landsat satellites are both polar and solar synchronous, at altitudes of 918 km for the first three satellites and 705 km for the other two. Each satellite scans the entire Earth every 18 days. The Landsat 4 and 5 satellites have two types of sensor. One, the Thematic Mapper (TM) sensor, registers seven channels in the visible and infrared spectra with an instantaneous field of view (IFOV) of 30 × 30 m, except for the thermal infrared channel, which has an IFOV of 120 × 120 m. The other, the Multi Spectral Scanner (MSS) sensor, registers four channels with an IFOV of 80 × 80 m. A single Landsat scene covers 185 × 185 km. Because they perceive reflected solar energy, almost all the SPOT and Landsat sensors fail to "see" through cloud cover or in darkness. The single exception is the Landsat mid-infrared spectral band sensor, which perceives reemitted energy; it is often referred to as *thermal infrared sensor.* See Figure 9.19 for a comparison of the characteristics of the Landsat 4.5 and SPOT satellites.

In recent years, several new satellites have been launched which supplement and partially replace data from Landsat and SPOT. The Russian Resurs satellite was launched in 1994. This has the same channels as Landsat TM but a geometric resolution which is no better than 170 m. An Indian satellite, the IRS-1C, has been operative since September 1996. The IRS-1C offers a spatial resolution of 5.8 m in panchromatic mode (0.50 to 0.75) and 23.5 m in multispectral mode (0.52 to 0.59, 0.62 to 0.68, 0.77 to 0.86, 1.55 to 1.75). The ground-swath width is 142 km.

There are plans to launch several new resource satellites with high-resolution sensors before the end of the decade, including the IKONOS, which will offer a panchromatic image of the earth's surface at 1-m resolution, and color image (blue, green, red and near infrared) of 4-m resolution; the ground-swath width is as small as 11 km. The Early Bird was planned with two sensors: one that collects 3-m resolution in panchromatic mode and another that collects 15-m resolution color images (green, red, and near infrared). Early Bird had an unsuccessful launching in December 1997. OrbImage has a more uncertain launch date but is planned with 1- and 2-m resolution in panchromatic, 4-m resolution in color and 8-km ground-swath width.

Sensor	Band	Wavelength (μm)	Resolution (m)	Ground–swath width (km)
Landsat (4,5)				185 × 185
MSS	4	0.5–0.6 (green)	80	
	5	0.6–0.7 (red)	80	
	6	0.7–0.8 (near IR)	80	
	7	0.8–1.1 (near IR)	80	
Landsat (4,5)				185 × 185
TM	1	0.45–0.52 (blue)	30	
	2	0.52–0.60 (green)	30	
	3	0.63–0.69 (red)	30	
	4	0.76–0.90 (near IR)	30	
	5	1.55–1.75 (mid IR)	30	
	6	10.4–12.5 (thermal IR)	120	
	7	2.08–2.35 (mid IR)	30	
SPOT	Multispectral	0.50–0.59 (green)	20	60 × 60
		0.61–0.68 (red)	20	
		0.79–0.89 (reflected IR)	20	
	Panchromatic	0.51–0.73 (green)	10	60 × 60

Figure 9.19
Characteristics of Landsat 4.5 and SPOT.

9.6.1 Correction

Radiometric correction

The measured magnitude, or radiometric value, of each pixel may be degraded by noise caused by:

- Sensor inaccuracy
- Atmospheric attenuation and scattering of reflected energy
- Variation of reflected energy as solar incidence varies with the curvature of the Earth, such as shadow effects
- Variation of reflected energy due to changes of surface slope

As most of these causes of pixel noise cannot be quantified directly, radiometric corrections are based on assumptions and models. The goal is to apply corrections so that the pixel values representing any one phenomenon will be the same, regardless of their locations in the overall image.

Geometric correction

Geometric correction is necessary if the data generated are to be used in mapping or for performing various analyses. The geometrical accuracy of data are influenced by:

- Rotation of the Earth during sensing
- Curvature of the Earth
- Variations in surface elevation
- Satellite instability in orbit

- Instability of recording devices
- Inaccuracies in the sensor projection system
- Atmospheric aberrations

Geometric correction of satellite data may be performed in various ways at two levels of precision:

- *Level 1:* systematic errors due to:
  - Rotation of the Earth
  - Instabilities of the satellite and its instruments (panorama effect, striping, angle of observation)
- *Level 2*
  - Internal image improvement
  - Transfer to chosen cartographic projection
  - Reduction to mean terrain elevation
  - Transformation based on map coordinates
  - Corrections for terrain variations

After all corrections have been made, geometrical accuracy may actually be better than resolution. For example, data with a resolution of 10 × 10 m may be geometrically accurate to ± 6 to ± 7 m, and data with a resolution of 80 × 80 m may be geometrically accurate to ± 50 to ± 70 m. This implies that these data are best used for overviews of areas that were poorly mapped previously, such as in developing countries.

9.6.2 Analytic methods

As a rule, satellite images must be analyzed and interpreted to yield the information desired. Satellite data are digital, so analyses and interpretations may be partially computerized. Three analytic methods are now common:

1. Color enhancement
   a. RGB color adapting
   b. HLS color adapting
2. Classification
   a. Supervised classification
   b. Unsupervised classification
3. Segmenting

In practice, these three methods are often used together.

Color enhancement

Satellite image data have no colors in the sense that the eye perceives color in the visible spectrum. All coloring of satellite data relayed to Earth is done after the data are compiled into images. Normally, the data are processed interactively on screen in order to compile color images that enhance the themes of interest in mapping, such as coniferous and deciduous forest, cultivated fields, and so on. As the colors do not need to be the same as those seen by the unaided eye in daylight, the images created need not resemble conventional color photographs. In fact, in most cases they do not.

Colors are assigned to the individual spectral bands perceived by a satellite sensor assemblage in order to yield a *false-color-enhanced image.* For instance, red, green, and blue (RGB) colors may be assigned to three LANDSAT spectral bands to produce an image in which deciduous trees appear orange, coniferous trees brownish, earth and rock blue, and water black.

The basic colors may be adjusted individually, although the disadvantage of the simplest approach—basic color adjustment—is that adjusting any one color affects the aggregate image. For instance, amplifying the red basic color in an image not only makes the red pixels redder, but also adds yellow to the green pixels. Often, a more suitable method of improving false-color-enhanced images is to adjust the color hue, lightness, and saturation (HLS). The final color-enhanced images may be interpreted in the same way as aerial photographs.

Classification

Classification is based on the assumption that like phenomena in the image belong to the same class on the ground and have similar characteristic spectra; that is, they have a *spectral fingerprint.* There are two approaches to classification that are distinguished primarily by their initial assumptions. In *supervised classification,* ground truth data from direct in-field observations are used to identify the initial parameters used in classification. *Unsupervised classification* needs no ground truth in its initial stages, but the final images and maps produced must almost always be verified in the field.

Supervised classification

Supervised classification begins with a compilation of *training data.* First, each class to be mapped, such as water, cultivated fields, pine trees, fir trees, and so on, is named and assigned a color. Next, a skilled analyst works on screen, circling areas known to contain the classes of objects named. Spectral band means and deviations are calculated for each encircled area. In other words, ground truth data are used to "supervise" subsequent manipulations.

Once the training data are compiled, the analyst initiates automatic classification of all classes with ground truth in the entire image. Each pixel is classified according to the greatest probability of its belonging to a particular class. Hence the entire image may be classified using relatively few in-field observations. Various statistical methods are used to assign pixels to classes, including contextual classification which takes account of the values of neighboring pixels. The pixels of the classified image displayed on screen are colored according to the class colors assigned in the training data.

Unsupervised classification

Supervised classification requires ground truth data when starting the classification. The classification is also limited to the classes with ground truth. Unsupervised classification is used to overcome these obstacles. Unsupervised classification starts with an *unsupervised cluster training procedure.* The first step is *clustering,* a way of ordering data by sorting pixels into classes according to their spectral values (Figure 9.20). In other words, clustering comprises classification by spectral characteristics rather than by relation to ground truth observations. The analytical procedure employed may be interactive and includes four main steps:

1. Each cluster is assigned a value corresponding to its center.
2. Every pixel in an image is placed in the nearest cluster most closely matching its value, according to a method of distance measuring.
3. When all pixels of an image have been assigned to clusters, new centers are calculated to form the means of the constituent pixels.
4. Repeat steps 2 and 3.

The process stops when the clusters no longer change between successive iterations. The clusters are then labeled by assigning colors and in-field observations used to liken the colored clusters to ground phenomena.

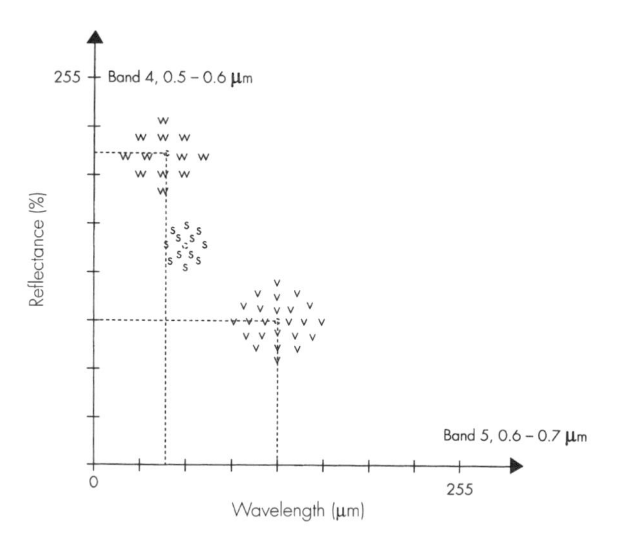

Figure 9.20
Clustering is a process whereby pixels that are adjacent in a spectrum are grouped together. Several parts of the spectral bands can be used simultaneously. w, water; s, soil; v, vegetation.

Segmenting

Segmenting is usually performed automatically by a computer program. It starts with the comparison of a pixel value with the values of its eight immediately neighboring pixels. Of these eight, those with values—within set limits—that are closest to that of the start pixel are selected. These pixels are then compared to their neighbors, which are also selected if their values are within the set limits. The pixels selected form a segment. The process is repeated until no more neighboring pixels fulfill the criterion for selection. At this stage, a new start pixel in a new area is chosen, and a new segment is built up successively. Segmenting thus takes account of both geometry and pixel values in grouping pixels.

9.6.3 Integrating satellite data in GIS

The data of a geometrically corrected satellite image that were acquired via a satellite spectral band may be regarded as a thematic layer of pixels. Multispectral data can therefore constitute a GIS overlay. Images taken at other times or by other satellites may be adapted to the same map projection, coordinate system and pixel size, and thereby form a new GIS overlay. Map overlays in raster form (geology, vegetation, etc.) may be integrated in the same system provided that their respective geometries and map references are the same as those currently used in the system.

The classified raster data may be converted to vector data by using suitable conversion functions. The data may then be entered in vector GIS. It is also possible to use data from different sources simultaneously. For example, SPOT PAN can be used together with Landsat TM to take advantage of SPOT's good geometric resolution and Landsat's many channels. To combine the two, data from Landsat with 30-m resolution is resampled to data with 10-m resolution. This results in better quality of the end products but is, of course, a more expensive solution.

CHAPTER 10

Data Collection II

10.1 Surveying

In surveying, measured angles at, and distances from, known points in a coordinate system are used to determine the positions of other points (Figure 10.1). Surveying field data are almost always set out in polar coordinates, which may be transformed to rectangular coordinates using the relationships

$$\Delta x = D \times \cos \varphi \qquad \Delta y = D \times \sin \varphi \tag{10.1a}$$

$$X_p = \Delta x + X_1 \qquad Y_p = \Delta y + Y_1 \tag{10.1b}$$

Heights are computed from horizontal or inclined distances (Figure 10.2). Vertical angles are computed from known points to new points, or vice versa. The corresponding transformations between polar and rectangular coordinates are

$$\Delta e = D \times \cos Z \tag{10.2a}$$

$$E_p = \Delta x + E_1 + i - s \tag{10.2b}$$

Surveying has traditionally produced numerical data in the form of angles measured by transits and theodolites and distances measured by tapes and chains. Starting in the 1950s, however, the traditional optical and mechanical tools of the surveying trade were gradually supplanted by more accurate electronic and electrooptical devices for measuring distances. Modern surveying is computerized: A *total station theodolite*, as illustrated in Figure 10.3, can now store and process digital measurement data either in the field or, subsequently, through an interface to a computer. Data may be coded or assigned themes in subsequent computer editing.

Surveying point location accuracy usually varies from ± 1 to ± 10 cm, depending on how measurements are made. Measurement speed depends on topographical variations and the degree of modeling desired. As a guideline, surveying of complete terrain details can cover about 1 to 2 decares per hour. Surveying is usually employed whenever up-to-date information is needed for reference marks as the basis for the measurement of terrain details, on property lines, buildings, manhole locations, and the like, as well as for smaller areas and for details that have to be measured accurately. Photogrammetry is usually employed in mapping larger areas.

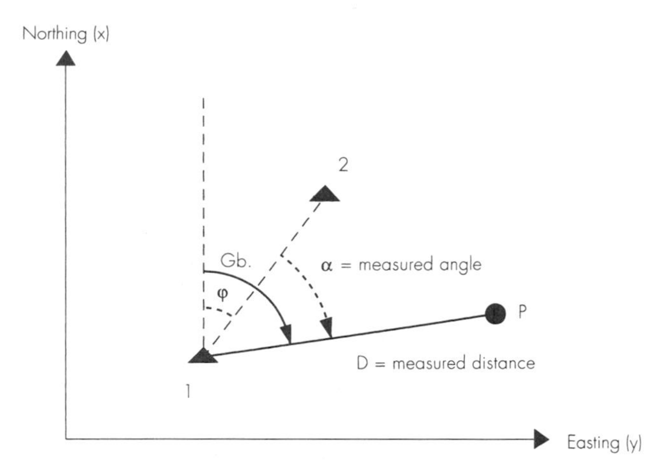

Figure 10.1
The geometrical principles of surveying (polar coordinates). Polar coordinates with direction and distance can be converted to right-angle coordinates.

Computed coordinates are most often used to supplement and update existing maps, and coordinate lists are compiled for data on survey control stations, properties, manholes, and other fixed objects. Normally, surveying requires two people, one to operate the instrument and one to walk the terrain and select survey points with a measuring prism. Other systems allow the instrument to be remote-controlled, and the same person can operate both instrument and prism. Together with a penpad, they form a complete one-person measuring station. The choice of survey point is decisive for how reality can be reproduced in GIS. In situations where a high degree of accuracy is required, special instructions and measuring procedures are often created. Measured data are coded directly under measuring, in relation to both connectivity and object type.

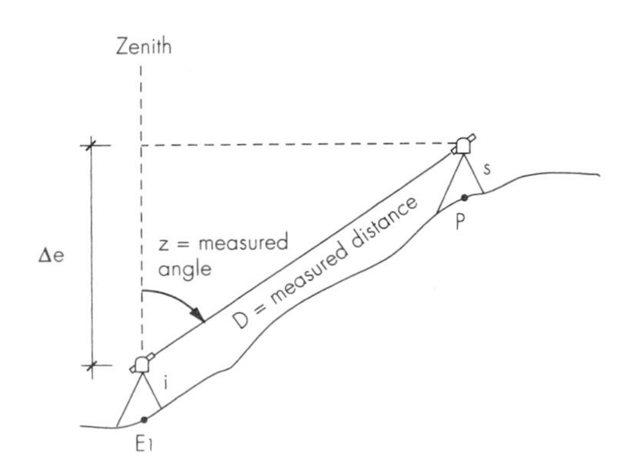

Figure 10.2
Surveying measurement of height. Height differences can be calculated from the measured zenith distance and the given or measured distance.

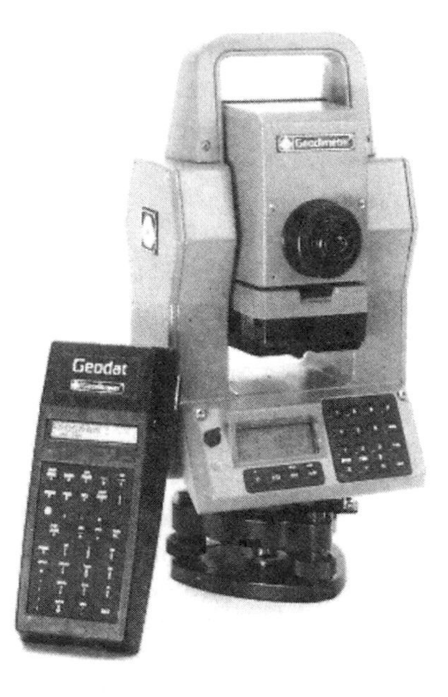

Figure 10.3
Most surveying is carried out using modern equipment known as *total stations.* With this equipment it is possible to carry out measurements in the field and register observations through an interface to larger computer equipment in the office. (Courtesy of Geodimeter.)

Entry of coordinates

Survey data may be entered in GIS whenever accurate map data are required, for example, when determining property borders for legal purposes. New data that are intended for EDB are transformed in digital form to GIS. However, larger amounts of older analog data are available where coordinates have to be keyed in manually, either directly to GIS or via a simple registration program. Data for each property should be verified, by identifying nodes and closing polygons, as entry progresses.

This process is time consuming—two or three times longer than the corresponding task using a digitizer, inclusive of all preparatory and editing tasks. However, the data produced are easier to apply than digitizer data.

10.2 Satellite Positioning System

Global positioning system

The global positioning system (GPS) is a military satellite-based navigational system developed by the Department of Defense in the United States. GPS controls 24 operational satellites (21 in operation); new satellites will be launched to replace older ones which do not function satisfactorily. Not only are their orbits highly predictable, but

GPS and electronic chart display and information system

The global positioning system (GPS) may be integrated with an electronic chart display and information system (ECDIS). As discussed in this section, the normal accuracy of GPS is ±100 m. Accuracy better than ±10 m, which may be needed in narrow channels, is available in differential GPS, which requires coastal reference stations. The accuracy provided by GPS and differential GPS is better than that of many older nautical charts. This means that accurate nautical charts are a prerequisite for the sensible use of GPS.

Updating is essential to the successful use of ECDIS, as navigational conditions continually change. National cartographic and hydrographic agencies regularly issue updating information, which may be entered into an ECDIS. Test are now being conducted on transmitting updating information to ships via the INMARSAT satellite communications system.

An ECDIS incorporates three information subsystems:

1. An electronic navigation chart (ENC) that combines nautical information, such as lighthouses and markers, and hydrographic information, such as coastlines and sea-bottom topography, for display on screen. The basic information is usually digitized from traditional nautical charts.
2. Position information from the ship's radar and positioning system.
3. Steering information from the ship's steering system.

ECDIS provides a complete electronic navigation and warning system that displays the ship's position and other information vital to navigation. The system also provides audible and visible warning signals in the event that:

- The course of a sailing or a planned sailing is charted through non-navigable waters (anti-aground).
- The ship deviates from its planned course.
- A planned change of course is not made.
- The ship does not deviate from a course of potential collision with other moving objects (anticollision).

If radar images are entered into the ECDIS, other vessels and their maneuvering in nearby waters may be displayed. If an autopilot is connected, the ECDIS can follow a planned course automatically. Changes of course are made after the navigator enters an acknowledgment of a warning of course change. As is the case in GIS, an ECDIS user can build up a database containing descriptions (attributes) of objects shown on digital maps.

the satellites themselves also carry atomic clocks, which enable them to transmit highly accurate radio signals. As illustrated in Figure 10.4, position on the ground is determined by using a small, portable receiver to receive and compare the signals from three satellites. For each signal received, the interpretation of codes or measurement permits computation of the satellite's position in space in relation to the

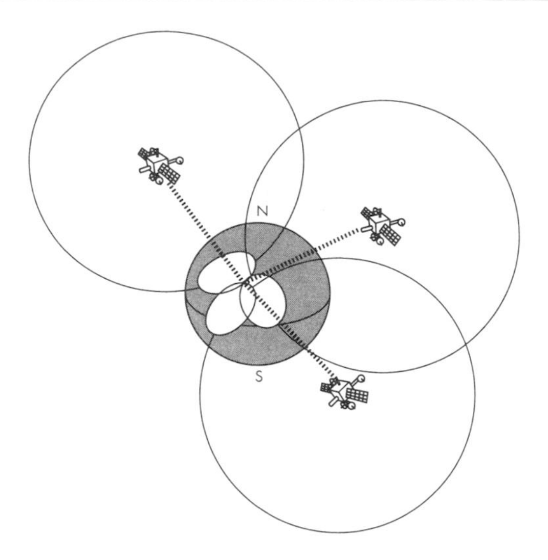

Figure 10.4
Global positioning system (GPS). GPS positions are determined from the intersection of cones.

earth's center and the distance between it and the receiver. With these data for each of three satellites, the receiver's position in *x*, *y*, and *z* orientation is computed from the unique intersection of three cones, for which the apexes are at the satellites. Positional accuracy may be improved by cross-checking the signals from four or more satellites.

Since we operate in three-dimensional space, it is correct to say that with GPS, vectors (linear sections in a given orientation) are measured instead of distances. Treatment of the measurement readings is thus based on calculation of the vector in a coordinate system with the earth's center as origo. GPS is based on the WGS84 as datum and geocentric coordinate system, which may be transformed to other coordinate systems. GPS allows the position of points throughout the world to be determined 24 hours a day in all kinds of weather conditions.

GPS now has two accuracy codes. The most accurate, P-code, is reserved for the military, while the C/A-code is available for civilian uses. Both codes were freely available during the test phase of GPS implementation, but the military code has not been available to civilian users since the system became fully operational in 1993. The military P-code gives an accuracy of approximately ± 30 m, while the C/A-code currently provides positions with an accuracy of approximately ± 100 m (± 150 m in elevation) with no further corrections.

Accuracy can be increased further by using data from two receivers simultaneously, provided that the position of one of them is known accurately (Figure 10.5). Known as differential GPS, this system involves either code measurements or phase measurements; the latter give the greatest accuracy. In addition to measuring methods, accuracy is dependent on the distance between stations, measuring time,

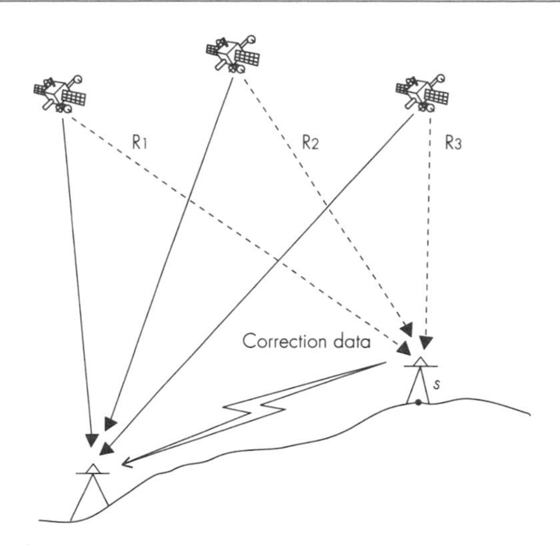

Figure 10.5
Differential GPS. GPS values are measured at two stations, one of which has known coordinates. Correction values can then be used to calculate the exact position of the unknown point, making it possible to calculate the exact position of survey control stations on land or of mobile objects such as boats and cars.

and the satellite's position in relation to the point to be determined. Normally, accuracy will be reduced by a distance of about 0.5 cm per kilometer from the reference station. The satellite geometry, *dilution of precision* (DOP), can be controlled by choosing an observation period where the geometry is good. With a distance between the two receivers of less than 10 km, a measuring time of 10–120 minutes, and good geometry, phase measurements can result in position accuracy of ±1 to ±2 cm. Once a particular point has been determined, the neighboring point can also be determined with a few seconds of measurement time, called the *stop-and-go function.* This means that GPS can also be used for collection of terrain details. In addition, GPS can be connected in the field to special data screens for sketching, etc.

Accuracy in centimeters is attainable by virtue of extensive, rapid computations of error corrections. In the field, relatively straightforward real-time differential measurement and computations may result in a position being determined to an accuracy of ±10 m, as for cars and ships in motion. In some countries a service has been established that broadcasts GPS corrections over the radio and phone network. This is particularly useful in boat and vehicle navigation.

One of the drawbacks of GPS is that it requires a direct line of sight between the receiver and each satellite accessed. In many instances, particularly for mobile users, the terrain, buildings, trees, and other objects may block the satellites from view or may disturb the signals.

As we have seen in Chapter 6, it is difficult to transform elevations measured in relation to the geocentric coordinate system (Figure 10.6)

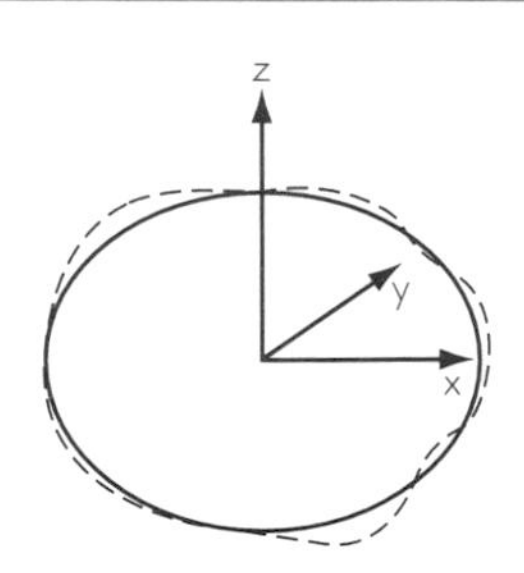

Figure 10.6
In GPS, the coordinates are expressed in relation to the Earth's center. Elevations are expressed in relation to the rotation ellipsoid, but these will not coincide with elevations stated in the national systems since these are based on the winding geoid.

to elevations measured in relation to the ellipsoid in the national elevation systems. Where the difference between geoid and ellipsoid in an area is unknown, the exact height to a GPS measured point can only be determined by leveling or by trigonometrical elevation measurement or by finding the transformation parameters by measuring the ellipsoid height (using GPS) at points with known geoid height.

GPS is most useful for:

- Locating new survey control stations and upgrading the accuracy of old station positions
- Measuring terrain features that are difficult to measure by conventional means
- Positioning offshore oil platforms
- Updating road data with a GPS receiver in a car
- Marine navigation, including integration with electronic charts
- Car navigation
- Determining camera-carrying aircraft positions to reduce reliance on fixed marks in aerial photography
- Determination of differences in elevation

GPS has been integrated directly with several GIS. In general, GPS data may be regarded in the same way as other digital map data. Coding of GPS data in GIS may be performed in the field to determine accurately the positions of intersections, road signs, the ends of bridges, road surfaces, and the like.

The cost of a GPS receiver depends in part on the accuracy it can achieve. High-precision receivers like those used in geodesy are the most expensive. At the time of writing, they are slightly more expensive than electronic theodolites. Low microprocessor costs and large-series production will certainly boost the market for GPS receivers and spawn new applications, not least systems that combine GPS positioning with GIS route finding for car navigation.

GLONASS

GLONASS is the Russian equivalent of GPS. GLONASS also consists of 24 satellites and is based on the same principles as GPS. With GLONASS there is no military interference with civilian signals, and the system has an accuracy of ± 25 m without use of reference stations. Receivers are available that can use both GPS and GLONASS simultaneously and thus obtain a better geometry in the measurements.

10.3 Photogrammetric Mapping

Photogrammetry means "picture measurement." The word describes a basic process first used in 1851 by A. Laussedat in France. Although Laussedat's work was original, it had little practical application because the lenses and photographic printing techniques of his time were too inaccurate for precision measurements. Toward the end of the nineteenth century, however, E. Deville, a Canadian, devised a stereoscopic machine to process photographs taken from theodolites. Subsequently, phototheodolites were developed specifically for that purpose. Phototheodolite techniques were eventually widely used, particularly for mapping the Alps of Central Europe. In 1909, the first instrument for map making was built. As in so many other fields, it was military necessity that accelerated development. During World War I, balloons and airplanes were found to be excellent platforms for photography. The airplane eventually proved to be the most successful because a single aircraft could fly along a prescribed course and take numerous overlapping photographs (Figure 10.7), which were then processed using various stereoscopic machines. The utility of the approach became so dominant that today photogrammetry is regarded as one of the techniques of aerial photography.

The stereo plotter is the instrument most commonly used to transfer aerial photographic information to planimetric and topographic maps which, in turn, may be the bases for vector data. In a stereo plotter, a photographic interpreter views two overlapping aerial photographs taken from different positions to form a three-dimensional image. Several systems for forming three-dimensional images are used. One of the earliest is the anaglyphic system, in which one photograph is projected in red and the other in green. When viewed through lenses with a red filter for one eye and a green for the other, the image acquires the third dimension of height. Another system (Figure 10.8) uses ray rods to reproduce (re-create) the passage of the ray and an optical system which ensures that each eye sees only one photograph.

By moving a floating point around in the three-dimensional image, the operator can draw in roads, rivers, contours, and other details. If the plotter has a digital transmitter, all movements used in drawing in the photogrammetric model can be numerically encoded and entered

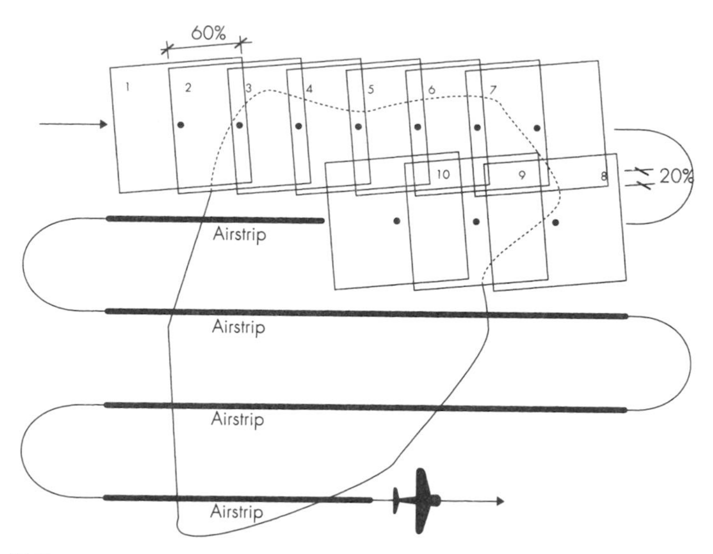

Figure 10.7
During aerial photography for mapping purposes, the entire area is covered by aerial photographs. There is usually a 60% overlap along each flight line and a 20% overlap between flight lines.

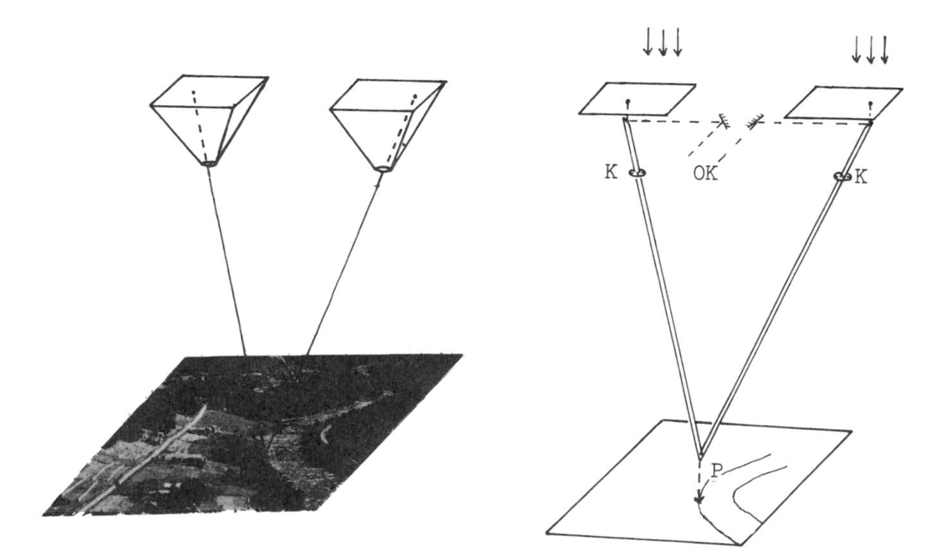

Figure 10.8
A stereo plotter; the system shown is the Wild autograph. Left: aerial photography. Right: a three-dimensional reconstruction of the terrain based on aerial photographs. The right image can be seen only with the right eye, the left image only with the left eye. When the images are positioned correctly in relation to each other, the brain will merge both image impressions into a three-dimensional model. The light beams in the model can be reconstructed in both analog and computerized forms (Ø Andersen 1980).

in computer storage. Many analog instruments, descendants of the original optomechanical photogrammetric instruments, are still used but are often rebuilt with digital encoder, although analytical instruments are the workhorses for GIS data capture.

In modern analytic plotters, software replaces the rods, and all movements of the instrument are digitally controlled. Stereo plotters of the latest generation use scanned digital images. The two overlapping images are shown on the same screen, with rapid exchanges of their polarized light. The operator uses glasses with synchronized polarized glass so that the right eye sees only the right image, the left eye only the left image. Computer programs now replace all the movable parts of previous instruments. This type of instrument uses a very powerful (and expensive) computer. A mouse, tracer ball, or similar is used to move a pointer around in the model. The digital images allow digital terrain models to be established very effectively through the pattern recognition for automatic determination of elevation points and profile lines. It is thus possible to produce orthophotos very efficiently using this type of instrument.

However, we have to be aware that automation cannot differentiate between the surface of the terrain and buildings, trees, and so on; thus the tops of buildings and trees will be part of the digital terrain model and may have to be corrected manually afterward. Automatic routines have yet to be developed for the recognition of buildings, roads, rivers, and other such details in the photos. If this can be realized, it will mean a true revolution in all map production.

Either stream encoding or individual point encoding may be used, as in the corresponding digitizer process. In digital instruments, transformation parameters are used to transform the internal system coordinates continuously to terrain coordinates. The transformation parameters are calculated from the photographic coordinates of points known from field survey control or triangulation. Unlike the digitization of existing maps, processing photogrammetric data directly encodes elevation information at any given point within the stereo model.

The three-dimensional transformations are

$$X = c_1 + m(a_{11}x + a_{12}y + a_{13}z) \tag{10.3a}$$

$$Y = c_2 + m(a_{21}x + a_{22}y + a_{23}z) \tag{10.3b}$$

$$Z = c_3 + m(a_{31}x + a_{32}y + a_{33}z) \tag{10.3c}$$

where X, Y, and Z are the terrain coordinates; x, y, and z are the system coordinates entered; and m, c_1 to c_3, and a_{11} to a_{33} are the transformation parameters.

In mapping, new thematic codes and ID codes are entered whenever the category of an object changes, for example, from residential to industrial use. Photogrammetric encoding requires both trained operators and their concentration. Experience indicates that ID codes are more efficiently entered on screen in the final editing phase.

The coordinates of such individual points as manholes, drains, and property boundaries may be registered in large-scale aerial photographs. So on the ground, such points are often marked with white paint or white paper prior to photographing to ease their identification in the resultant photographs. The size of these markings normally varies from 60 × 60 cm for aerial photo scales of 1:15,000 down to 30 × 30 cm for scales of 1:6000. Aerial triangulation, a special technique for point location based on measurements in the stereoscopic model, may be used to register marked points with an accuracy of ±5 to ±10 μm on photographs, corresponding to 3 to 6 cm on ground in an image scale of 1:6000. For other details, photogrammetric accuracy is 10 to 20 μm.

The product of photogrammetric mapping must be edited to ensure that map data correspond to the map image desired. Map editing programs are used either during the mapping itself or during the subsequent editing phase. Again, experience indicates that editing can be as time consuming as traditional manual redrawing. A 60- × 80-cm map in a scale of 1:1000 can involve 1 to 3 MB of data, depending on the complexity of the map.

When producing base maps with details and topography, map data can normally be entered at a rate corresponding to the registration of 10 to 12 decares an hour in 1:6000 scale images, while editing may proceed at a rate of 20 decares an hour. For 1:15,000 images, the corresponding rates are 20 and 40 decares an hour, respectively. GIS is unable to distinguish between a digitized map and a digitized stereo model. However, a map is inherently generalized, whereas a photo is not, especially in respect of elevation data.

10.4 Collection of Attribute Data

Just as digitizing work may be divided into three phases, so the acquisition of attribute data may be divided into:

- Preparation
- Entering
- Editing and quality enhancement

Preparation

Data of interest are often recorded on paper and then scattered among a multitude of different conventional registers and files. Consequently, the initial acquisition of data may require a major organizational effort involving many different agencies. For instance, registering all the undeveloped land in a county may require data from all the agencies in all the towns in the area.

In some cases, registration has been organized in accordance with statutes and regulations. For example, many countries have their own regulations governing the registration of fixed property. In other cases, it may be necessary to supplement new field registration with local knowledge. Attribute data may be collected effectively only if a suitable collection structure is available or is specified as an initial step. A common approach is to complete various forms, either on paper for manual registration and subsequent entry via a PC, or to hold them in files for direct keying-in on a laptop or PC.

The data fields of the forms should, of course, correspond to the data fields of the database. Registration should be unambiguous and provide for all data variations to ensure that data always locate in the correct data fields. Consider, for example, setting up a GIS for a water supply system. Such a system will require a systematic structuring of data.

Piping

1. Water mains
2. Supply mains
3. Service pipes
4. Sewer pipes
5. Venting
6. Manholes
7. Drains
8. Inoperative piping
9. Inoperative piping not to be removed
10. Basins
11. Booster stations
12. Pump stations
13. Pressure zones
14. Stop valves
15. Joints and joining methods
16. Nonreturn valves
17. Pressure reduction valves
18. Separators
19. Vent valves
20. Fire hydrants, above ground
21. Fire hydrants, below ground

Service pipes

1. Type
2. Tapping
3. Sprinklers
4. Stop valves
5. Water meters

Manholes

1. Type
2. Piping accessed

3. Status (operational, planned, abandoned)
4. Location
5. Construction

Miscellaneous

1. Units of length
2. Dimensions
3. Model numbers
4. Valve numbers
5. Supply areas
6. Soil conditions
7. Road traffic conditions
8. Damage
9. Repairs
10. Service interruptions

These data may be divided roughly into three categories: locations, components, and network parameters.

The acquisition of attribute data often requires verification in the field. Verification may entail going down a manhole (Figure 10.9) to sketch or photograph details, rodding pipes to ascertain direction and junctions, measuring and leveling, poring over old files and drawings to locate details of dimensions and materials, interviewing retired plumbers, and so on. It is thus apparent that it requires a great deal of work to produce attribute data of satisfactory quality. As far as piping data are concerned, a rule of thumb is that 30 minutes are used per manhole and additional time is required to measure the route of the service pipes. It is claimed that 800 worker-years would be needed to

Figure 10.9
A manhole. A large number of the details in this picture will have to be registered for entering in a GIS.

convert the sewerage network in England to digital format, including in-field registrations. The establishment of an EDB-based national property register (3 million properties) in Norway required 150 worker-years. Once these data have been collected and structured in lists, on forms, and/or on maps, entering data in GIS may begin.

Entering data

A convenient way to enter data in GIS is to use a database application program on a laptop or PC for initial entry, subsequently exporting the database file to GIS. PC database application programs support forms that can be designed and filled out on screen, allowing a specific form to be designed for each type of object.

There should be space on the form to include verification of formats in all data fields to ensure, for instance, that text is entered into a text data field while only figures are entered in a numerical data field, and that values entered are restricted within set extremes. It is occasionally advantageous to call up a graphic display of map data on screen as the attribute data are entered. It might also prove useful to microfilm information contained in old files on site. Processed microfilm may be viewed on a microfilm reader placed beside a PC or terminal keyboard, so data can be keyed in as the microfilmed pages are viewed. Whenever data *are* available in computerized form, they may be entered directly.

Editing

Automatic verification functions reveal only formal errors. Incorrect information, such as a wrong name entered in a name field, can be spotted only by manual proofreading. Major errors and meaningless text are usually easy to spot, but incorrect spellings, inversions, omissions, and other less obvious errors are more difficult to find. Consequently, it is better to verify attribute data by printing out sorted lists which may then be checked line by line by a copy reader.

It is important to eliminate errors in entered data, although it is invariably time consuming because it requires mostly manual work.

Once the digital map data and attribute data are verified, corrected, and entered in GIS, the identifiers linking the databases should be checked and all missing identifiers marked. Checking in GIS will not identify incorrect links due to map data code errors or mistakes (in legal fields) in attribute data entered.

10.5 Text Data

Public administration agencies, public utilities, and other organizations dealing with georeferenced data may frequently need to integrate text data in GIS. All kinds of documentation may be involved—

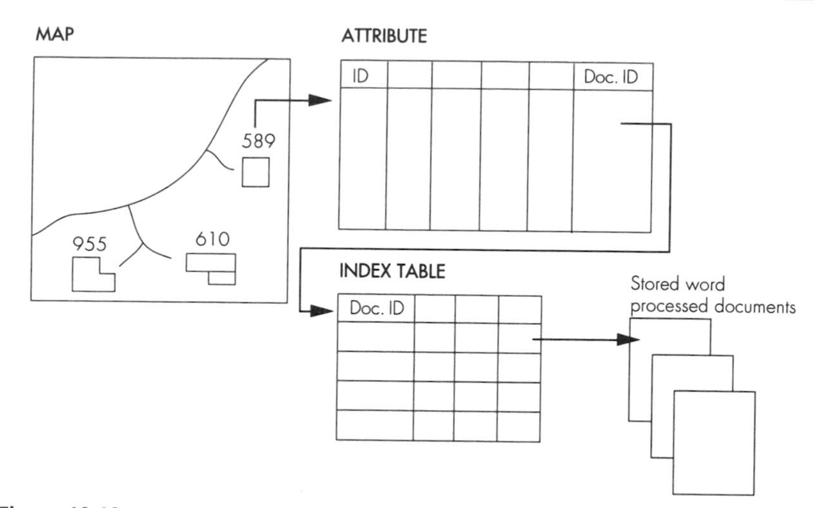

Figure 10.10
Documents in a word processing system can be integrated into GIS via an index table between the attribute table and the word processing system.

from letters and property deeds to zoning directives and service overviews. Text data from word processing systems may be integrated into GIS via a table of object identifiers and pointers to text file addresses (Figure 10.10). A GIS may also conduct searches of text documents. In practice, this requires careful organization in the naming and filing of text data. GIS graphic map images may be imported in word processing and incorporated in texts using programs such as Microsoft Word or PostScript. Text with maps may then be printed out on a laser printer.

CHAPTER 11

Data Quality

11.1 Selection Criteria

The use of analog maps leads us to assume that all objects on a map are of similar quality. This is not the case with GIS, where data of differing origin and accuracy may well be mixed. In addition, the introduction of GIS into mapping has produced a completely new (and perhaps less professional) user group compared with traditional map users. In such a situation, an understanding of the concept of quality will be very important, and the marking of data according to its degree of quality a necessity. Users frequently expect digital data to be of higher quality than either conventional map data or data in manual registers. This is because digital data have a far broader range of application than analog data and also because users presume that technological advances enhance the quality of data.

The applications of digital data are indeed many and varied, but digital data systems are not inherently more accurate than the analog systems they supplant. In general, digital systems are capable of processing data more precisely than are analog systems, but their overall accuracy still depends on the accuracy of their source data, which in most cases remain analog.

To see why this is so, let us consider time, perhaps the most accurately measured quantity in everyday life. In the 1950s, extremely accurate ships' chronometers and other timepieces became available, at a high price; ordinary wristwatches and clocks were accurate to only a few minutes a day. In most applications, clock accuracy was a question of how much a user was willing to spend for a finely made mechanical timepiece. National standards of time were kept by using large electronic clocks which derived their signals from the oscillations of quartz crystals housed in temperature-controlled ovens. These clocks were completely analog and accurate to the order of 1 part in 10^9, or about 1 second in 10 years. Thanks to microchips and digital techniques, similar levels of accuracy are now common in everyday wristwatches and clocks. The accuracy is still due to quartz crystals, although it is the digital techniques (and their ultracompact realizations using microchips) that have made such accuracy available in small timepieces. In terms of overall accuracy, the quality of a timepiece now depends both on the quality of the crystal oscillator and on the precision with which its signals are processed to give the final

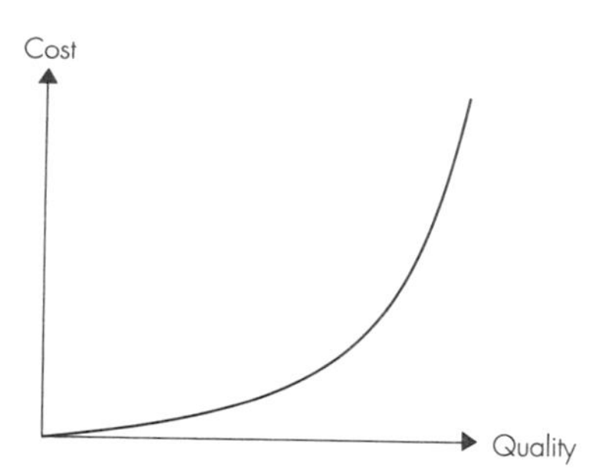

Figure 11.1
Relationship between quality and cost. There is a strong correlation between data quality and cost—costs increase with quality. Quality is therefore often determined on the basis of cost.

display. In technical terms (marketing oversell of expensive timepieces notwithstanding), user needs determine accuracy, and accuracy follows price because greater accuracy requires a higher-quality crystal (or atomic) source and more complex processing of its signals (Figure 11.1).

In GIS, the situation is similar: Final accuracy depends on source accuracy. In principle, the user dictates the need for greater accuracy. This is provided to some degree by improved technology, which is purchased or developed in response to the original need. While technical achievements are virtually limitless from a technical viewpoint, the actual level of achievement, and thus the available technology, is determined—and consequently limited—by costs. The final accuracy of data depends on the qualities of the original input data and on the precision with which input data are processed. Higher accuracy entails higher initial data quality and more precise processing, both of which increase system costs. Furthermore, the requirements for up-to-dateness affect the measure of data quality and hence system cost (Figure 11.2). In general, the costs of updating may strongly influence the determination of data quality. In theory, therefore, GIS data quality is a compromise between needs and costs. In practice, however, the choice is often a question of what is currently available or can be acquired within a reasonable amount of time.

Delays may be due to a variety of causes. Where snow falls in winter, users of aerial photographs must wait until the snows melt before they can proceed; similarly, users of satellite images may have to wait for cloud cover to disperse over an area of interest; users of map digitization produced by national mapping agencies may be assigned to backorder queues because the speed of agency work is often constrained by budget allocations. Finally, users may suffer delays in resolving copyright or other legal difficulties concerning the ownership of data. Any of these factors may cause data to be of lower quality than originally envisioned.

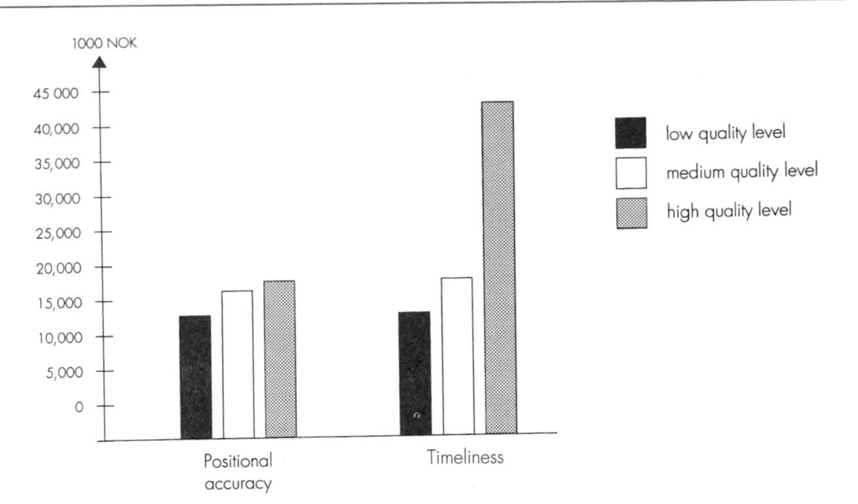

Figure 11.2
Simulation of cost sensitivity in relation to various quality standards for the Norwegian national road database. The bar chart shows clearly that currentness is here much more susceptible to quality choice than to positional accuracy (see Figure 17.16).

In summary, four aspects of data acquisition comprise the criteria for selecting data accuracy:

1. Needs
2. Costs
3. Accessibility
4. Time frame

Quality can be defined as one or more characteristics of geographical data that describe the extent to which it is fit for use. The most important measures of data quality are:

1. Data accuracy
   a. Positional data accuracy
   b. Attribute data accuracy
2. Data consistency
3. Data completeness
4. Data timeliness

Additional measures often used in GIS as concepts of accuracy are:

- Lineage
- Accessibility

These two terms are generally considered as the indirect goals of accuracy.

These characteristics are important because they encourage multiple use of data both within an organization and between organizations.

The users can judge whether they can use the particular data and whether they have gotten what they expected, while the producer can judge whether their products meet market expectations.

11.2 Accuracy

Inaccuracy in the data basis can be caused by any of the following, singly or in combination: random error, systematic error, and/or gross errors. Statistical methods are generally used to specify accuracy requirements and total accuracy. As in GIS, statistics is a tool used to describe reality; since reality cannot be described exactly, the methods used will always contain an element of uncertainty.

Accuracy versus precision

Although commonly considered to be identical, accuracy and precision express different aspects of measurements and should be carefully distinguished. *Accuracy* expresses how closely a measurement represents the quantity measured. *Precision* expresses the repeatability of the measurement. For the measurements normally of interest in GIS, precision is limited only by the instruments and methods used. In physical measurements, precision is customarily indicated by the number of significant figures carried in a quantity expressed. For instance, a figure of 101.23 cm indicates that the distance is measured with a precision of a fraction of a millimeter. The last decimal place may be uncertain.

Accuracy should be stated in terms of an interval in which a true value is assumed to lie. For instance, 101.21 ± 0.04 cm indicates that the true value is assumed to lie between 101.17 and 101.25 cm. However, some textbook authors may use a notation with estimated value ± standard deviation. Care should be taken to avoid misunderstandings.

Measurements of accuracy

Standard deviation (σ) is the square root of the variance. For a series of n observations x_j, where j takes the integer values 1, 2, up to n, with known expectation μ, the variance Var(x) and the standard deviation σ are given by

$$\mathrm{Var}(x) = \sigma^2 = \lim_{n \to \infty} \sum_{j=1}^{n} \frac{(X_j - \mu)^2}{n} \tag{11.1}$$

The estimate of σ from a limited number of observations (n is finite) is often denoted s. Hence

$$s^2 = \sum_{j=1}^{n} \frac{(X_j - \mu)^2}{n} \tag{11.2}$$

$$s = \pm\sqrt{s^2} \qquad \text{(standard deviation)} \tag{11.3}$$

If the expectation for X_j (denoted μ) is not known, the expectation may be substituted by the mean of all observations and the number of redundant observations reduced by one:

$$s^2 = \sum_{j=1}^{n} \frac{(X_j - \overline{X})^2}{n - 1} \tag{11.4}$$

$$\overline{X} = \frac{\sum_{j=1}^{n} X_j}{n} \tag{11.5}$$

The standard deviation takes the number of redundant observations into account [Eq. (11.4)]. The root mean square is always divided by the number of observations. [Eq. (11.2)]. Thus, if the difference between the number of redundant observations and the total number of observations is relatively small (e.g., the number of observations is large), the difference between the root mean square and the standard deviation may be negligible. The standard deviation is the measure of how close a measurement/observation lies to the value expected.

In surveying, error ellipse (two-dimensional) and error ellipsoid (three-dimensional) are utilized to visualize the uncertainty of single points (Figure 11.3). The standard error ellipse has semimajor and semiminor axes equal to standard deviation in each direction. The probability (approx. 60%) of the point being outside the standard error ellipse is much larger than the probability of it being inside. Therefore, it is better to visualize with confidence error ellipses or ellipsoids scaled to give, for instance, 95% probability of the point falling inside.

The precision of observations may be visualized with colors or with error ellipses or ellipsoids, depending on their dimensions. For one-dimensional observations, different colors are used to indicate different classes of precision when the observations are shown on the screen. In two dimensions, error ellipses are used, and for a vector in three dimensions, error ellipsoids. The likelihood of observations/measurements lying within given boundary limits is known as *confidence*

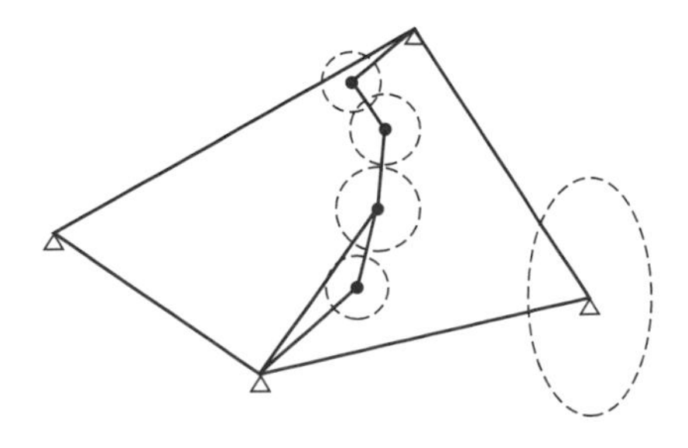

Figure 11.3
The error ellipses visualize the uncertainty in the points with regard to both size and direction.

and is expressed as a percentage. For example, the requirement could be that observations should have a confidence of 95%.

Deviations may be random or systematic. The systematic errors may be computed according to the equation

$$m = \sum_{j=1} \frac{X_j - \mu}{n} \tag{11.6}$$

The systematic component may be relatively unimportant when data originating from the same source are used, but it can be disruptive in data originating from dissimilar sources.

Serious errors (outliers, gross blunders) can always occur in the data and have to be removed. The definition of what constitutes gross blunders is a matter of opinion, but the limit is often set as three times the standard deviation for the data in question. It is difficult to spot serious errors with only a few redundant observations. There is a certain likelihood that geometrical data will contain serious errors that are not detected and removed. For example, these can arise in connection with in-field measurements, calculations or manual entry of coordinates into the database.

Redundancy is used to remove blunders. *Outliers,* which are probable blunders, are detected through statistical testing. A statistical test will reveal larger outliers, while smaller outliers may remain. *Internal reliability* is for each observation the largest possible remaining outlier after the test. The impact of this largest remaining outlier on the coordinates calculated and similar factors is called *external reliability.* External reliability is a worst-case scenario.

Data accuracy may be verified through reference to other, more accurate data that have been designated as error-free for the purpose. These reference data may be compiled from survey measurements in the field, from maps of considerably larger scale, and from more precise instruments for the measurement of attributes such as wind and temperature.

Errors may accumulate in data processing. For independent error sources, the standard deviation of a multiplication $C = A \times B$ is as given by

$$s_C = \sqrt{S_A^2 + S_B^2} \tag{11.7}$$

11.3 Lineage

Origin comprises the description of the source material, including date, original application, data capture method, data producer, and treatment method (e.g., transformation algorithms). Digital map data can thus be established on the basis of photogrammetric measurements or by digitizing existing maps. Attribute data can be taken from public registers or from direct in-field registrations.

It is not sufficient to know the original data source; we also need to know something about how the data are collected. For example, what type of instruments are used (e.g., type of surveying instrument, type of aerial camera). Data are often treated before they are stored in the databases. The treatment method is important not only for the accuracy of the end result, but also because knowledge of the method makes it possible to reconstruct original data and, possibly, to carry out new (and better) calculations. For example, documentation of a transformation algorithm should be available on request by a user obtaining digital data.

Digital map data generated directly from aerial photos or entered during surveying entail fewer steps and are consequently less subject to error than data from digitized maps. From the standpoint of accuracy, original sources are *always* preferable to maps.

11.4 Positional Accuracy

The goal of accurate georeferencing is to locate objects exactly as they are located on the ground, as related to a common coordinate system. In ordinary mapping, accuracy is inversely proportional to map scale: a map to scale 1:1000 is more accurate than one to scale 1:100,000. The corresponding case for digital map data is rather more complex. Dissimilar data are stored in the form of terrain coordinates, regardless of their respective sources and accuracy. In other words, all data are stored to a scale of 1:1 and may then be used to generate maps of any scale.

The use of insufficiently accurate data may eventually prove more expensive than the original acquisition of the data. A GIS can plot smooth lines on a map and perform convincing computations and analyses, all on the bases of its stored data and regardless of the accuracy of the original data entered. Moreover, a GIS can perform calculations of cut-and-fill works from data accurate to ± 2 m as comprehensively as it can from data accurate to ± 0.2 m, and plot equally fine profiles from both. However, the potential consequences of using the less accurate set of data are severe: the results may be totally useless or, at best, cause serious miscalculation of construction costs.

Whenever original data are based on digitization of existing maps, the cartographic adaptation may introduce error. For example, roads are not defined on a map by their edges but by solid lines. A line representing a railway may have been moved slightly so that both it and a second line representing a road, when both lie close to a coast, can be shown clearly. Maps of scale 1:50,000 are more generalized than maps of scale 1:5000 and are as incompatible for GIS uses as they are for viewing together. Cartographic adaptation can even mar the accuracy of a large-scale map, such as a detailed map of an electrical supply network, where the lines drawn may be moved in order to draw all the electrical cables connected to a substation.

Geometrical accuracy includes:

- Uncertainty
- Resolution

All observations contain elements of uncertainty. The *uncertainty* of positional data is best described by use of the statistical error theory [e.g., expressed either by standard deviation (*root-mean-square error*—RMSE) or confidence interval and confidence error ellipse].

The accuracy of georeferencing in digital data also depends on *resolution,* which specifies the densities of points used to represent objects, such as the interpoint interval selected in digitizing or the size of the cells in satellite data. If the interpoint interval is relatively large and the lines digitized have sharp curvatures, accuracy will be less than if the interpoint interval is smaller.

Since ordinary maps comprise the most common source of digital map data, they are also the source of inaccuracies stemming from the original surveying, photogrammetric mapmaking, map drawing, and digitizing. These errors can be cumulative [see Eq. (11.7)] This can have an adverse affect on the final data.

Measurement of positional accuracy may be obtained by one of the following methods, based on the U.S. Geological Survey Spatial Data Transfer Standard:

1. *Deductive estimate.* Any deductive statement based on knowledge of errors in each production step should include reference to complete calibration tests and also describe assumptions concerning error propagation. Results from deductive estimates should be distinguished from results of other tests.
2. *Internal evidence.* National procedures should be used for tests based on repeated measurement and redundancy such as closure of traverse or residuals from an adjustment.
3. *Comparison of source.* When using graphic inspection of results *(check plots),* the geometric tolerance applied should be reported and the method of registration should be described. Use of check plots should be included in the lineage portion.
4. *Independent source of higher accuracy.* The preferred test for positional accuracy is a comparison with an independent source of higher accuracy. When the dates of testing and source differ, one must ensure that the results relate to positional errors and not to temporal effects.

All geometrical data should carry accuracy designations so that GIS users may assess the consequences of the various sources of error involved. Ideally, accuracy codes should be assigned either to various objects to designate RMSE or through a quality overlay.

11.5 Attribute Data Accuracy

Considerations of accuracy usually focus on the accuracy of geometrical data, although in practice the accuracy of attribute data is equally important. Ill-defined attribute data can introduce errors in the final data, as when incomplete definitions of object types result in objects being wrongly classified. Thus a residential building could be classified as a commercial building if the demarcation between the two classifications is unclear.

Quantitative data may be defined in nonnumerical terms *(ordinal data)* either as discrete variables divided into classes *(interval data)* or as continuous variables without numerical limits *(ratio data).* Regarding accuracy, continuous variables may be treated as georeferenced data [see Eq. (11.1).] For discrete data, the situation is different. If, for example, discrete data classes are defined incompletely, the final data may contain errors arising from unintentional misclassification. In some cases, misclassification may involve simple sorting errors: An object of type A is put in class B. In other cases, the class structure may be faulty, as when there is no class C for objects containing elements of both A and B.

In practice, boundaries between areas of differing characteristics, say between different types of vegetation, are often inaccurate because no two persons interpreting the original data will draw exactly the same boundaries. Fragmented boundaries, as so often occur in nature, cannot be measured accurately, regardless of the instruments used or the skill of the operators (this may be why fractals so befuddle GIS experts). Operators encoding data may be unequally skilled and may therefore introduce systematic error, and whenever borders are ill defined, the information of the polygons they delineate may be questionable.

Furthermore, measurement instruments may be poorly adjusted, thus introducing systematic errors; or they may be subject to random inaccuracies, as for instruments measuring acidity and alkalinity in pH, or for satellite sensors viewing reflected and reemitted solar radiation. The methods used to collect attributes also influence accuracy of data. Here, examples include collections based on statistical distribution hypotheses or other assumptions that fail to agree with the real world, as when observations are excessively disperse. This is obvious in the choice of cell size in raster GIS, as shown in Figure 11.4, where the density of observations (cell size) is too small to register the phenomena desired.

Coding errors arising when manual register data are entered in GIS may also diminish accuracy. As for geometrical data, the accuracy of attribute value includes:

- Uncertainty
- Resolution

As discussed previously, uncertainty and reliability are expressed with statistical measurement values (RMSE, etc.). Resolution denotes the density of observations: the cell size in raster GIS, the size of a polygon in vector GIS, the priority of measurement points, and so on (see Figure 11.4). The respective qualities of completeness and resolution are often interrelated. For example, the specification may state that all properties with an area of half a hectare or more must be registered. Some areas may be completely covered, yet the coverage of all properties within the boundaries of the area may still be incomplete. In this case, the observation density is not sufficient to depict each property. The accuracy of classified data is best expressed with a percentage correctly classified; probable misclassification can be expressed with a matrix; deductive estimates are expressed by standard deviation.

Generally speaking, the accuracy of numerical attributes is verified by comparing data with randomly assigned true values and is expressed in terms of standard deviations and systematic errors. Errors can accumulate even as data are being processed (Figure 11.5). For example, when two overlays of a raster GIS, $A = 10$ and $B = 15$, with standard deviations $s_A = \pm 1$ and $s_B = \pm 2$, are multiplied to a resultant C, the standard deviation is as given in Eq. (11.7), provided that A and B are uncorrelated.

$$C = 10 \times 15 = 150$$

$$s_C = \pm\sqrt{1^2 + 2^2} = \pm 2.2$$

The many probable sources of attribute data error can impose constraints that restrict the validity of data and limit the accuracy of final analyses. A suitable expression of classified attribute data accuracy is

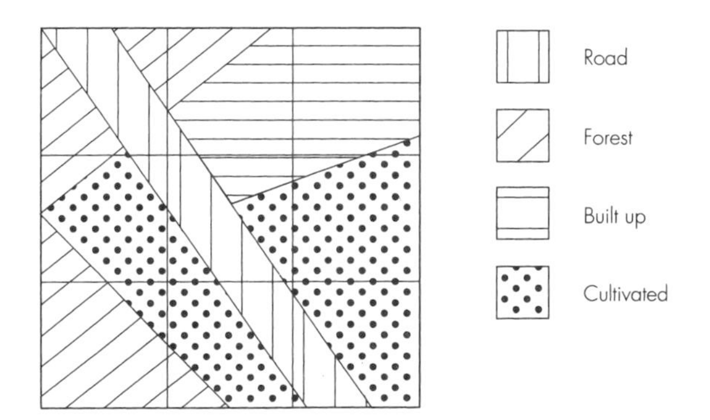

Figure 11.4
The observation unit (cell size) selected here is too large to register all the objects desired.

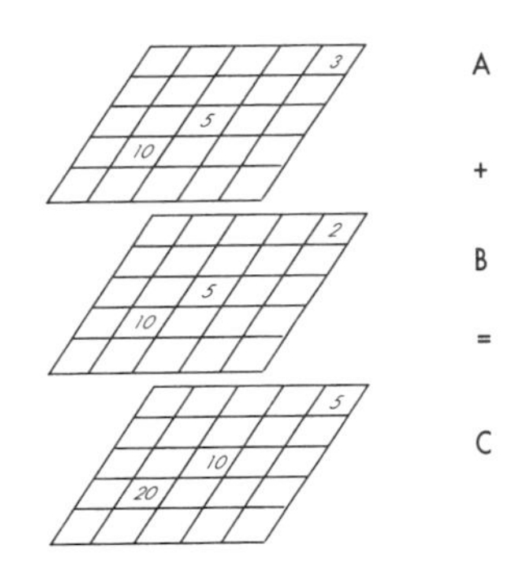

Figure 11.5
Errors in addition of the attribute values in a raster layer are transmitted to the total.

to state a percentage of correct or erroneous classifications, such as "99% of all objects correctly classified."

Accuracy tests for categorical attributes may be performed by one of the following methods, based on the USGS Spatial Data Transfer Standard:

1. *Deductive estimate.* Any estimate, even a guess based on experience, should be permitted. Terms such as *good* and *poor* should be explained in as quantitative a manner as possible.
2. *Test based on independent samples.* For independent samples, a misclassification matrix should be reported as counts of sample units cross-tabulating the categories of the sample with the test result.

All methods should refer to map scale in interclassification. Ideally, the accuracy of all attribute data should also be accuracy coded as geometrical data. This is seldom done, however, either because the accuracy remains unknown, the cost of assessing it exceeds the benefit of the knowledge it provides, or the user fails to appreciate the need for accuracy.

Data storage of both geometry and attributes can also affect data accuracy whenever entries comprise many significant figures. Storage with *single precision,* in which a single computer word (string of bits) is used to represent a number, results in precision to six or seven significant figures. Storage with *double precision,* in which double the usual number of bits are used to represent a number, results in precision of up to 14 significant figures. If data are transferred from a single-precision system to a double-precision system, digits after the sixth or seventh are unreliable. Single-precision systems may also introduce noticeable errors in rounding off, as when computing nodes.

11.6 Logical Consistency

Data *consistency* includes the limitations and rules that apply for the data as they are specified. This can include rules for:

- Connection
- Homogeneity
- Level of generalization

The logical relationships of data must satisfy the requirements imposed on relationships between objects, which, in turn, are based on the tasks to be performed. A test of valid values (e.g., to detect gross blunders) should be applied but does not ensure all aspects of logical consistency. A database that contains lines may have typical faults, such as:

- Overshoot and undershoot
- Polygons too small
- Polygons not closed
- Double lines
- Incorrect intersection of lines
- Lines too close

Various tests may be applied to address these questions, and the tests should be described. Specific topological tests could be carried out to verify the following conditions:

- All links intersect at nodes
- The cycle of chain and nodes is consistent around the polygons or nodes.
- Inner rings embed consistently in a closing polygon.

The consistency of logical relationships can be difficult to measure. In any case, quality depends on the use of a checking program as data are entered or processed through topology building; data of good quality are used to correct data entered to form topologies.

Logical connections in map data are essential for GIS users, although they are less important for others. Users of map-editing systems, for example, may manage perfectly well with spaghetti data, but poor-quality topology complicates or introduces error in many GIS functions, not least in overlayering and network analysis.

Consequently, data with unverified topology may cost a user dearly in terms of verification and correction work, or in terms of degraded quality of the final product. Topology consistency is best expressed as a percentage of junctions correctly formed, percentage of polygons correctly formed, and/or a tolerance used for merging adjacent areas.

Validity of values can be expressed as percentage conformance. The relations between attributes are often expressed (e.g., a building is located on a particular property, a road borders particular properties). Relational consistency is an important data quality and is best expressed as a percentage of correct relationships. Redundancies are also to be found in data of poor consistency. It is therefore normal to avoid redundancy. Whenever small-scale maps are used to generate digital map data, the generalization imposed by the constraints of scale may degrade data coverage. So although the houses of a group may be shown individually on a large-scale map, a single symbol may represent them all on a small-scale map with less precise resolution. Consequently, data consistency should also be recorded for data entered and stored, to inform users of the limitations entailed.

11.7 Completeness

Data *completeness* describes how fully data on an object type have been entered in relation to a specification: for example, whether attribute data as well as digital map data have been entered for all properties, all roads, all buildings, and so on, in an area. Completeness may be verified by comparing the number of objects for which data are entered against the number of real objects specified in the area involved. The result of the comparison is then expressed in terms of a percentage. In general, the more attributes that are entered, the better the description, yet there are no objective measures that express how completely an object should be described. So verification is relative and expressed only in terms of a specification compiled for the purpose. Relative to the specification, the completeness can be both under- and overcomplete.

Completeness is essential for some purposes. For example, local or national regulations may require that neighbors be given notice whenever a property is divided, and GIS may be used to locate those neighbors. Should just one person be overlooked because of database incompleteness, the entire notification procedure may have to be repeated. Where a map is the data source, account should be taken of the fact that generalizations and selections made during map production can affect completeness.

Data completeness also describes the completeness of coding. Linking geometry and attributes correctly depends on the same ID codes being used in both sets of data. Consider, for example, a situation in which road geometry data are stored in one database and road attribute data in another. The roads are divided into numbered segments, and the numbers serve as common ID codes. When new roads are built, new ID codes are added in the databases, which should be updated simultaneously. If they are not, link inconsistencies may arise, diminishing the quality of the data stored jointly in the two databases. Such problems arise most readily whenever several agencies or users enter data into common databases at different times.

Consequently, data completeness should also be recorded for data entered and stored, to inform users of the limitations entailed. It is also advisable, whenever geometry and attribute data are to be linked, to compile tables of any missing links as data are entered. The consistency of linking is customarily stated in terms of a percentage (e.g., "99% successful links").

11.8 Data Timeliness

Two situations may affect the timeliness of data:

1. When the geometry and attributes of existing objects have been altered
2. When new objects, with geometry and attributes, have appeared

The degree of timeliness required depends on object type and data application. Public works agencies and utilities often require the most up-to-date information on property boundaries, buildings, roads, manholes, and other works. It is therefore advisable to update the data for such objects every two to three weeks.

Zoning boundaries are typical of data that change more slowly and thus require less frequent updating. However, attributes may change more rapidly than geometry. Examples of this include new definitions, new registration methods, and changes of vegetation during a growing season. Another common case is that of property data, where ownership changes more often than the boundaries.

Lack of current information on maps has always troubled users. Studies (Bernhardsen 1985) have shown that the use of outdated 1:1000 and 1:5000 maps can cost each professional user an amount corresponding to one-third of the annual salary of a technical staff member. This is because map data must be supplemented, often with in-field measurements. Given the number of professional users in the country as a whole, the aggregate cost of outdated map information could clearly be enormous.

Timeliness of data depends on the frequency of changes and updating and becomes particularly important whenever tasks involve data from various sources. It is also important to state how long the data are valid—*temporal validity;* without this, it is easy for data to be used incorrectly. Timeliness is normally indicated by the date of registry, such as the dates on aerial photographs, photogrammetric data, and satellite data, and the dates of surveys, accidents, water leakage, and the like.

11.9 Accessibility

Accessibility describes where data for a particular area can be found, who has the original rights, and what other conditions and limitations (price, form of delivery, format, etc.) apply to the acquisition and use of the data. These factors are closely linked to standard formats, metadatabases, and use of the Internet. We should also reemphasize how important the quality marking of data is, with respect to lineage, positional accuracy, attribute accuracy, logical consistency, completeness, data timeliness, and accessibility.

11.10 Probable Sources of Error

Error is inevitable in all measurements, either quantitative or qualitative, and may range from the serious to the negligible. Yet neither error occurrence nor magnitude is known in advance. Errors and omissions degrade the accuracy of georeferencing, attribute data, logical consistency, linking, coverage and resolution, and timeliness.

Error may be introduced in any phase of data processing. The following conditions (adopted from Burrough 1986) may affect data quality.

A. Original error sources independent of GIS processing
   1. All surveys and field registrations. Errors may be ascribed to:
      - Instrument inaccuracies
      - Satellite sensor systems
      - Aerial cameras
      - GPS
      - Surveying instruments
      - Various instruments for measuring attribute values
   2. Data processing, as when producing maps used as GIS sources. Errors may be ascribed to:
      - Mapmaking
      - Errors in computations and geodetic networks
      - Mapping instrument inaccuracy
      - Map drawing inaccuracies
      - Data editing
      - Computations
      - Enlarging/reducing and redrawing
   3. Errors due to changes in the field:
      - Registered objects change character.
      - New phenomena arise.
   4. Errors introduced due to lack of coverage or resolution.

B. Errors introduced in GIS data processing
   5. Errors in data entry:
      - Inaccuracy in digitization
      - Equipment error
      - Operator error
      - Inaccuracies in entering attribute data
      - Human error (lack of verification routines)

6. Errors in storing data:
    - Insufficient computer numerical precision
    - Storage medium errors
7. Errors in data manipulation:
    - Raster to vector
    - Vector to raster
    - Generalization and thinning
    - Combining classes
    - Overlayering
    - Interpolation (contours, elevations, etc.)
    - Analyses of satellite and other data
8. Errors introduced in data presentation:
    - Plotter inaccuracy
    - Paper and other presentation media inaccuracies

C. Errors in methods

9. Errors in methods used to collect data:
    - Insufficient observation density
    - Ill-defined objects and classes
    - Insufficient expertise of data compilers
    - Uncertain borders between areas

11.11 A Few Rules

A few simple rules may help to reduce the problems of inaccuracies:

- Employ verification routines to ensure quality.
- Verify data as early as possible.
- Verify data at several stages of their manipulation.
- Avoid combining data of high and low quality in the same base.
- Know the nature of the data, be it geometry data or attribute data.
- Be critical in all data uses.
- Apply processing results carefully.
- State inaccuracies associated with results and analyses.

Without control measures, there is no guarantee that specified quality requirements are met. Checks should certainly be carried out during production, but they can also be necessary when the data are first being put into use by the end user. Even though the data meet specific accuracy requirements, they can still be used incorrectly because they are not understood properly or because users lack the requisite expertise. Even hypothetically error-free data may be used incorrectly. Good data descriptions and specifications of accuracy, as well as user expertise in the relevant application, are the best assurances of correct data use.

CHAPTER 12

Database Implementation

12.1 Database

Within the framework of GIS, data are divided logically into two categories: geometric data and attribute data. This division may extend to a similar division of physical storage, although the relationships between the two categories of data must be preserved regardless of whether the division is logical or physical. In current GIS, this condition is satisfied through one of four approaches to the overall database storage involved:

1. Two separate database systems, one for geometrical data and one for attribute data
2. A single database system that stores both geometrical data and attribute data (Figure 12.1)
3. One database for geometrical data connected to several different databases for attribute data
4. Several databases for geometrical data and attributes joined into one system—distributed database solution

Systems that store geometrical data and attribute data in two separate databases are known as *hybrid systems* (Figure 12.2). The databases are linked by unique identifiers. Systems with separate databases normally employ relational DBMSs that are commercially available for attribute data and tailor-made file-based systems for geometrical data. Generally, it is possible to achieve a more rapid response time by using file-based systems than by using relational DBMSs. This is particularly important for geometrical data where large volumes of data are involved and users are dependent on rapid search and recording on the screen. The hybrid systems include both vector topological data and raster data. To increase the search and record speed, the geometrical data are often split into different layers and areal units (i.e., map sheets, etc.) for storage. This division will normally be invisible to the user; he or she will experience the database as seamless. Quad-tree-based techniques are also used in some hybrid systems for storage of the geometrical data.

Storage of attribute and geometrical data in the same base is known as an *integrated system* (Figure 12.3). The DBMS tools for databases storing both geometrical and attribute data tend to be specialized. Coordinates and topological information are stored in relational tables,

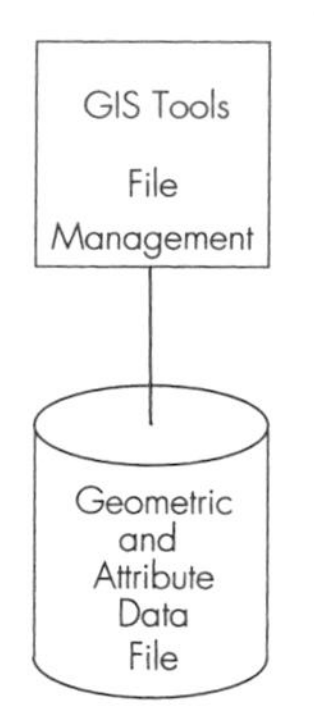

Figure 12.1
Tailor-made file-management system for geometric and attribute data both within the same database.

while attribute data are stored in separate tables. Systems that combine geometrical data and attribute data in a single database most often employ a relational database with a geometry of nodes and links that is structured topologically.

Hierarchical databases, network databases, or combinations of network and relational databases are used whenever the data to be stored are restricted to geometrical data alone. Relational or network databases are employed whenever attribute data alone are stored. The current trend in GIS follows the general information systems' propensity

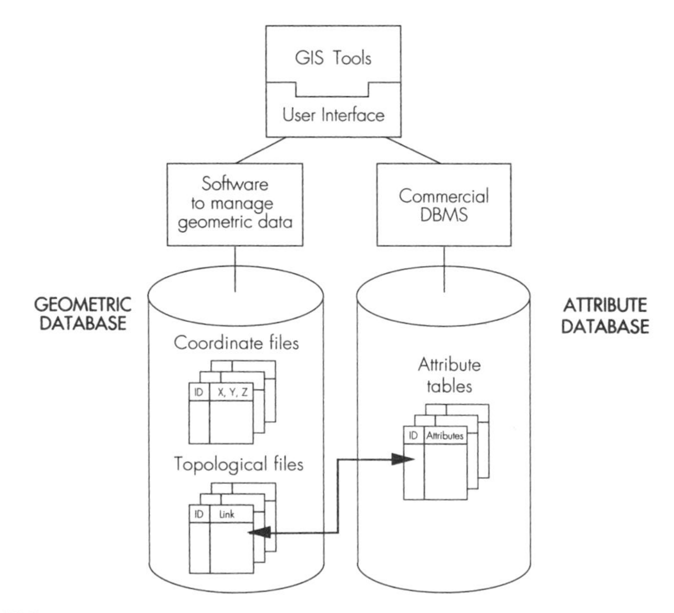

Figure 12.2
Hybrid database model, where geometric and attribute data are stored in separate databases.

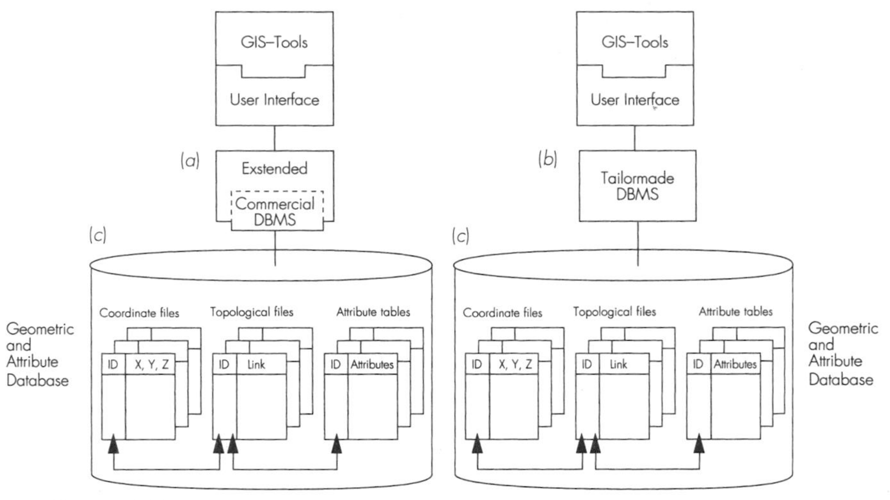

Figure 12.3
Integrated GIS model, with alternative solutions for a database management system.

toward generalized, flexible programs and tools. Some systems feature special functions dedicated to geometrical data, which enhance yields in manipulation and presentation. Such special functions, however, impose constraints that may complicate their use with general programs. Systems that employ a single database or commercially available DBMS tools are often more easily adapted to work with newer database systems.

Normally, basic database design may be divided into three categories: stand-alone, giant, and joint (Figure 12.4). The choice between the three depends on design goals. The joint database design is the

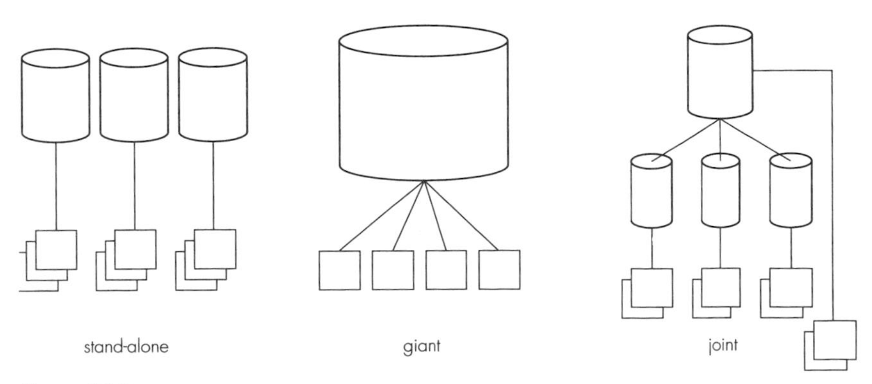

Figure 12.4
Basic database design may be divided into three categories: stand-alone, giant, and joint. The choice among the three depends on design parameters. The joint database design provides the greatest benefit but is the most complex.

most complex, but provides the greatest benefit. Today, the trend is to use standard database management systems such as Oracle, Informinx, DB2, and SQL Server for storage of common data, both attribute data and geometrical data. Often, the purpose is to store data in the simplest possible structured form. By using simple storage structures in standard DBMS, the basic data model and applications become less dependent on each other.

12.2 Distributed Databases

The modern data network has made possible decentralized database solutions. Distributed database systems are a special decentralized solution (Figure 12.5). A system with a distributed database comprises several databases on different computers closely integrated with the assistance of a network and treated as one unit. Users experience this as if they are working against one database.

Subdatabases may have different database management systems, copies of the same data can be found in several of these subdatabases, and the data can be stored in different formats. This level of complexity makes particularly high demands on control routines in the joint database management system. Distributed solutions have special requirements with respect to common data models, routines for maintaining and checking data quality, routines for access control, and so on. All this can result in the system being relatively expensive to run. However, there are a number of advantages, such as increased total capacity, increased reliability and access to data, increased efficiency and flexibility, as well as the advantages of pooling (sharing) resources.

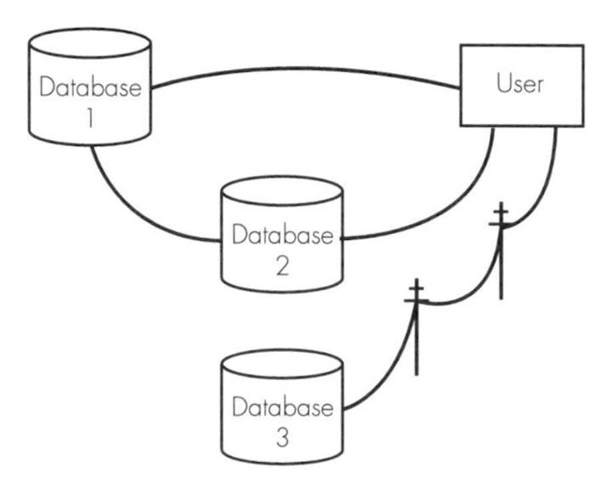

Figure 12.5
Storage can be divided among several producers or users. The manipulation (search, selection, transport, and organization) of common data is carried out by a DBMS in a data network. Distributed databases are managed as a logical database, which is stored at several locations.

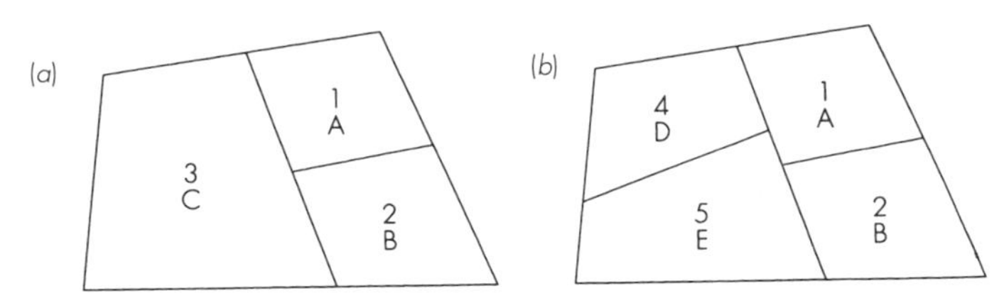

Figure 12.6
Updating is complex. Property 1 has been divided and has two new owners, D and E. Property 3 has to be deleted from the database and the two new properties, 4 and 5, registered. Property 2 now has property 5 instead of property 3 as its neighbor. Property 1 has lost one neighbor (3) and gained two new ones (4 and 5).

12.3 Databases for Map Data

Compared to network databases, relational databases for digital map data have some drawbacks and limitations that restrict their applicability in specific searches and prolong their response time. Thus, in editing and updating, the system must update all relations and pointers continuously (Figure 12.6), a time-consuming exercise.

Databases for digital map data should be able to manipulate records of varying length efficiently. For example, line length may vary considerably, resulting in a corresponding variation in the number of coordinates entered. Standard databases cope poorly with such variations, which is why many GIS suppliers have developed their own databases for digital map data. These database systems reflect geographical reality by such means as requiring that data on objects of the same type, such as the lines forming a property boundary, be stored in close proximity in the database, to speed up the response. This feature is not possible in hierarchical or network systems which lack linkage between objects of the same group.

Tailormade database systems using relatively simple file structures are often used for geometrical data. These databases are usually based to a large extent on network or relational databases, or a combination thereof. Map databases for vector data store:

- Point counts
- Codes (such as ID codes)
- Coordinates
- Plotting routines (how outputs are drawn)
- Search data (pointers)

12.4 Partitioning and Indexing

We have ascertained that spaghetti data require a long search time since the data are stored in a relatively casual and unconnected sequence in the file. The time used to search for and retrieve topological data is also governed by the way in which the data are structured for storage. A rational data structure will reduce the storage volume. Special techniques have therefore been developed for dividing and structuring data.

Generally, map data are stored in map sheets or other geographical units, but storing map sheet data in single sequential files lengthens the response time. This has resulted in some GISs employing indexing to speed up the searching process, and enabling current map sheets to appear on screen almost immediately. Indexing specifies locations, so map sheets are divided into sections which are distributed in such a manner as to accelerate the search. For example, *zooming* focuses on data in those sections relevant to a selected area and ignores the remainder of the map sheet.

Hashing

The underlying concept for hashing is that a key is used to locate the data physically on a disk. Each map section is assigned a number, which is listed in a table of pointers to storage locations (Figure 12.7). The map section numbers, and hence the entries in the index table, may be calculated from Cartesian coordinates.

The following is an example of the use of index tables to search in a point cloud. A single function, known as the *hash function,* can be set up based on dividing up the map area into equally large rectangles. The entry value for a point in the index table can be calculated as follows:

$$Hx(x) = \frac{X - X_{min}}{X_{max} - X_{min}} N \qquad (12.1a)$$

$$Hy(y) = \frac{Y - Y_{min}}{Y_{max} - Y_{min}} M \qquad (12.1b)$$

where N equals the number of grids in the x orientation and M equals the number of grids in the y orientation. The totals should be rounded down to the nearest whole number. To be able to divide the area into grids, the largest and smallest of the coordinate values must be specified. As a rule, the map area is divided into square grids, but rectangular grids allow for a more uniform division of the keys in the index table.

The time required to find all the points within a given window and bring data on screen increases as the amount of information in the map sections grows. In other words, the smaller the sections, the more rapid the search. But smaller sections mean more pointers, which also slow retrieval, so compromises are often necessary.

The method of dividing a map area into regular units takes its strength from its simplicity. The weak point of this method is that information is not available on the data content in each grid and that all

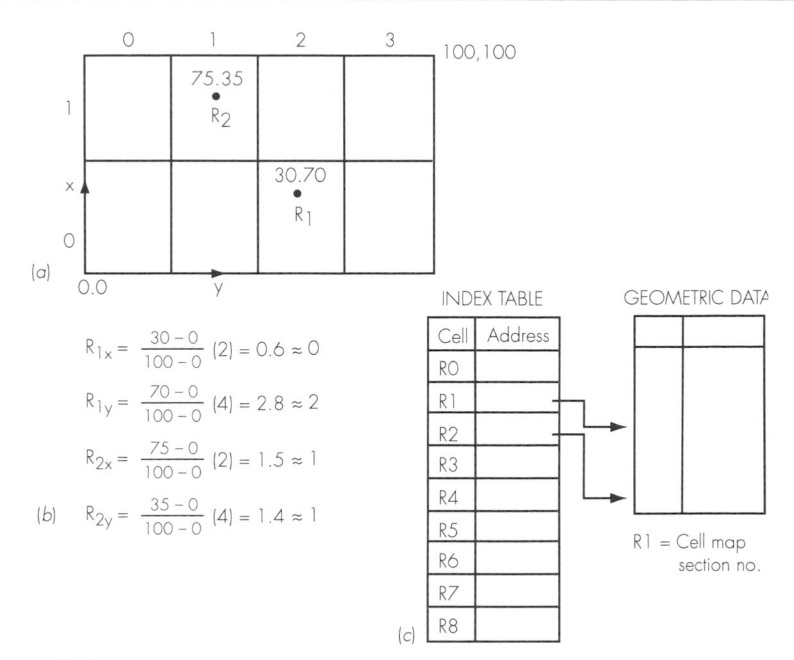

Figure 12.7
The manipulation of digital map data is made more efficient by storage in small squares according to a special addressing principle. The index table enables the system to access the data quickly once it has established in which square the data are located.

grids must be the same size, regardless of whether they contain a small or a large quantity of data.

Trees

Should one abandon the principle of constant grid size, the grid's addresses cannot be calculated by the simple method described above. The alternative will then be to choose a tree structure. In this section we deal only with the indexing method. (Quad-tree has been discussed in Chapter 4.) Each square may be assigned a number according to the *Morton index,* which consists of a successive number of squares as they are quartered (Figure 12.8). Numbers of successive quarterings are combined, so that the index number indicates the size of the square (a low number indicates a large square) as well as the relative geographical location and hierarchical location in the pointer table. Larger areas will have pointers located higher up in the table so that their data may be retrieved rapidly from the table. The attributes of neighboring squares on the ground are stored as neighbors in the table. In general, the search efficiency for trees is inversely proportional to the number of levels in the tree, since that is the number of disk accesses required to locate a given key.

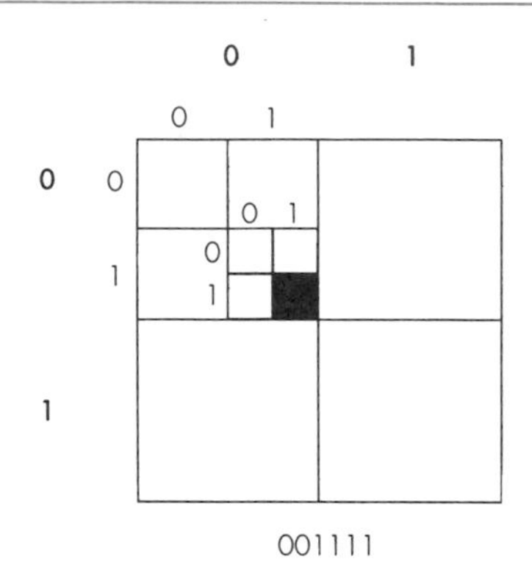

Figure 12.8
Morton index with numbered cells. Black = 001111.

Conclusions

The use of traditional hashing techniques and trees makes it very difficult to handle divided areas that overlap. However, routines have been developed that can handle overlapping data relatively efficiently. These have also been implemented for object-oriented solutions. In recent years, more powerful and rapid hardware has made it easier to use simple data structures for storage, but many GISs still use different "smart" solutions to obtain rapid access to data stored on the disk.

No current database system or structure completely fulfils the needs of database applications. There are grounds for suspecting that the excessively complex and voluminous data collections of many GISs may be ascribed to the databases employed. It goes without saying, then, that further database development is in order. One goal might be to develop better object-oriented database systems.

12.5 Structured Query Language

The simple structures of relational database systems have permitted the development of standard query languages, one of which is Structured Query Language (SQL). SQL gives users access to data in relational DBMSs by describing the data they may wish to see. SQL also

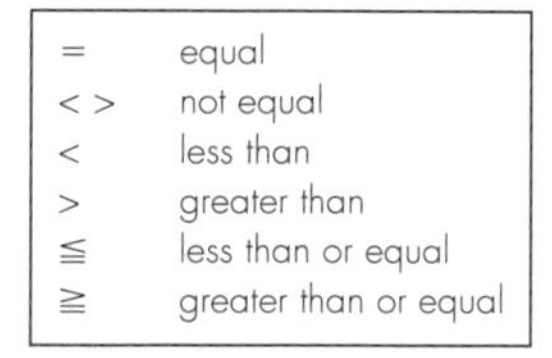

=	equal
< >	not equal
<	less than
>	greater than
≦	less than or equal
≧	greater than or equal

Figure 12.9
The logical operations in SQL.

allows users to define data in a database and to manipulate those data. Additional functions that SQL supplies to relational databases are very useful for many GIS applications.

Relational algebra may be performed using two classes of storage and retrieval operations. The *set operations* include union, intersection, difference, and product. The *relational operations* include selection (accessing rows), projection (accessing columns), joining, and dividing. Relational joining links tables and creates a new table from data retrieved from various tables. The new table need not be stored physically in the database.

There are six logical operations in SQL:

1. $=$ Equal
2. $<>$ Not equal
3. $<$ Less than
4. $>$ Greater than
5. $\leq$ Less than or equal
6. $\geq$ Greater than or equal

There are five aggregate functions:

1. The total of all the rows, satisfying any conditions, of the given column, where the given column is numerical
2. The average of the given column
3. The largest figure in the given column
4. The smallest figure in the given column
5. The number of rows satisfying the conditions

Most GIS users have developed application programs with various human–machine interfaces. SQL is used most frequently in searching, although other query procedures are also used. Complex GIS functions such as data search within specified rectangles or circles, creation of buffer zones, and overlay require operations that are not implemented in standard SQL. However, several suppliers of GIS software have developed special spatial SQL dialects. This applies in particular to systems that use relational databases for storage of both geometry and attributes.

12.6 Database Design

To solve typical tasks efficiently, GIS databases have to be designed in an appropriate way. Normally, database design comprises three main elements:

1. *Conceptual design* is hardware and software independent, and specifies the end user's needs. Usually, conceptual design

includes the same elements as those included in data structuring/data modeling.
2. *Logical design* is software dependent but hardware independent, and sets out the logical structure of the database elements determined by the different GIS software.
3. *Physical design* is hardware dependent and includes file structure, memory and disk space, access, and speed, among other attributes.

Conceptual and logical design are described in the following section, while physical design is discussed further in Chapter 18.

12.6.1 Conceptual database design

Even though good DBMSs have been developed for GIS, the usefulness of the system will be no better than the data entered into the database. To be useful, data must be organized. Disorganized data in DBMS are unproductive: in other words, garbage in, garbage out. For example, if the intention is to show all water pipes in a network that are 3 inches in diameter, the result will be of doubtful value if the pipe data are entered sometimes as inside diameters and at other times as outside diameters. A transport analysis can be performed only if elements of the road network—its bridges, tunnels, and ferry routes—are entered as secondary parts of the primary network, not as independent elements. If all the properties on a specified part of a street are to be identified, the data must contain stable, common identifiers between streets and properties. Similarly, if all the private roads of a township are to be identified, the definition of what constitutes a private road must be stated clearly.

Data collected and entered in a database must be processed and arranged to provide information that is meaningful for current use as well as for tasks still to be defined. This information might include the answers to questions relating to:

- The location of a particular object
- Attributes of a particular object
- Attributes of all objects in a region
- The identification of regions containing objects with attributes *A* and *B*
- The quickest route between points *A* and *B*
- The identification of all objects of type *A* contiguous with objects of type *B*

Clearly, then, data planning and structuring are essential to successful use in GIS, and almost all GIS projects should begin with careful analyses of how data are to be entered. The data model and its related structuring must be independent of the software and hardware chosen. In many ways, the problem is the converse of selecting software and hardware to suit the data model.

Data may be structured using the following procedure:

1. *Defining areas of application.* First, the real-world situation must be limited to the area of application (e.g., the operation and maintenance of water supply and sewage networks, property management, transportation planning). Second, if the database is to contain several areas of application, each must initially be treated separately.

2. *Constraints.* The tasks involved and the products to be produced are then identified for each area of application. Examples of constraints include:

   a. *Tasks*
      - Administrative and operational tasks
      - Statutory public tasks
      - Planning
      - Design tasks
      - Dissemination of information

   b. *Requirements*
      - Requirements and needs for standardized map products
      - Requirements and needs for standardized reports

3. *Criteria for assigning priorities.* As each constraint identifies a number of object types with their attendant attributes, only relevant data should be entered in the database and the numbers of object types must be constrained. Thus the data on a road surface are important, but not as important as the width of the road, which, in turn, is not as important as its geometry and location. Objective criteria must then be chosen to determine the objects included. Here, cost-benefit analyses are often a good starting point. In practice, however, the problem is often the converse: How can the necessary data be acquired?

4. *Descriptions of object types and attributes.* The object types chosen and their attributes must be described. Each type of object comprises a collection of individual entities to be treated equally, given names, and defined. For example:
   - Object type: road
   - Definition: All public and private roads passable for cars, excluding detached roads and roads shorter than 50 m

   Attributes are unique descriptions of objects. Each attribute has a definition and a permitted range of values. For example, the attribute description for the object *road* may be:
   - Attribute: load capacity
   - Permitted range: minimum 0 kg, maximum 20 tons

5. *Coordinating terms and definitions.* The various object terms and definitions must be coordinated, particularly for common data. Disparities, of course, are inevitable. For example, while parts of roads may be regarded as property boundaries in property administration applications, they would certainly be defined as roads in road and transport applications.

   Attribute overviews must be uniform. All attributes for a single object type, such as for roads, should be assembled and globally assigned permitted values. Uniform terminology may be difficult to achieve. For example, does a pedestrian area that is accessible to emergency vehicles classify as a road?

6. *Geometric representation.* Rules must be established to determine how objects are to be represented geometrically and which basic geometric elements are to be used. In principle, the choice is between:
   - Vector representation (points, lines, areas)
   - Raster representation

   Entities comprising several objects must be allowed.

   In principle, the areas of application determine how objects are to be represented, but other elements, such as costs and ease of updating, must also be considered.

7. *Relations.* Relations between objects must be defined before tasks can be addressed. Relations may include:
   - Composition of objects (e.g., a country comprises counties, a county comprises townships, and a township comprises statistical units)
   - Locations of objects (e.g., a particular building is located on a particular property, or a planned development is in a recreation area)
   - Affiliations of objects (e.g., which addresses are particular to a street, which stop valves are parts of which water pipes)
   - Neighbors of objects (e.g., which properties border on a particular property, which type of land use borders on a preservation area)

   Some relations may be computed from digital map data, as when forming topology, while others must be entered as attributes. Sometimes it may also be necessary to distinguish between actual relations and potential relations. Concrete user needs should dictate the selection of the relations described. Relations that are not used need not be described.

8. *Quality requirements.* The data must be subjected to quality requirements, including:
   - Geometrical accuracy
   - Attribute accuracy

- Geometrical resolution
- Consistency of linking between geometric data and attribute data
- Timeliness
- Comprehensiveness with respect to content and geographical coverage

9. *Coding*
   - Code lists of the geometric object type designations must be compiled. The code system should be based on the data structure defined in steps 4, 5, and 8.
   - Code lists may also be compiled for attributes and for descriptions of relations.
   - The identifiers between geometry and attributes must be designated, and if possible, their coding delineated.

Completion of steps 1 to 9 results in an object catalog that contains descriptions and codes for all objects to be entered in the database. Parts of the data structuring (steps 4 to 8) can be perceived as data modeling and are shown by the use of graphic diagrams. One of the most widely used approaches to forming a conceptual model is known as the *entity-relationship* (ER) *model.* The object-oriented model is an extension of the ER model. To illustrate the ER model, one can use the example of the object type "built-up area" (see Figure 12.10).

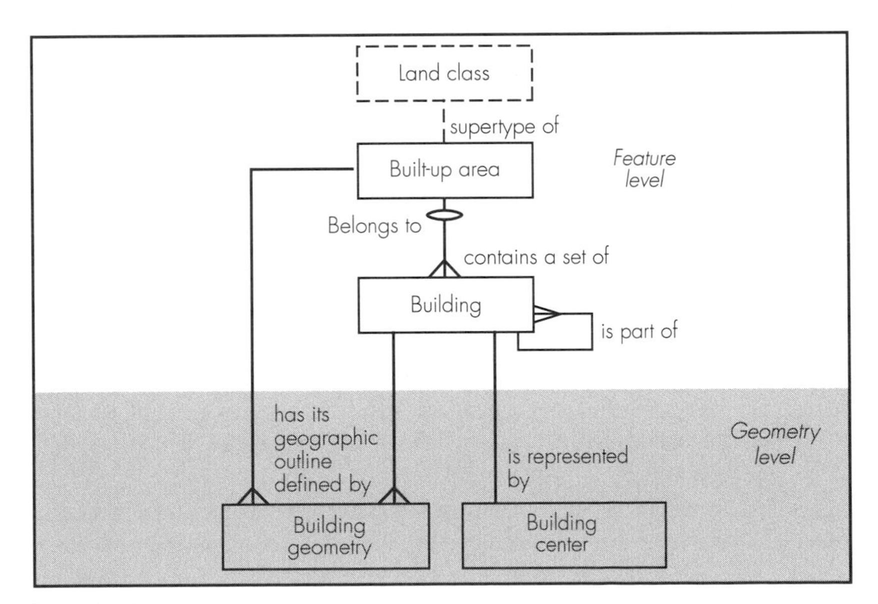

Figure 12.10
Data model for buildings, based on the entity-relationship (ER) method. (From National Standard Zimbabwe, Digital Geographical Information.)

Each element of an ER diagram has to be described. The following descriptions apply to Figure 12.10:

1. *Definitions*
   a. *Built-up area:* a chosen set of buildings spatially described by a generalized area outline (a subset of the "land" class).
   b. *Building:* any manufactured construction except those related to transport and utilities, which are described under the object type "transport and utilities."
   c. *Building geometry:* a set of geometry primitives (polygon) describing the outline of the building.
   d. *Building center:* a point representing the spatial position of a building.

2. *Description of the model.* A building will normally be described by its roof outline. A small building can be represented by a center point that will represent the spatial position of the building. Built-up areas are used for representing groups of buildings where the density at the actual scale does not make it possible to represent each building.

Once the data have been structured in a theoretical model, it remains to:

- Realize the logical structure in a GIS with a concrete database application system—*logical design.*
- Choose the physical files to be structured on disks for optimum utility—*physical design.*

Experience indicates that structuring entails problems, of which the most common are that:

- Existing maps are not always good data models.
- Terminology is imprecise and not based on common concepts.

12.6.2 Logical database design

Logical database design is software dependent. In most GIS database systems, data are organized in geographical units as map sheets, administrative areas, or other delineated areas. The sizes of the areas involved may vary considerably within a single database. However, from the user's viewpoint the data involved appear seamless; that is, the user does not see the borders between adjoining areas or sheets.

The size and shape of storage units affect data storage efficiency and accessibility and the ease with which data may be updated. Fixed unit size is most efficient whenever data are spread evenly over the entire region covered by a database. Small units are an advantage whenever data density is large. The scale of presentation and the sizes of areas that may be processed simultaneously also influence the choice of unit size. In practice, units are often chosen to correspond to the map sheets of the region covered by a database. The geographical storage structure should be chosen with care, as established database structures are difficult to modify.

In most GISs, data are also organized in thematic layers, which correspond to the overlays of traditional map production. The choice of topics for thematic layers determines how efficiently their data may be used. Each thematic layer comprises types of objects that are to be processed together, so data use determines the allocation of objects to layers. For example, one application may benefit most from having all linear objects (e.g., roads, rivers, coastlines) in a single thematic layer, while another application might separate rivers and coastlines into separate thematic layers.

The greater the variety of uses of the data in thematic layers the greater the number of thematic layers. However, data may be collected in a common thematic layer if they are:

- Always used together
- Logically related
- Updated simultaneously
- Valid for several functions (e.g., lines indicating roads, property boundaries, land-use boundaries)
- Used together in presentation (e.g., as all 0.3-mm-wide lines, all areas colored red, all place names)

Logical database design also includes specification of the coordinate system and various tolerance definitions, such as those which coordinate movement or the presence of weeds.

The uses of a database may change with time, so databases should be designed to be modified to incorporate new object types and attribute types, as well as to allow for enlargement.

12.6.3 Physical design

Physical database design is hardware dependent. Database design depends on computer hardware with respect to:

- Data volume
- Storage medium
- Storage capacity
- Accessibility and speed
- Load of data (e.g., large and dense maps)
- Single or multiple users
- Number of workstations

- On-line
- Off-line
- Storage medium
- Network

Different systems employ different approaches to data storage, and storage methods often may be modified or expanded. For example, at the time of writing, the storage capacity of stand-alone PCs is limited by the storage capacities of hard disks. Future memory capacities may be considerably greater, as CD-ROM, optical disks, and other storage media are employed in PCs. Even with current technology, effective PC storage capacity may be increased considerably by connecting a PC to a local-area network that has a larger server with greater storage capacity. In many cases, the constraints imposed by data storage facilities are more administrative than technical, as organizations implementing new systems frequently wish to employ existing data processing infrastructure. Physical design is discussed further in Chapter 18.

CHAPTER 13

Housekeeping Tools

13.1 Introduction

From the user viewpoint, the ideal GIS should include enough functions to perform all conceivable manipulations of geographical data. In practice, user needs comprise various tasks. So, as listed in Figure 13.1, an overview of user tasks define the overall GIS requirement (Larsen 1991). The first two tasks in the table are customarily associated specifically with GIS and hence are often incorporated into infrastructures under the jurisdiction of national map agencies, military agencies, or public utilities. The remaining tasks are supported to a greater or lesser degree by the various GISs now on the market.

As a software system, a GIS usually comprises a set of software tools, including:

- Database
- Database management system (DBMS)
- User interface
- Query language (QL)
- Application functions and programs

Databases, DBMS, user interfaces, and QL have been discussed previously. The basic application functions employed in GISs are discussed in this chapter; the more specialized analysis functions are the subject of Chapters 14 and 15.

The basic application functions often employ standard functions supported by the DBMS, the QL, and the user interface. The wide range of needs encountered requires numerous special GIS functions. Only a few GISs have all possible functions. In all cases, each function may be executed by one or more commands. The basic GIS functions may be divided into four main categories:

1. Functions for storing, registering, and entering data
   - Entry (including format transformation and similar operations)
   - Organization of storage operations
   - Registration and verification
2. Functions for correcting and adapting data for further use
   - Geometrical manipulation
   - Editing

User need (task)	Data manipulation required
Access catalog	Locate relevant data (e.g., metadatabase)
Requisition	Send data requested
Enter	Enter and/or register in the user system data selected
View	Study and evaluate data received or registered
Select	Choose data required
Correct	Rectify errors and omissions
Compile	Compile dissimilar data to a model or image of the real world
Overview	Use model or image to provide a general overview of an area
Monitor	Use model or image to monitor occurrences in an area
Navigate	Use model to locate position of vehicle or ship and to determine quickest route between designated points
Analyze	Evaluate connections, possibilities, conflicts, and consequences
Make decision	Ensure that data and data processing support choices of alternative actions
Present	Present overviews, analyses, and decisions in graphic form (maps and/or reports)

Figure 13.1
User tasks required by GIS.

- Edge matching
- Map projection transformations
- Coordinate system transformations
- Attribute editing

3. Functions for data presentation
   - Use of symbols (cartographic variables)
   - Text insertion
   - Perspectives and other drawings
4. Presentation

The functions for data presentation are dealt with in Chapter 16.

The four main categories above are based primarily on the various phases of most GIS projects, but in practice the categories may be mixed. For example, functions for editing geometry may be employed both in editing and in presenting data.

TOOL KIT
A: Registrering, entering and storage B: Correcting and adapting C: Processing and analysing D: Presentation

13.2 Organization of Data Storage Operations

Software systems often organize data to ensure effective use. Such organization may involve various logical paradigms concerning the grouping of object types and the divisions of geographical areas. The physical limitations of system file capacities may also be a practical reason for thematic and geographic divisions. A list of all maps in a system, organized by location and theme, forms a map library, from which the user can select the map he or she needs and store it in the workspace of the computer.

Thematic layers

Data in most GIS are organized in layers (levels), much like the overlays of conventional mapmaking (Figure 13.2). Similarly, individual data layers are stored in individual data files. These layers may contain object types intended to be processed together, such as points in one layer, lines in another, and polygons in a third. Alternatively, the individual data layers may be organized by theme, perhaps one layer for topography, another for property boundaries, others for roads or types of land use, and so on. Furthermore, each layer may contain subsidiary layers in a hierarchy. Thus a layer for roads might encompass subsidiary layers for national, county, urban, and private roads.

Collecting logically similar objects can reduce the amount of data required to describe an individual object. Objects that represent several themes, such as lines that are simultaneously roads and property or land-use area boundaries, may be collected in one layer. The line geometry of that layer may then be transferred to other layers as needed. Objects that are updated frequently or from the same source

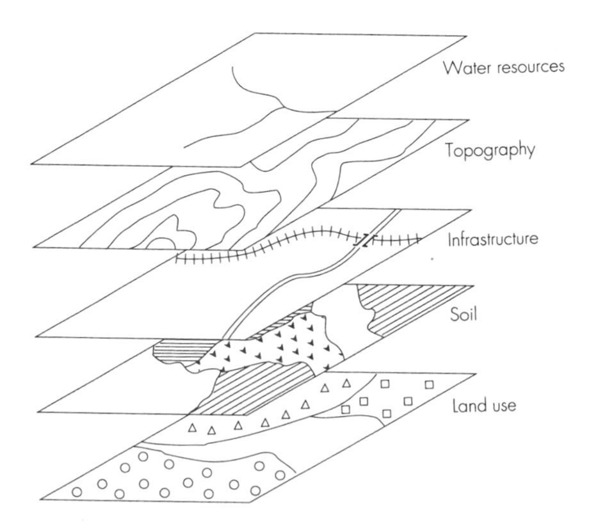

Figure 13.2
How data are organized. Many systems organize data into different layers (files). Their efficient use often depends on the well-planned organization of data into a layer system.

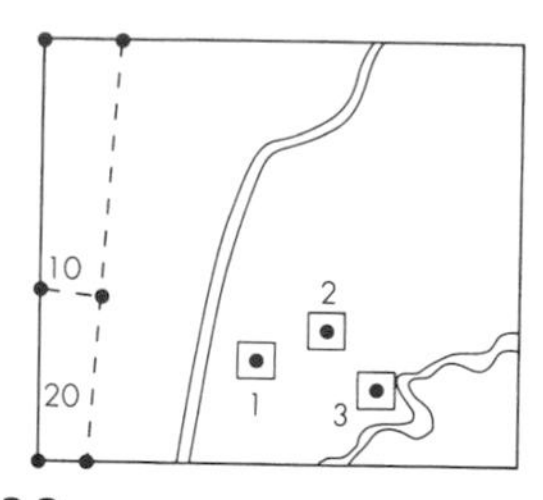

BR 012-5-2	BS 012-5-1	BS 012-5-2
BR 012-5-4	BS 012-5-3	BS 012-5-4
BR 011-5-2	BS 011-5-2	BS 011-5-2

Figure 13.3
How data are organized into separate map sheets. Many systems organize data into surface units such as map sheets. Careful consideration has to be given to which storage element is most suitable, since restructuring to other elements is difficult once the data have been organized in a specific way.

of information may also be collected in a single layer to facilitate updating work. The cartographic effects of plotting are frequently dependent on the sequential plotting of layers containing like objects.

The separation of data into layers may seem analogous to the traditional separation of map information into overlays, and therefore not always a realistic data model of reality. One of the reasons for this layered storage is that many earlier systems have not been able to store overlapping polygons in the same layer. Today, however, there are systems that can handle this problem. These GISs circumvent "map overlay thinking" by being more object oriented; that is, each object is manipulated as an independent entity with regard to both its geometry and its attributes.

Partitioning the area

Many GISs have facilities that will divide surfaces to promote efficient storage, use, and updating of data. Individual surface segments are then stored in individual files, division by map sheets being the most common. Some GISs support divisions of data structures into projects, each of which may then be further divided into subprojects.

The manipulation of data for a larger area often involves combining data for their constituent segments. This is done either manually by the user or automatically by the system. As many GISs are seamless (i.e., data need not be regarded as belonging to fixed map sheets), users are presented with data representing contiguous surfaces, even though stored data may be divided into map sheets, which in turn may be divided into cells in grids (Figure 13.3).

Users must select the most suitable elements for storing data, such as the map sheet sizes and area divisions. Choice is vital for two reasons. First, the organization of data storage elements can have a considerable influence on the efficiency with which data are used. Second, once the storage elements are chosen and data have been stored, restructuring to other storage elements is extremely complicated.

13.3 Data-Entry Functions

A GIS may be implemented with software supporting various forms of data entry, including entry from:

- Digitizers
- Photogrammetric instruments
- Surveying stations
- GPS
- Scanning and pattern recognition
- Miscellaneous entry program

The relevant programs include facilities for expressing original data in ground coordinates, for entering codes, and for performing quality checks. The software systems also have flexible modules for entering assorted types of attribute data, including the presentation of various screen displays.

13.4 Importing Existing Digital Data

All GISs contain software that permits the import of existing digital data. The relevant processes entail transforming various exported formats to the GIS internal file format. Many GIS programs can read national or international standard formats such as ASCII, DIGEST, SIF, DXF, and HPGL for vector data, and TIFF and CGM for raster data. Some systems have routines for reading both vector and raster data. The ability to read different formats can often significantly enhance a program's user friendliness, since most GIS users will find that they have to collect data from many different sources. Suppliers of GIS software have started to take this into account, and today it is possible to find solutions that are able to handle simultaneously many types of data warehouses with different formats.

13.5 Functions for Correcting and Adapting Geometric Data for Further Use

Data are ready for use only after they have been verified and, if necessary, corrected. All GIS contain programs for adapting data. These include:

- General utility functions
- Editing and correction of errors
- The ability to create topology
- Transformations to a common map projection

- Transformations to a common coordinate system
- Adjustments at map edges or between other storage elements
- Coordinate thinning and line smoothing

13.5.1 General utility functions

Like all computer systems, GISs have collections of programs that provide such useful functions as file manipulation, editing, and program cross-referencing. GISs incorporate other utility functions specific to GIS tasks. Zoom, a utility for changing the scale of screen *images*, is an example. Zoom may be activated either in the main GIS software or in the screen software. *Screen zoom,* which allows either direct enlargement or reduction, is the more rapid of the two. Although slower, GIS *program zoom* can include various intelligent functions that alter texts, line widths, and other features in proportions differing from those of the overall scale magnification or reduction. In most systems there is a limit to how much an image may be enlarged. Another GIS utility can highlight selected themes or objects, thus easing on-screen work with complex maps comprising many elements.

13.5.2 Editing and correcting errors and omissions

Most GISs support geometric map editing functions, including:

- Supplementing
- Copying
- Deleting
- Moving
- Rotating
- Dividing lines
- Joining lines
- Altering form

Raw digitizing data always contain some errors and omissions, such as lines crossing erroneously, unintentional wiggles in lines, missing points and lines, and so on (Figure 13.4). Omissions are corrected most easily by entering data directly from a digitizer. Errors are corrected most efficiently using a keyboard or a mouse.

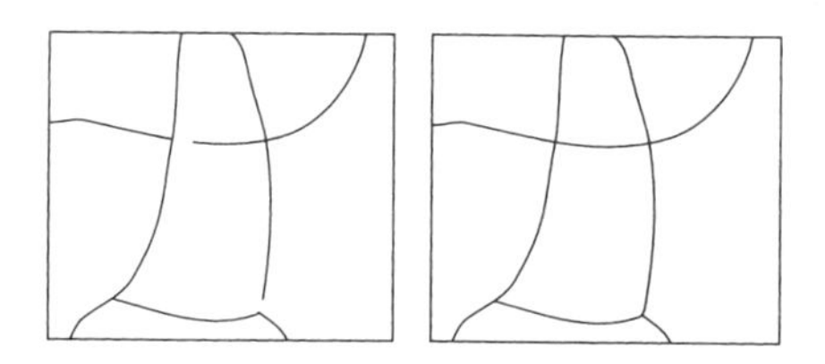

Figure 13.4
Polygons that are not closed can be corrected by joining lines automatically where the gap is below a given parameter. It should be noted, however, that automatic closing can have undesirable consequences should the closing parameter be set too high.

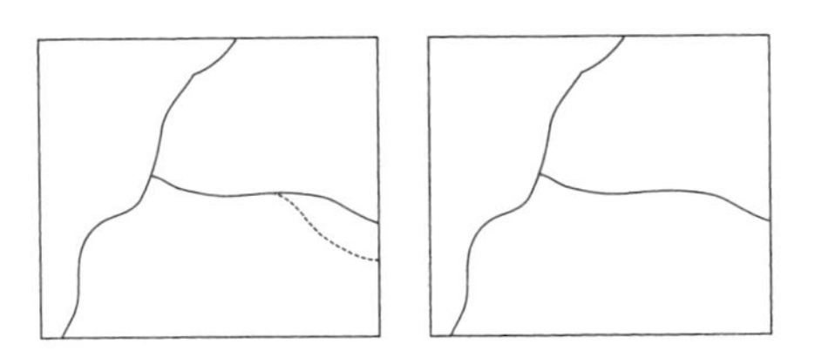

Figure 13.5
Digital map data can be edited by, for example, the addition of new data and removal of old data, so that objects retain their correct shape.

Lines may be divided or line segments deleted to allow text or symbols to be placed on maps. Lines and composite map objects may be moved whenever objects are digitized incorrectly. Line segments to be moved or deleted are marked, as well as any new positions (Figure 13.5). In some cases, objects are rotated through prescribed angles.

Other utilities support text editing for altering or amending place names, road names, and other names. Many systems support the direct printout, for example, of such attribute values as addresses and property registry numbers on maps (Figure 13.6). Some systems treat all text as attributes. Raster-based systems have programs with editors dedicated to editing raster data. Hence systems that can manipulate both raster data and vector data have separate raster and vector editors. Coordinate values may be modified using an ordinary text editor.

13.5.3 Creating topology

Corrected data may result in an unacceptable map image, but even when such data are verified, the image may contain errors that are not evident until logical connections are established. Consequently, some GIS have functions that compute nodes and links automatically and compile topology tables (Figure 13.7). By creating topology it is possible to identify such errors as:

- Polygons that are not closed
- Disconnected lines (dangling ends)
- Missing or repeated ID codes
- Polygons lacking, or having too many, ID points (in some systems. All polygons are assigned ID points with codes and coordinates)
- One or more missing polygons

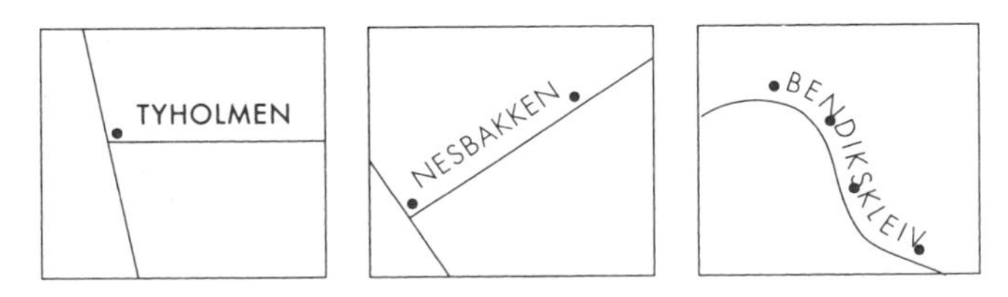

Figure 13.6
Normally, text can be freely placed with regard to beginning and end, rotation, following lines, and so on. Font types are generally system defined.

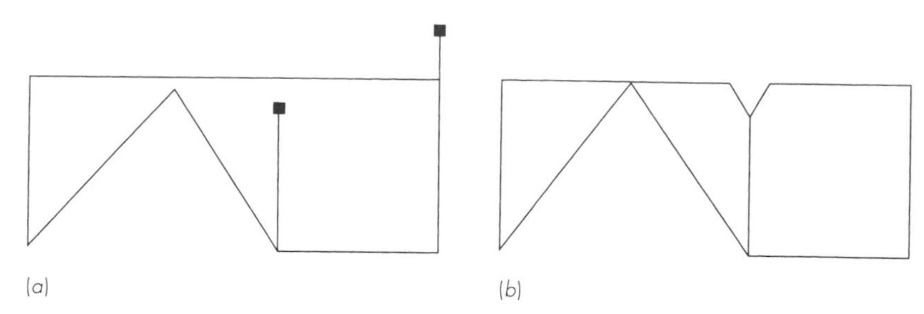

Figure 13.7
(*a*) Various types of data error can often be identified during topology creation.
(*b*) Automatic error correction can be risky since it can introduce completely new, highly undesirable errors.

Errors such as these may be reported by a dedicated error reporter or marked with specific symbols on screen plots; such errors may be corrected manually or automatically. Errors in ID codes and points are corrected manually via a keyboard.

Small dangling ends can be deleted automatically by setting a parameter that stipulates the sizes of the ends to be deleted. Gaps may also be closed automatically. However, if the gap-closing parameter is set too high, automatic gap closing may introduce errors by connecting lines that ought not to be connected. It is always advantageous, therefore, to digitize line intersections in order to reaffirm line connections. If the gap-closing parameters are set too low, it will not be possible to close all the faulty gaps.

Small, meaningless polygons can also be removed automatically by using the area size. Here it is also possible to introduce errors. Should the test value be set too high, polygons that should have been included will have to be removed. Generally speaking, the automatic correction functions should be used with great care.

13.5.4 Transformations

Transformations to a common map projection

Data from a variety of sources are useful only if referenced to a common map projection. However, data from differing sources are often referenced to differing map projections. One common disparity is that survey points are entered in coordinates of latitude and longitude in an azimuth projection, while other map data are entered with reference to a national rectangular coordinate system based on a cylindrical projection. Consequently, most GISs support various transformations for converting data from one projection to another.

Mapping hemispheres and other major land areas usually entails selection of the map projection that most closely suits the data presented. In some cases, this can mean that entire sets of data must be transformed.

These transformations are based on the mathematical relationships that describe the various map projections. Thus, a transformation con-

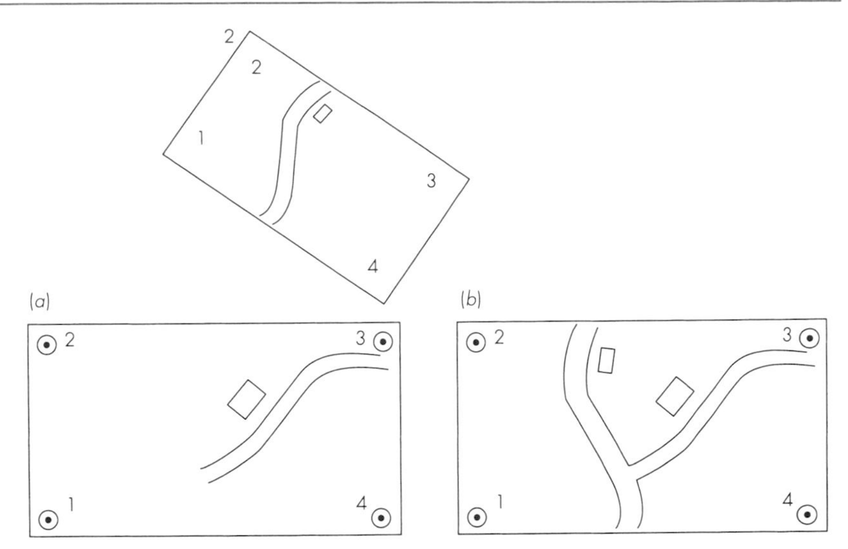

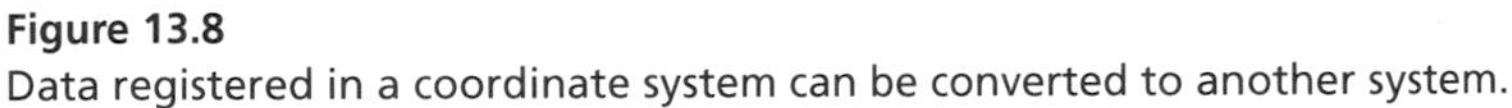

Figure 13.8
Data registered in a coordinate system can be converted to another system.

verts the coordinates of one system to the coordinates of another. As the transformations themselves are digital, ancillary operations are necessary whenever positional information is not decimal, as when the coordinates are in degrees, minutes, and seconds.

Transformations to a common coordinate system

Data from various sources, as from map digitizing and field surveying, may be used together only if referenced to a common coordinate system (Figure 13.8). Sometimes systematic errors in such themes as displacements, rotations, and scale errors can easily be compensated if the data involved are transformed to an error-free base. In other cases, the data of a particular thematic layer must be transformed to be compatible with the data of other thematic layers. Consequently, most GISs support a variety of coordinate system transformations.

Coordinate transformations involve mathematical functions that relate coordinate geometries to each other. A conversion function also contains parameters based on a knowledge of the respective coordinates of a number of points that are common to the initial and final systems of the transformation involved.

13.5.5 Adjustments at map edges and between neighboring areas

As anyone who has glued neighboring map sheets together to make a larger map knows, lines that *should* meet at the joined map edges often do not. Digitalization errors, scanning errors, and nonuniform paper shrinkage are often the causes of this common and vexing problem. The problem is compounded whenever map sheets are digitized because digitization may also introduce error. Most GISs therefore support func-

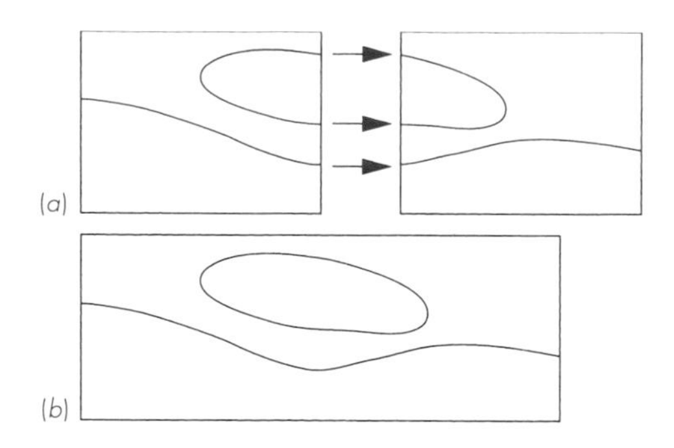

Figure 13.9
Map sheets can be aligned automatically and combined into a new, seamless unit.

tions for automatic adjustment of disparities along neighboring map sheet edges (Figure 13.9). This is a time-consuming process because for each disparity along the meeting edges, a decision must be made to move either one or both lines. When the process has been completed for two adjoining edges, the polygon representing the edge may be deleted.

13.5.6 Thinning coordinates and smoothing lines

Thinning coordinates

Digitization often produces more data than are needed for the immediate task, particularly with linear objects, such as contour lines, borders for types of soil and vegetation, and so on. Consequently, most GISs have routines for *thinning,* deleting superfluous data. Thinning increases data efficiency, reduces the storage capacity required, and increases the plotting speed. In some cases, data volume may be reduced by 60 to 95% without reducing the data quality.

Various methods are available for thinning points on lines. The simplest approach is to delete every *n*th point—every other point, every third point, and so on—up to the number *n* chosen (Figure 13.10). A more advanced approach entails moving a small, variable-width corridor along the lines. The number of points needed to describe a line may be reduced by moving forward a corridor of given width until it touches the digitized line. All the points on the line that lie within the corridor, apart from the first and last, are thereby deleted. This process is repeated until the entire line has been trimmed. Lines may also be replaced by polynomial spline functions. Good thinning routines remove points without having any noticeable effect on the geometric accuracy.

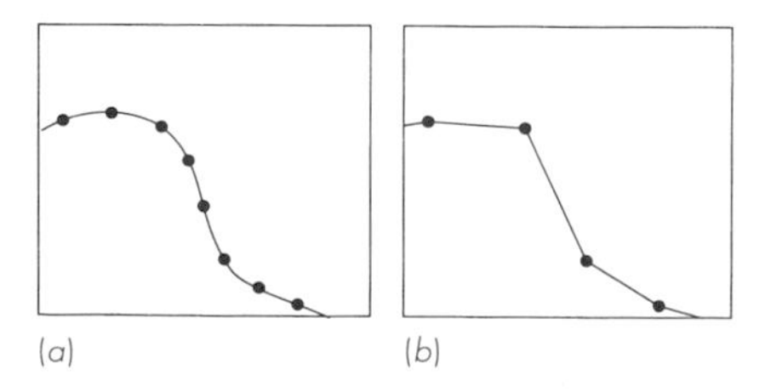

Figure 13.10
Lines can be thinned by removing points based on certain functions.

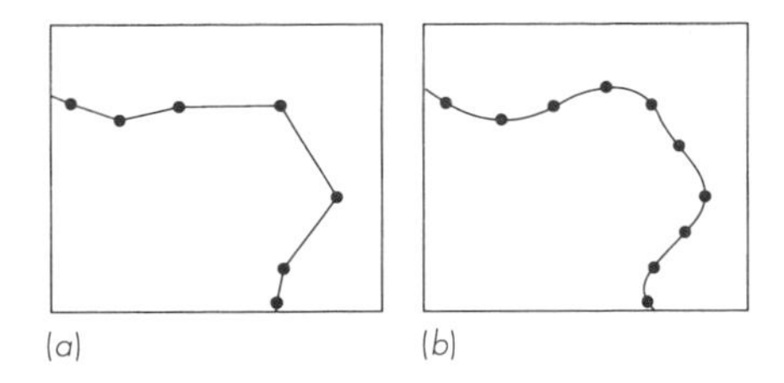

Figure 13.11
Height contours and other linear elements can be smoothed by the automatic addition of new points, based on certain functions.

Line smoothing
Lines delineated by points are never completely smooth, so line smoothing is often used to improve the graphic quality of lines. Smoothing functions are usually associated only with data presentation. Line smoothing means increasing the total number of points delineating a line. Several smoothing methods are available, but those used most frequently are based on curve fitting, using third-degree polynomials, as illustrated in Figure 13.11.

Smoothing affects appearance only. Although a smoothed line may seem more accurate, it is no more accurate than the original data for the line. Since smoothing increases the number of points, this function is often linked to the plotting itself and thus does not affect the storage volume.

13.6 Editing Attribute Data

Like digital map data, attribute data must be edited and corrected. These operations include error correction, updating, and amending. The editing tasks may be carried out by using standard editing tools, such as those available in word processing, or specific GIS commands. Some GISs use SQL (Structured Query Language) to manipulate attribute data in relational databases. Specific GIS commands include commands for changing object thematic codes and switching codes between objects, as well as for editing thematic codes containing texts.

The guidelines for entering data may change with time, mandating changes in the codes of older data. Usually, common mathematical signs such as +, −, *, and / are used for this purpose. The currencies in which prices in attributes are expressed may be changed [e.g., from U.S. dollars ($) to pounds sterling (£)] by entering an exchange rate. Relational and other databases used to store attribute data usually incorporate effective editing tools. These permit a variety of operations, including searching for members of a prescribed class and then editing one by one, or assigning new values to an entire class using a single command.

CHAPTER 14

Basic Spatial Analysis

14.1 Analysis of Spatial Information

Even though most of the real world can be observed with the naked eye, it is often difficult to interpret and systematize what is observed. It becomes even more difficult when the image of that reality is stored in digital form as map data and attribute tables. To bring out the patterns, connections, and possibly, the causes of variations in the data, various computer-based techniques are applied to analyze the data. Spatial analysis techniques are an attempt to imitate what concerns the human brain (i.e., to create an understandable image of reality). These are the techniques discussed in this and the following chapter.

Technology can still only help to a limited extent, however. Stating problems and delineating the approaches to solutions together comprise one of the most difficult steps in GIS analyzing. This has to be solved by the individual user or operator before the technology can be put to use, based on his or her professional knowledge in the fields of agriculture, environmental protection, planning, and so on, supplemented with knowledge of GIS.

Analyzing data normally comprises two principal phases:

1. Choice of data
2. Analyses of the data chosen

All GISs provide functions for analyses of data chosen and for storing the results of such analyses. Data may be selected according to:

- Geographical location
- Thematic content

Most GISs permit defining the criteria for selection. These are often based on SQL or are in menus with provisions for generating SQL queries. Some GISs provide predefined selection criteria; other systems use a macro language to set up the selection criteria. Specific systems have predefined menus dedicated to the relevant applications. In most systems, selection criteria may be stored for subsequent use.

Data may be analyzed at various levels:

1. Data in attribute tables are sorted for presentation in reports or for use in other computer systems.
2. Operations are performed on geometric data, either in search mode or for computational purposes.
3. Arithmetic, Boolean, and statistical operations are performed in attribute tables.
4. Geometry and attribute tables are used jointly to:
   a. Compile new sets of data, based on original and derived attributes.
   b. Compile new sets of data based on geographical relationships.

Within each of these levels, the operations used may be logical, arithmetic, geometric, statistical, or a combination of two or more of these four types. Operations may be performed on individual points or on areas, and may involve considerations of proximity or of changes over time. Numerous operations may be performed on line networks. The more important commands are discussed below. The functions implemented vary from one GIS to another, and some GISs contain functions not discussed here.

14.2 Logic Operations

Logical searches in databases normally employ set algebra or Boolean algebra. Set algebra uses the three operators equal to, greater than, and less than, and combinations thereof:

$$=, >, <, \geq, \leq, <>$$

These operators are included under SQL.

Practical applications include:

- Identifying extrema, such as finding attribute minima or maxima within various polygons and, as a result, delineating a new thematic layer (new row in the attribute table)
- Selection or isolation, where particular values are selected for subsequent ranking in a new thematic layer

Boolean algebra uses the AND, OR, NOR, and NOT operators to test whether a statement is true or false. AND, OR, and NOT are used in SQL. For two items, A and B, we might have any of the following statements:

A AND B, A OR B, A NOR B, A NOT B

Such statements may be illustrated in a Venn diagram, which is a schematic representation of a set in which magnitudes illustrated by surfaces are superimposed, as shown in Figure 14.1. The shaded areas

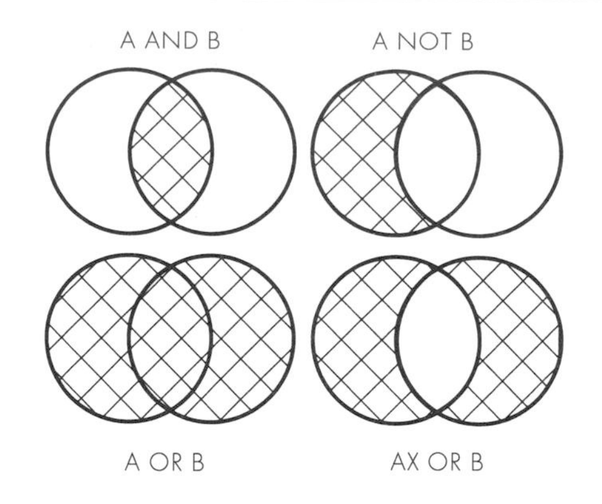

Figure 14.1
Logical operations shown in a Venn diagram. Similar operations can be carried out on geographical data. The shaded areas represent true statements.

represent true statements. This technique is well suited to analyzing geographical data. For example, potential conflicts between forestry and cattle farming can be illuminated by assigning A to potentially productive forest tracts and B to known grazing areas. The tests A AND B on the two operands will identify conflict areas that can be assigned special symbols and drawn on maps.

Logic operations are particularly powerful when the relationships are complex. In GIS, logic operations may be performed simultaneously on more than two themes and involve several operators.

14.3 General Arithmetic Operations

Arithmetic operations are performed on both attribute and geometric data. All GISs support the customary arithmetic operations of addition, subtraction, multiplication, division, exponential, square root, and the trigonometric functions:

$+, -, \times, /, ^n, \sqrt{}, \sin, \cos, \tan$

These operators may be used for many purposes, including assigning new thematic codes. Typical examples include:

- Reclassification of soil types, in which areas are to be converted from decares to hectares by dividing all area figures by 10.
- Conversion of distances along roads to driving times, by dividing all distances by a specified average vehicle speed. The result is a new set of attributes that are useful in transportation planning.

Arithmetic functions are used in all geometric computations involving coordinates, as in calculating distances, areas, volumes, and directions.

14.4 General Statistical Operations

Statistical operations are performed primarily on attribute data, but may also be effected on some types of geometric data. Most GISs support a range of statistical operations, including sum, maxima, minima, average, weighted average, frequency distribution, bidirectional comparison, standard deviation, multivariate, and others. The computation of averages requires averaging two or more attribute values and stating the result as a new attribute. Frequency distributions are used to compile *histograms,* charts comprising rectangles whose areas are proportional to relative frequencies and whose widths are proportional to class intervals. The data used to draw a histogram may also be employed to plot a curve. Other statistical operations in common use include least-squares computations of transformation parameters from regression models, with the standard deviation as an expression of accuracy. Bidirectional comparison involves point-by-point correlation of two themes to produce a new statistical thematic layer and hence a new attribute.

Satellite data are usually analyzed statistically in dedicated image processing systems, which are often connected to GIS facilities. Some vector GISs support image processing. Multivariate operations, such as cluster analyses, are vital in image analyses. These operations assign new classes to entities on the basis of statistical selection criteria. Pattern recognition based on statistical models is incorporated in some GISs.

14.5 Geometric Operations

Operations on geometric data involve the customary arithmetic operations in computations of distances, areas, volumes, and directions. In principle, distances should be measured using a uniform unit and a fixed procedure. In a raster GIS, the unit may be cell width or diagonal. In transportation analyses, results may best be stated in travel time or travel costs rather than in meters or other units of length.

In many GISs, the periphery, area, and centroid are computed automatically for each polygon and connected to objects as attributes for topological purposes. With distance operations, points within a prescribed distance must be identified by comparing all distances computed with the distance prescribed. Volumes and directions are calculated in models manipulating terrain, as in determining excavation volumes, terrain slopes, exposures, and the like. Flow analyses calculate geometric computations to find the shortest routes between designated points.

14.6 Report Generation from Attribute Data

Inquiry in attribute data may be the primary GIS task in daily practical applications, particularly in the operational, maintenance, management, and service sectors. Inquiry calls for a human–machine dialogue in which the user employs various input/output devices and their supporting software. Tailor-made software is employed in some GISs. In others, standard routines such as SQL are employed, particularly in those that access relational databases.

Inquiry of and search for data may be based on logical and arithmetic operations and specific relational database functions, as with projection, selection, and joining. The query criteria may be complex and cover several attribute tables. In practice, responses to inquiries should be displayed in a comprehensible manner, not simply as brief on-screen information. This is particularly true when attribute data are used as a reference source. Consequently, many GISs support a standard report format with fixed headings, column locations, and other layout features. In addition, users may define their own report formats.

Some GISs support report formats that are tailor-made to each application, while others use the report function and report form storage facilities of the report generator supported by the DBMS in use. In many cases, reports may be simple. For example, an inquiry may be for the name and address of the owner of a property at a particular address, or for a list of all the properties registered on a specific street. Reports in various formats may also be stored in separate files for use with other system files or for export to other systems for processing, as in statistical analyses.

14.7 Map Data Retrieval and Search

Map data may be searched using specified criteria, using a dedicated query language or SQL. Searches start from coordinate values, and in principle, any part of a database may be searched. Seamless databases permit searches anywhere within a geographical area. Databases based on map sheets, however, limit searches to one map sheet at a time, although contiguous map sheets may be combined for search purposes. In this case, the combined map sheets function as a partial copy of the database as a whole. For example, a search may be conducted within an area specified by the coordinates of its lower left and upper right corners, without involving the rest of the database.

Map data usually carry thematic codes for various types of objects, such as roads, administrative borders, water supply systems, buildings, properties, and so on. In searching these codes, the various map themes may be turned off and on, or shown in different colors, for the different types of objects.

A typical task may require the display of all map details within the limits of x minimum = 10,000 to x maximum = 12,000 and y minimum = 5000 to y maximum = 7000. The query program then reads and tests all coordinate values in the database against these constraints and retains those that lie within the limits stated. Proximity operations often necessitate searches in map data.

14.8 Complex Operations of Attribute Data

Mathematical, logical, and statistical operations may be performed on attribute data (Figures 14.2 to 14.5). The mathematical operations include addition, subtraction, multiplication, division, exponential, square root, and the trigonometric functions:

$$+, -, \times, /, {}^{n}, \sqrt{}, \sin, \cos, \tan$$

The logical operations require the operators of set algebra: equal to, greater than, and less than, and combinations thereof:

$$=, >, <, \geq, \leq, <>$$

as well as those of Boolean algebra:

AND, OR, NOR, and NOT

The statistical operations include sum, maxima, minima, average, weighted average, frequency distribution, bidirectional comparison, standard deviation, multivariate, and others.

Attribute data may be either quantitative and expressed numerically or qualitative without numerical magnitudes. Numerical operations may be performed only on numerical data and hence mainly on

ATTRIBUTE TABLE

ID	Area value $
1	7.24
2	10.13
3	19.05
4	1.58
5	22.01

(a)

ATTRIBUTE TABLE

ID	Area value £
1	10.86
2	15.20
3	28.58
4	2.37
5	33.02

(b)

Figure 14.2
Arithmetic operation on attribute values. Here U.S. dollars (*a*) are converted to British pounds (*b*) by multiplying by the exchange rate (1.50).

ATTRIBUTE TABLE

ID	Area value $
1	7.24
2	10.13
3	19.05
4	1.58
5	22.01

(a)

ATTRIBUTE TABLE

ID	Area value $
1	7.24
4	1.58

(b)

Figure 14.3
Logical operation on attribute values (a). All values below 10 are selected (*b*).

	Attributes			
ID	A	B	C	D
1	10	10.1	0.3	55
2	11		0.7	55
3		98	0.8	51
4	6	116		50

(a)

	Attributes			
ID	A	B	C	D
1	10	10.1	0.3	55
4	6	116		50

(b)

Figure 14.4
Boolean algebra on attribute values (*a*) identifies all objects that contain both attributes A and B (*b*).

quantitative attribute data. The numerical treatment of qualitative attribute data is limited to counting operations, as in determining the number of classes into which data may be ranked or how many observations fall within each class. For example, four classes—A, B, C, and D—may be identified and found to contain differing numbers of observations: 12 in A, 23 in B, 2 in C, and 9 in D. Attribute data are frequently processed to discern new patterns in the data.

ID	Attributes
1	7.24
2	10.13
3	19.05
4	1.58
5	22.01

(a)

Sum $	59.91
Average $	11.98
Max. $	22.01
Min. $	1.58

(b)

Figure 14.5
Statistical operations on attribute (a) values make it possible to find the sum and the average, maximum, and minimum values in the table (b).

14.9 Classification and Reclassification

In the classification process, attributes are grouped according to limits set by the user. For example, three classes may be set for the attribute "year": A = before 1970; B = 1971–80; C = 1981–90. Each object with a year attribute, such as a water supply pipe, is then assigned a new year-class attribute A, B, or C. Plotting the classes in distinctive symbols or colors may reveal new patterns: showing, for instance, that water supply pipes in class B are those most subject to leakage.

Reclassification involves changing attribute values without altering geometries (Figure 14.6). Arithmetic and some statistical operations are used to assign new attribute values. In many ways, reclassification may be compared to changing colors on a map, in that the reclassified attributes may plot out in new colors. Reclassification may be used to isolate object types. For example, in a raster GIS, a "built-up area" characteristic may be isolated by assigning all other areas a value of zero. When the data are plotted out, only the built-up areas will then be shown. The boundaries between polygons of the same type may be deleted to combine the polygons as larger units. Polygons of dissimilar types must be reclassified prior to being combined (Figure 14.7).

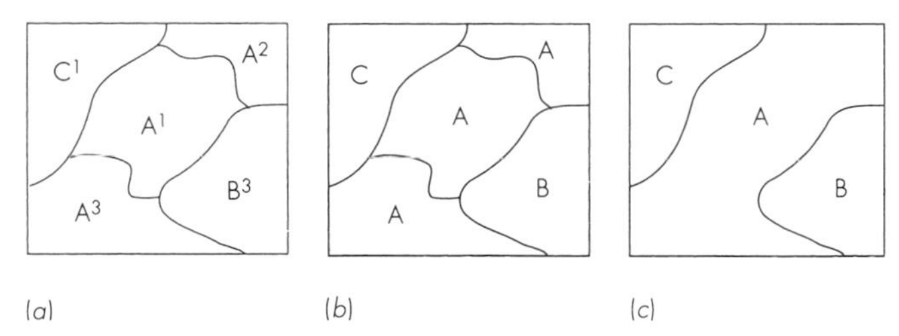

Figure 14.6
Reclassification of polygons (*a* and *b*) and combining polygons (*c*).

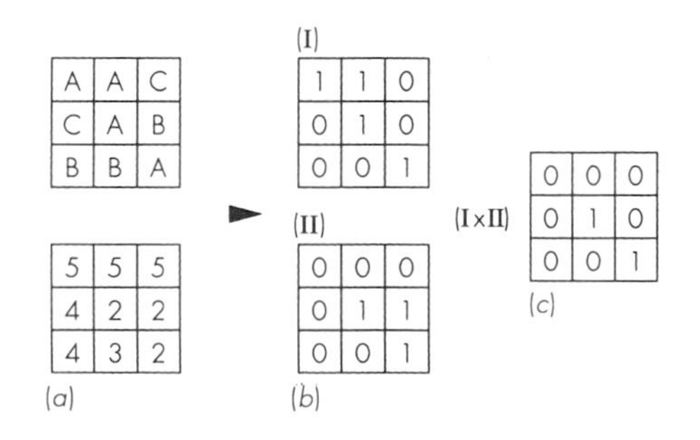

Figure 14.7
Operations on raster data: finding cells in the two raster layers I and II which contain both A and 1. The cells are reclassified (*b*), and I multiplied by II gives a new layer (*c*).

14.10 Integrated Processing of Geometry and Attributes

One of the simpler forms of integrated processing of geometry and attributes is to point to the location of a building displayed on screen and request retrieval of all information stored on the building. On receiving the query, the GIS searches the map database to find the building corresponding to the coordinates that have been pinpointed. Using the building ID number stored with the coordinates, the system then searches the attribute database for all available information, which can then be displayed or printed out.

More advanced integrated processing is also based on the condition that each object type (cultivated land, deciduous forest, protected area, etc.) is represented both in geometry and in an attribute table. The geometry concerned may be likened to a single thematic map. Single maps may then be superimposed to integrate with each other and thus produce a multithematic map containing information from each of the initial thematic maps. The integrated map comprises comparable units [integrated terrain units (ITUs)] and a new attribute table is compiled, as illustrated in Figure 14.8. Arithmetic, logical, and statistical operations may be performed in the new attribute table. The geometry and the attributes may then be used to compile a new thematic map.

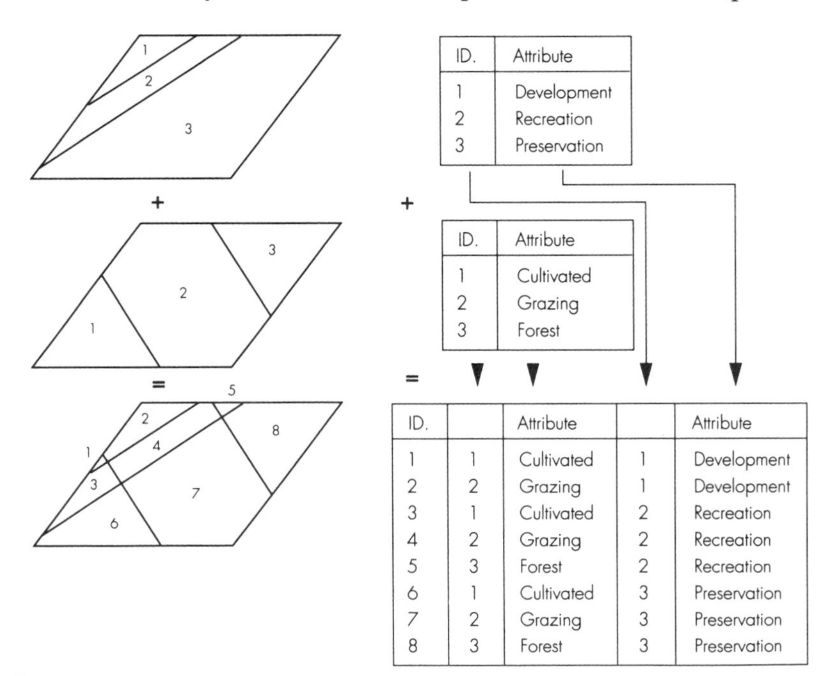

ID.	Attribute
1	Development
2	Recreation
3	Preservation

\+

ID.	Attribute
1	Cultivated
2	Grazing
3	Forest

=

ID.		Attribute		Attribute
1	1	Cultivated	1	Development
2	2	Grazing	1	Development
3	1	Cultivated	2	Recreation
4	2	Grazing	2	Recreation
5	3	Forest	2	Recreation
6	1	Cultivated	3	Preservation
7	2	Grazing	3	Preservation
8	3	Forest	3	Preservation

Figure 14.8
A computerized superimposition can, in many ways, be compared to a series of map overlays. However, this also leads to an expansion of the attribute table, in addition to the geometric changes. Arithmetic, logistical, and statistical operations can be carried out in the expanded attribute table.

14.11 Overlay

Overlay is used in data integration and is a technical process, the results of which can be used in realistic forms of spatial analysis.

14.11.1 Polygon overlay

Polygon overlay is a spatial operation in which a first thematic layer containing polygons is superimposed onto another to form a new thematic layer with new polygons. This technique may be likened to placing map overlays on top of each other on a light table (Figure 14.8). The corners of each new polygon are at the intersections of the borders of the original polygons; hence computing the coordinates of border intersections is a vital function in polygon overlay. The computations are relatively straightforward, but they must be able to cope with all conceivable geometric situations, including vertical lines, parallel lines, and so on. Computing the intersections of a large number of polygons can be very time consuming.

If areas are stored as links in a topological model, fewer intersections need to be computed, thus reducing the computing time. The new intersections are identified as nodes and the lines between the nodes as links. The new nodes and links then comprise a new topological structure. Let us take as an example polygon C4 in Figure 14.9, which is a combination of polygons C and 4. The system cannot associate attributes with C4 unless topology is associated with either the original or the new data. With topology for the new data, the system recognizes

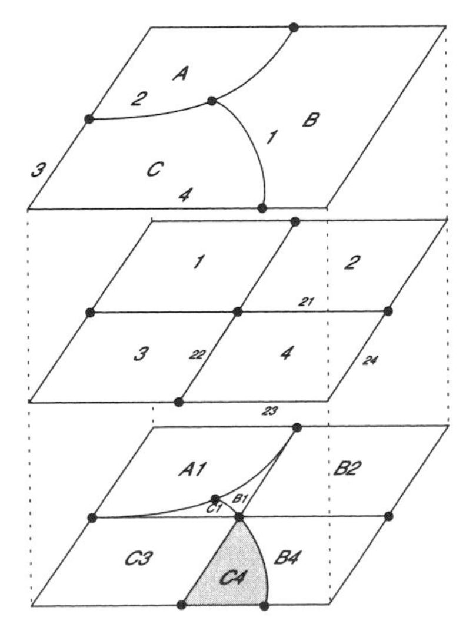

Figure 14.9
The data topology has to be corrected prior to overlaying (see Section 14.11). (From ESRI 1995.)

that polygon C4 comprises line 22 and parts of lines 23 and 1. Furthermore, the system is aware that polygon 4 is on the right side of line 22 and on the left side of line 23, and that polygon C is on the left side of line 1. Hence the system now knows that polygon C4 is a composite of polygons 4 and C both geometrically and in terms of attributes.

Each new polygon is a new object that is represented by a row in the attribute table. Each object has a new attribute, which is represented by a column in the attribute table. Superimposing and comparing two geometrical data sets of differing origin and accuracy often give rise to a large number of small polygons. For example, two polygons representing land areas may have slightly differing geometric borders on a lake, yet on a piecemeal basis the borders may coincide. Superimposing the two polygons can then produce an unduly large number of smaller polygons (Figure 14.10). This proliferation of smaller polygons may be counteracted automatically by laying a small zone around each line. If two zones intersect when the polygons are superimposed, the lines they surround combine into one line. Smaller polygons may also be removed later, using area size, shape, and other criteria. In practice, however, it is difficult to set limits that reduce the number of undesirable small polygons while retaining smaller polygons that are useful.

In addition to performing the overlay computations, the system can present a new image of the new structure, borders between polygons of like identity being removed to form joint polygons. This combination process may be either automatic or controlled by the user.

The overall procedure for polygon overlay is to:

1. Compute intersection points.
2. Form nodes and links.
3. Establish topology and hence new objects.
4. Remove excessive numbers of small polygons where necessary and join like polygons.
5. Compile new attributes and additions to the attribute table.

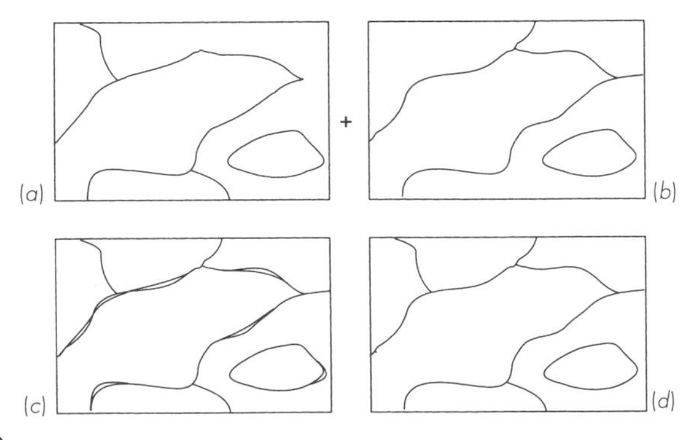

Figure 14.10
The overlaying technique (*a* and *b*) can easily lead to the creation of many small meaningless figures—slivers (*c*). These can be removed automatically (*d*).

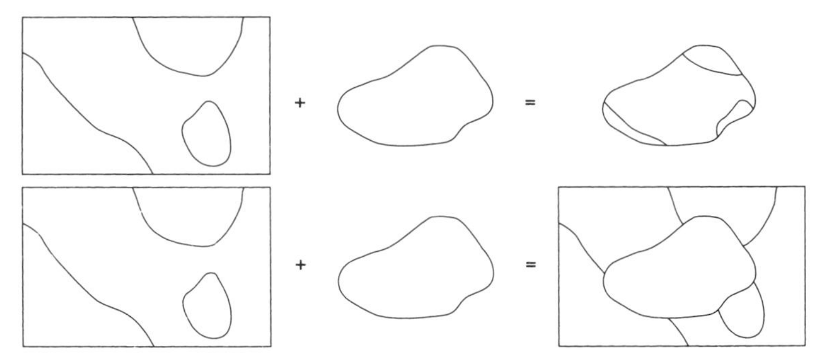

Figure 14.11
Various overlay operations can be used to clip out geometric windows.

Polygon overlay is a sizable operation for which even on the most powerful computers may require relatively long processing time. The process may be used to clip geographical windows in a database. For example, a township border in one thematic layer may be used to clip all other thematic layers in order to produce a collection of data for that township only (Figure 14.11).

Overlay produces a thematic map comprising thematically comparable homogeneous units [integrated terrain units (ITUs)] and an expanded attribute table. Arithmetic, logical, and statistical operations may be performed on the attributes, for example, when simulating alternatives and studying consequences. The advantages that GIS offers in these analyses are that the number of alternatives considered are not limited by system capacity, that all alternatives can be based on the same data, and that all available information may be used.

14.11.2 Points on polygons

Just as polygons may be superimposed on other polygons, so may points be superimposed on polygons (Figure 14.12). The points are then assigned the attributes of the polygons upon which they are

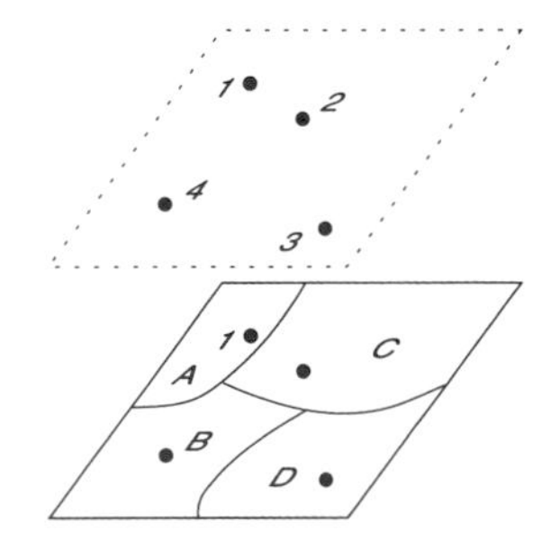

ID	Building no.	Polygon	Property
1	660	A	44/110
2	659	C	44/95
3	610	D	44/121
4	665	B	44/81

Figure 14.12
Superimposing points on polygons.

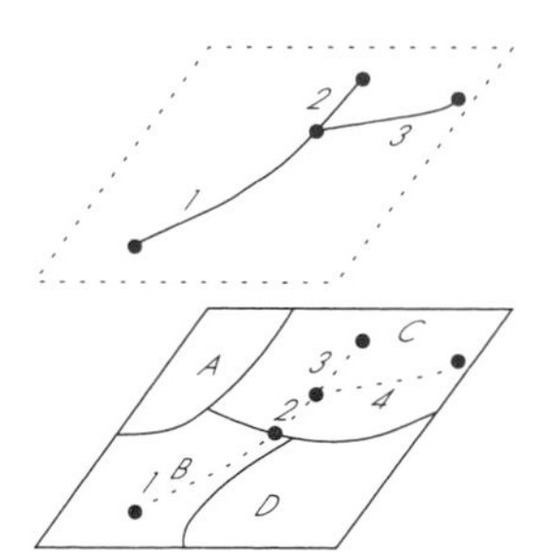

Attributes

ID	Line	Road no.	Polygon	County
1	1	Rv. 410	B	Akershus
2	1	Rv 410	C	Oslo
3	2	Rv. 9	C	Oslo
4	3	E 18	C	Oslo

Figure 14.13
Superimposing lines on polygons.

superimposed. The relevant geometric operation means that points must be associated within polygons. One approach requires computing the intersection of a polygon border with parallel lines through points. Attribute tables are updated after all points are associated with polygons.

14.11.3 Lines on polygons

Lines may also be superimposed on polygons (Figure 14.13), with the result that a new set of lines contains attributes of both the original lines and the polygons. These particular computations are similar to those used in polygon overlay: intersections are computed, nodes and links are formed, topology is established, and attribute tables are updated.

14.12 Buffer Zones

Buffer zones are used to define spatial proximity (Figure 14.14). These comprise one or more polygons of a prescribed extent around points, lines, or areas. The new polygons have the attributes of the original

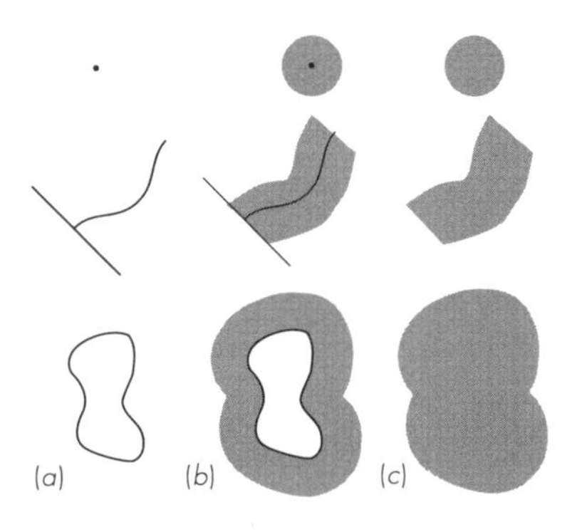

Figure 14.14
Buffer zones can be established around points, lines, and polygons (*b*). The buffer zones are the basis for the creation of new polygons (c) that have the attributes of the original objects, to be used in further analyses.

objects. Many GISs support the automatic compilation of buffer zones. Here, the operator interaction usually consists of keying in specific zone parameters, such as stipulating a 50-m zone width or either side of a road. The creation of a buffer zone is not in itself an analysis, but the new polygons that are created can be used in analysis. Buffer zone polygons may be processed in the same way as polygons generated during operations such as overlay, arithmetic, logical, and statistical computations in which attribute values come within the respective zones.

14.13 Raster Data Overlay

Raster data may also be overlaid. Indeed, raster overlay is often more efficient than vector overlay (Figure 14.15). The positions of the overlaid thematic layers need be tested only to see whether or not they contain cell values. The resultant cell-to-cell comparison presupposes that all cells in each thematic layer are queried, regardless of their values. The total number of cells therefore has an effect on processing time. The new composite cells are assigned attributes composed of those from the original cells. These new cells are registered as a new thematic layer.

Raster data consist of equally spaced cells of equal size (assuming that the various thematic layers cover the same area or have been modified to do so). Consequently, there is no formation of smaller erroneous polygons as with vector data overlays, and there is no need to distinguish between polygons, lines, and points, because all raster data comprise cells. In raster data, attributes are not usually listed in tables as in vector data, but are represented by thematic layers. Therefore, arithmetic operations and some logical and statistical operations may be performed directly during the overlay process; two thematic layers may be combined, subtracted, multiplied, and so on. If, for example, an attribute of volume in liters is to be modified to deciliters,

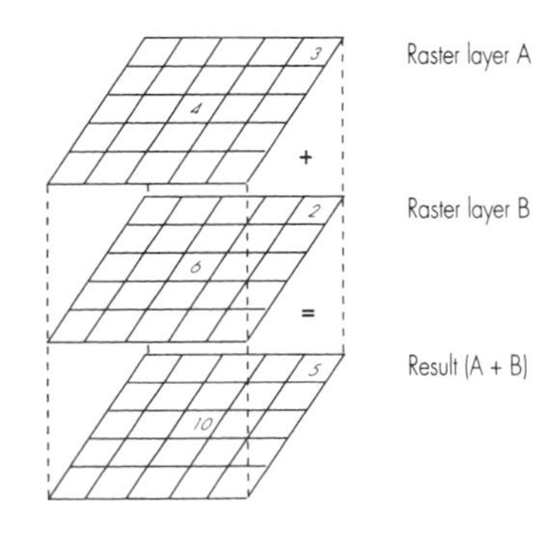

Figure 14.15
Raster data overlay is simpler than vector data overlay and can be carried out directly on the attribute or cell values.

the thematic layer of volumes merely needs to be multiplied by 10 in each cell of an ancillary thematic layer.

The arithmetic operations on two thematic layers, A and B, produce a new thematic layer, C, through the operations

$$C = A + B, C = A - B, C = A/B, C = A \times B$$

Typical logical operations might be

if $A > 100$, $C = 10$; otherwise $C = 0$
or
$C =$ max. (or min.) of A or B

Some GISs support logical operations in the resultant C layer but not directly in the original A and B layers.

As raster overlay is far more efficient than vector overlay, many GISs support functions for manipulating both raster and vector data. Vector data may be converted to raster data in order that overlaying can be performed, and the results can then be converted back to vector form.

14.14 Procedures in Integrated Data Analyses

We have already seen that overlay and the creation of buffer zones are mechanical operations and should be looked upon as a link in an integrated data analysis. Integrated data analyses may be relatively complex, so fixed procedures should be employed to rationalize the various tasks, as follows:

1. State the problem.
2. Adapt the data for geometric operations.
3. Perform the geometric operations.
4. Adapt attributes for analysis.
5. Perform attribute analysis.
6. Evaluate the results.
7. Redefine and instigate new analyses if needed.

Stating the problem

Stating problems and delineating the approaches to solutions normally comprise the most difficult steps in using GIS analytic tools. Consider, for example, a typical assignment: the location of a recreational area that can provide a wilderness experience. The selection criteria may include:

- Remoteness, at a specified distance from manufactured facilities
- Reasonable accessibility
- Lakes and streams

- Varied topography
- A variety of vegetation

Each criterion may be analyzed using layering, buffer zones, and other GIS analytical functions, so before the analyses begin, the criteria to be used must be chosen carefully. Particular regard must be given to the potentials and limitations of the original data: data quality, object definitions, the bases for divisions into classes, and so on:

Adapting the data for geometric operations

As a rule, data available from a database must almost always be modified before they can be used for geometric operations. Map data, attribute data, or both may, for example, be modified by clipping a selected area out of a map database, modifying the attribute statements of the area from hectares to square kilometers, or removing borders to join polygons in order to reduce the number of land-use classes involved.

Performing the geometric operations

Geometric operations, employed to sort out the objects to be analyzed, include specification of buffer zones, overlaying, search and retrieval, joining polygons, and other tasks. Each operation results in new data for further processing. For example, in the location of a recreational area, broad buffer zones may be set up around roads and other manufactured facilities. The untouched areas outside these zones may then be sorted out and overlaid with vegetation, hydrologic, and other data relevant to the selection. For each task, the geometric operations must be defined on the basis of the analytic criteria involved. In the process, new polygons are formed, net topologies are established, and new attributes are added.

Adapting attributes for analysis

Most analyses require prepared sets of geometric and attribute data, and just as geometric data must be adapted for processing, so must attribute data be prepared for the arithmetic, logical, and statistical operations that need to be performed. Attribute tables must contain an adequate number of empty rows and columns for new entries. For example, in selecting a recreational area, a new attribute labeled "suitable characteristic" may be compiled to hold codes that indicate the degree of suitability of the various combinations of thematic layers.

Performing attribute analyses

Arithmetic, logical, and statistical operations are performed on attribute data associated with the objects chosen in the geometric selection phase. For example, for the recreational area discussed above, an analysis may be performed to identify all areas outside the buffer zones around manufactured facilities. Another analysis might aim to select areas classified as moderately hilly. Yet another might try to identify minor and major streams and rivers. In all cases, data may be sorted with respect to extrema stipulated. The final results of the operations are identifications of georeferenced characteristics that satisfy the selection criteria.

Evaluating the results

The results of analyses must always be evaluated with respect to accuracy and content. In short, do the results seem trustworthy, and do they make sense? These evaluations are made most easily using maps and written reports.

Redefinitions and instigating new analyses

Unacceptable results must be modified or the analyses that produced them improved and performed again. In some cases, new data may be required and the initial criteria may need to be changed. Similarly, new analyses may be required to provide results with a greater selection of alternatives. New analyses usually start at a suitable stage in the overall analytical process. Here again, GIS provides an efficient tool because GIS systems can process large quantities of data and perform complex computations rapidly to arrive at and compare alternatives.

Presenting the final results

Analytical results are best presented in easily read maps and written reports.

CHAPTER 15

Advanced Analysis

In this chapter we have collected a number of special GIS functions and complex operations when functions are selected according to certain rules. The term *advanced* will not always be adequate, because although individual functions might be relatively simple, they can be used in a complex context.

15.1 Network and Raster Connectivity Operations

Connectivity operations, which utilize sets of functions that exploit the connectivity in data, include:

- Network operations
- Contiguity
- Spread functions
- Streaming

Network operations are performed on vector data. Spread functions, streaming, and some connection operations are performed primarily on raster data.

15.1.1 Network operations

Many GISs support analyses in networks, which are systems of connected lines represented in vector data (Figure 15.1). In the real world, networks consist of road systems, power grids, water supply, sewerage systems, and the like, all of which transport movable resources. Network operations are based on:

- Continuous, connected networks
- Rules for displacements in a network
- Definitions of units of measure
- Accumulations of attribute values due to displacements
- Rules for manipulating attribute values

GIS network operations usually include:

- Displacement of resources from one place to another (route optimization; Figure 15.2)
- Allocation of resources from/to a center.

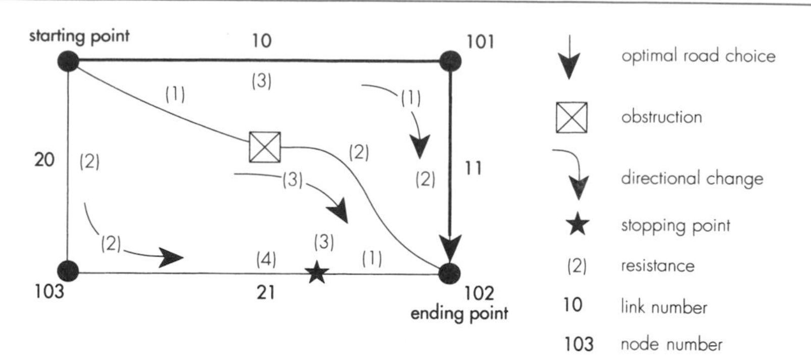

Figure 15.1
Network analyses can enable a system to find the fastest route between start and end points. Resistance (time delay) is entered as an attribute and the route with least resistance is selected. (Adapted from ESRI 1995.)

These operations are best described through the example of transport analyses.

As is the case for other GIS analytical tasks, network analyses depend on the existence of topology in the data. The system must, for example, know which roads can accept traffic flow at intersections. Every link and every node in a network must have a unique identity, and the program must contain specifications of where roads begin and end.

An analysis may include the simulation of moving resources—cars, people, refuse, and so on—along the lines representing roads. The resources moved along these lines encounter resistance in the form of speed limits, road work, peak traffic, barriers, one-way streets, weight limits, traffic lights, bus stops, and sharp curves, to name only a few.

The network system elements comprise links, barriers, stopping points, and centers. Assigning attributes to such elements permits the

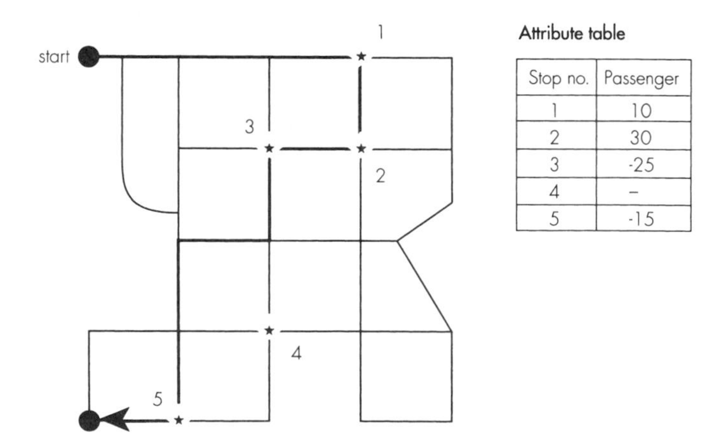

Attribute table

Stop no.	Passenger
1	10
2	30
3	-25
4	–
5	-15

Figure 15.2
The figure simulates a given bus route. At stop 1, 10 passengers board the bus; at stop 2, 30 passengers get on board; at stop 3, 25 people dismount; stop 4 is omitted; at stop 5 the remaining passengers get off and the bus carries on to the depot (end point). (Adapted from ESRI 1995.)

simulation of realistic situations. For example, attributes representing resistance in the network may be expressed in seconds or minutes—perhaps a delay of 20 seconds at each traffic light or of 1 minute at each bus stop. Resistance in the network may also be computed on the basis of the accumulated distance and stipulated speed limits. A section of a road that is 1 km long and has a posted speed limit of 60 km/h represents a delay of 1 minute.

Once all the attributes have been allocated, the system can assess the movement of resources through the network. A unit of measure (distance, time, monetary units, etc.) is then employed so that the system can evaluate alternatives and select the route of least resistance. In this manner, transport analyses may be conducted to optimize driving time, and minimize travel distance and costs. In practice, analyses usually result in several alternatives that illustrate the consequences of the foregoing simulations.

Other common network analyses include the distribution of resources from one or more centers, or the assignment of resources from an area to one or more centers. Such resources, flowing through a network to or from centers, must be associated with network lines or points. For example, all pupils living in a school district are associated with their respective home addresses.

The computational processes necessary to move resources out from a center continue to a prescribed limit. A limit of 10 minutes from a center in a street network will define those streets as having a total resistance of 10 minutes from the center. The result may be a plotted map of the streets involved (Figure 15.3). Resources displaced from or to a center may move along nonbilateral lines, perhaps one offering a

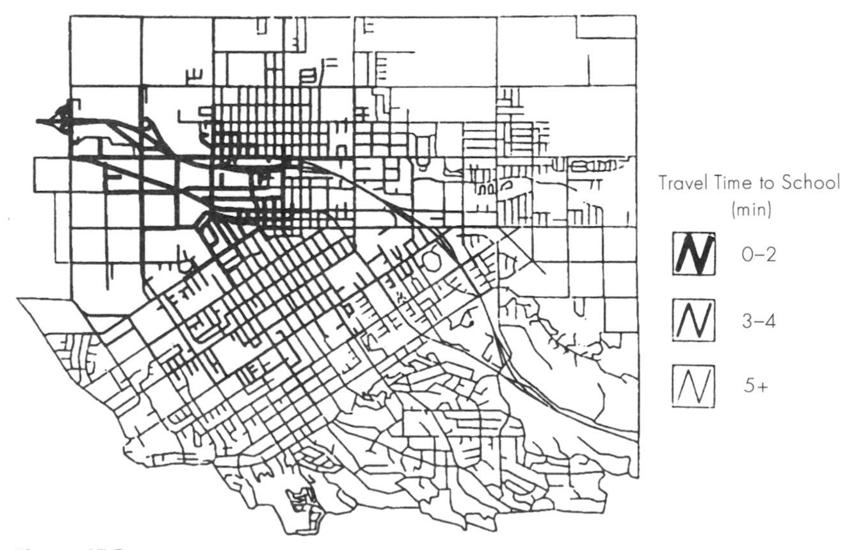

Figure 15.3
Displacement of resources (travel time) from a center. (From ESRI 1995.)

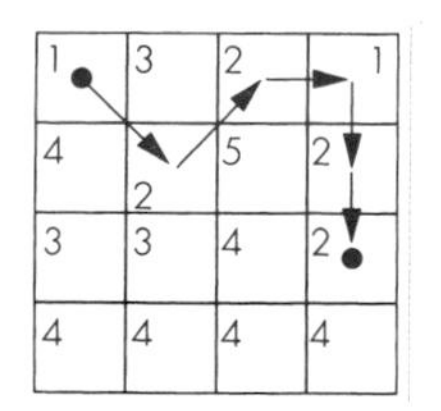

Figure 15.4
Optimized route location on raster data.

resistance of 10 seconds in one direction and 15 seconds in the other; several centers may be involved in the same computational system.

Centers can delineate areas even further. An example might be the area within 5 minutes driving time of a school. Thereafter, the area delineated may be employed either in a proximity analysis of the objects it encompasses, or in an overlaying function.

15.1.2 Connectivity operations on raster data

A few GISs support connectivity operations in raster data, a process that requires discrete cell-by-cell displacements, originating from a single starting point (Figure 15.4). The process paradigm involves the choice of units of measure and of other measurement parameters. The attribute values of the cells are accumulated during the transit of each cell. Connective operations may be used to determine travel distances in a road network, to identify areas of given shapes and sizes, and to compute precipitation runoff from a terrain surface. The accumulated cost surface can also be used in dispersion modeling and least cost analyses.

15.2 Proximity Operations and Spatial Interpolation

All GISs support some form of proximity operation, in which attribute values are assigned to new points on the basis of the values of existing neighboring points or observations. In ordinary circumstances, proximity manipulations (which include topographic functions associated with manipulations in digital terrain models) involve routines for area search and interpolation.

Search operations take place within a specified area, usually from a defined position to a new point, as within a circle around a point (Figure 15.5). Existing attribute values within the area are treated in accordance with the criteria for determining the values of new points.

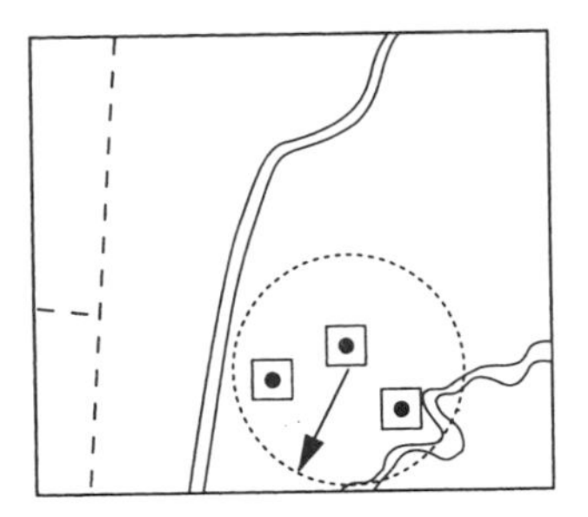

Figure 15.5
An area search operation can be carried out from a given point and within a circle of given radius.

The procedure for proximity operations usually comprises four main steps:

1. Identify a base point.
2. Define or compute the search area.
3. Select or search for objects.
4. Manipulate the attribute data in accordance with the selection criteria.

The manipulations of attribute data may comprise statistical computations, including:

- The sum of all values
- The average of values
- The greatest or least value
- Interpolation of values from those of neighboring objects
- Identification of the most frequently occurring value
- Statistical distribution of values

Area search and the subsequent computation of elevations of grid points from elevations of points within grid cells are used to compile digital terrain grid models. The search criteria for attribute data may be stipulated at the same time as the delineation of the search window, which may simply be a specification of the coordinates of the upper left and lower right corners. For example, in a search covering properties, criterion may be all residences on lots larger than 5 decares. Such searches often combine data from several thematic layers, thus necessitating searches among them. Another area search criterion might be a random polygon, perhaps all the cultivated fields within a given window that correspond exactly, within a given polygon. Hence, in many respects, proximity operations resemble overlaying.

Interpolation and extrapolation

There are many situations where the available observations are fewer than are needed to cover an area satisfactorily, whether it be attribute

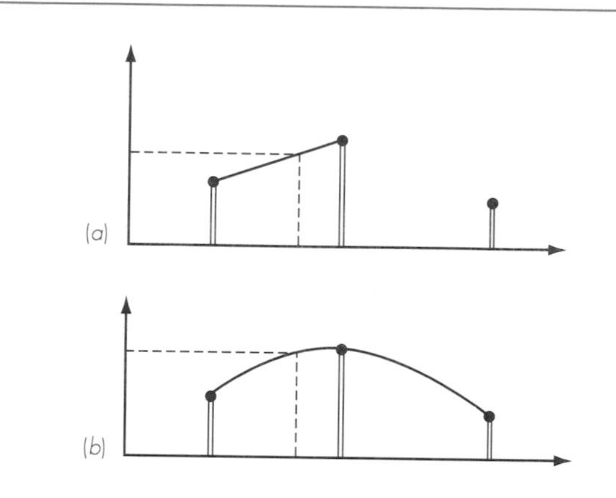

Figure 15.6
Interpolation in data: linear (*a*) and spline (*b*).

data or geometric data. With the use of both interpolation and extrapolation techniques (Figures 15.6 and 15.7), it is possible to calculate new values for points lying within or outside the applicable data area. The task is to calculate the most likely value of the new point based on available observations. The most usual techniques are:

- *Closest-point inter-/extrapolation.* The new point is given the same value as the closest known point.
- *Linear inter-/extrapolation.* The new point is given a value calculated from straight line between the two closest known points.
- *Spline inter-/extrapolation.* The new point is given a value calculated from a curve between the three closest known points.

Different variations of these techniques are available. For example, several known points can be used and the known points can be given different weight in the calculations (Figure 15.8). It is often the case

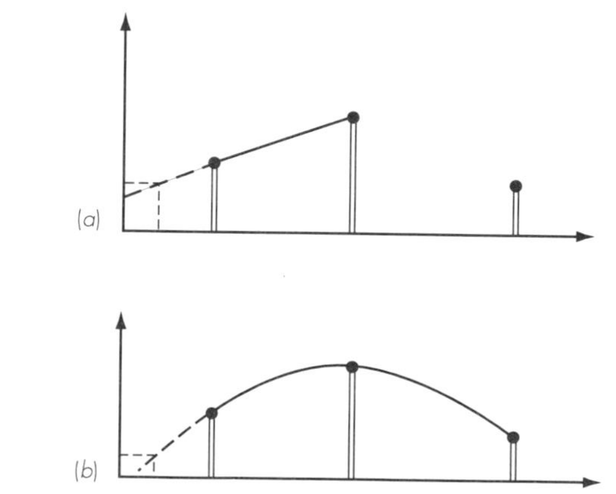

Figure 15.7
Extrapolation of data: linear (*a*) and spline (*b*).

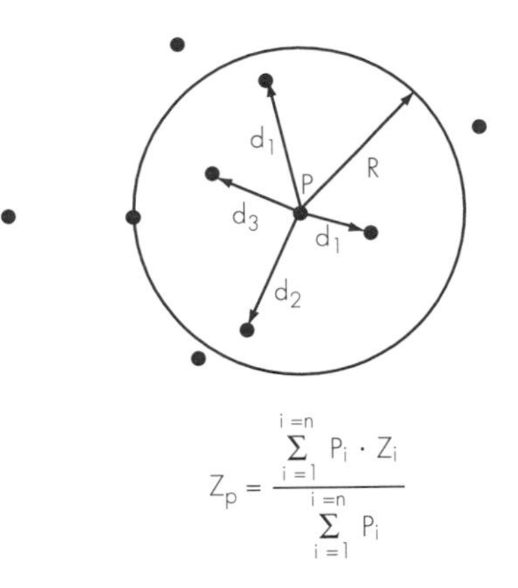

Figure 15.8
Weighted average: where the value (Z_p) of the new point (*P*) is based on the value (Z_i) of all known points within the circle. The known points that lie closest are given the highest weight (*P*).

that the closest known points represent the new point better than points located at a distance; thus close points can be given greater weight in the calculations than distant points. One variation of this is known as *kriging* after a South Africa mining geologist, D. G. Krig. Kriging is based on the theory that the same pattern of variation can be observed at all locations and uses a mathematical function to model the spatial variation in values within the input sample points. The variation is measured using the semivariance, which is half the average squared difference in values between pairs of input sample points; it is used to determine the optimal weight for interpolation. Borders may also be interpolated around points to form surfaces, which themselves may be formed using Thiessen polygons, in which the borders around a point are delineated equally distant from all neighboring points (Figure 15.9).

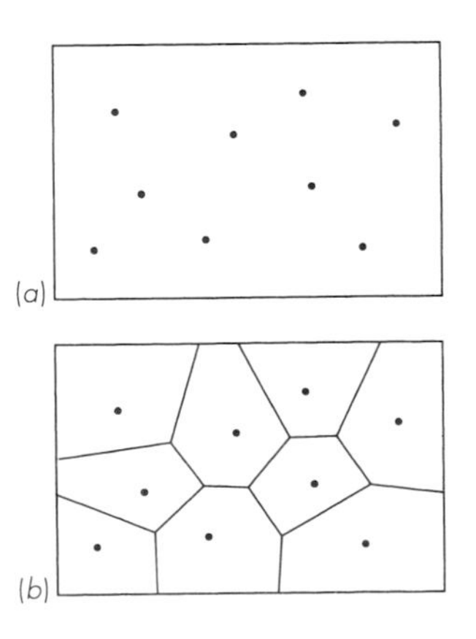

Figure 15.9
The Thiessen polygon is created around a point in such a way that the polygon's border will be located equidistant from each adjacent point.

3 x 3

1	2	2	2	2	2	2	2
1	1	2	2	2	2	2	2
1	1	3	3	3	2	2	3
1	1	3	3	3	2	3	3
4	4	4	4	4	3	3	1
4	4	4	4	3	3	1	1
4	4	4	3	3	1	1	1
4	4	3	3	1	1	1	1

Figure 15.10
A contiguity operation using step-by-step window scanning of raster data. The value of the cells within the window can be manipulated in various ways.

Proximity operations on raster data

Raster data are well suited for contiguity operations. A raster GIS can assign new cell values that are independent of the values of the neighboring cells of the original layer. This may be accomplished, for example, by step-by-step scanning using a 3 × 3 cell window (Figure 15.10). The value of the cell in the middle of the window is computed from the average of the values of all nine cells viewed. Proximity operations may also be used in networks, for example by determining areas to be searched on the basis of stipulated driving times from specified points in a road network.

15.3 Fuzzy Analysis

In the traditional discrete model, objects are defined as being either outside or within specific classes. Should areas with a population density of more than 1000 persons per square kilometer be defined as high density, areas with a density of 999 persons per square kilometer should be classified as medium density even though the real difference is very slight. These problems can be solved by using *fuzzy set theory,* which allows an object to belong only partially to a class. Each object can be given one or more additional values between 0 (nonmember) and 10 (full member); for example, 7 might be used to specify how certainly an object can be said to belong to a certain class. Should there be several attributes within one area, a corresponding number of additional values can be used. In an analysis context, such as overlay, fuzzy values are handled equally with other attributes that are linked to the actual areas.

15.4 GIS Analytic Models

Several theoretical models have been evolved to exploit GIS analytic functions. The three prime examples are cartographic algebra, expert systems, and linear combinations. The models must be regarded as aids to expert judgment, because the drawback in all of them is that they are based on theoretical assumptions that cannot cover all practical situations.

15.4.1 Cartographic algebra

Cartographic algebra is based on the assumption that a set of simple operations can be found and joined sequentially to comprise relatively complex modeling (Figure 15.11). This process starts with an existing set of attribute tables, which are then processed in a sequence of operations, each of which produces a new column in a table. Finally, the attribute sought is compiled. It may be represented by a flowchart in which the thematic map is processed successively until the final version is produced (Strand 1991a).

Cartographic algebra may require the following operations:

- Reclassification
- Averaging
- Maximizing
- Subtracting
- Adding
- Scattering
- Streaming

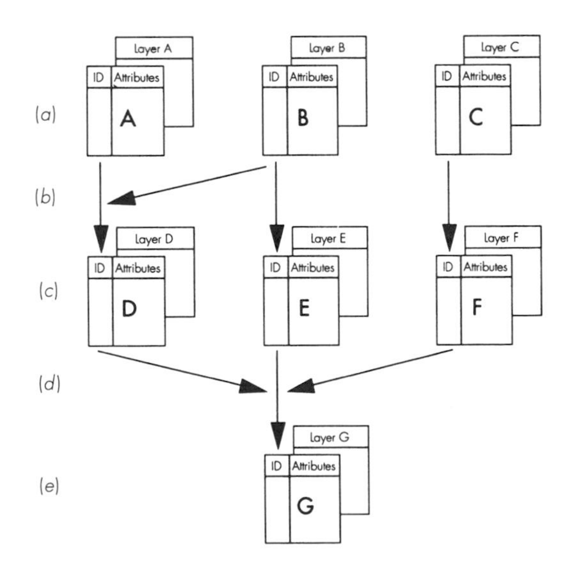

Figure 15.11
Cartographic algebra. Thematic layers A, B, and C with their own attribute tables (*a*) are manipulated by simple operations (*b*) to a new thematic layer and attribute tables (*c*) which are, again, by a simple operation, manipulated to one result layer (*d*) and a result attribute table (*e*).

The success of the model is contingent on the manipulated data being of a form suitable to the operations concerned. For example, attribute values arranged in classes cannot be added or subtracted and are therefore unsuitable for cartographic algebra.

15.4.2 Expert systems

GIS are increasingly implemented with ever more "intelligence." One intelligent approach is that of the *expert system,* in which analytical results are assessed automatically in terms of criteria entered as ancillary information, or attributes. On the basis of these criteria, the system assigns priorities to various combinations of attributes and then provides an output comprising recommended choices and courses of action (Figure 15.12).

In GIS, each attribute type may comprise a number of attribute values. "Soil quality" may have three quality classes, "conservation" may have two conservation classes, and "Quaternary geology" may comprise three types of geologic deposits. The attribute values of these three attribute types may be combined to $3 \times 2 \times 3 = 18$ possible combinations. One or more experts may then rank the 18 combinations with regard to a particular goal, such as development. All combinations, including conservation classes, are ill suited to development; some may be unworkable for technical or other reasons, and hence must be discarded.

(a)

Attribute	Attribute	Priority
Cultivated	Development	9
Cultivated	Recreation	6
Cultivated	Preservation	7
Grazing	Development	3
Grazing	Recreation	1
Grazing	Preservation	8
Forest	Development	4
Forest	Recreation	2

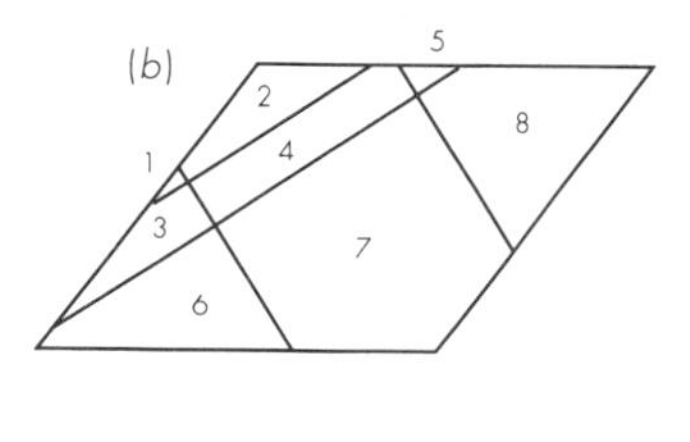

ID		Attribute		Attribute	Priority
1	1	Cultivated	1	Development	9
2	2	Grazing	1	Development	3
3	1	Cultivated	2	Recreation	6
4	2	Grazing	2	Recreation	1
5	3	Forest	2	Recreation	2
6	1	Cultivated	3	Preservation	7
7	2	Grazing	3	Preservation	8
8	3	Forest	3	Preservation	5

Figure 15.12
Expert systems. The expert's criteria in table (*a*) are combined with attribute table (*b*) for an automatic evaluation of the result of an overlay operation (see Figure 14.8).

The evaluations of the experts are then arranged in a table in which the 18 combinations are ranked qualitatively: "very suitable," "suitable," "moderately suitable," "moderately unsuitable," and "very unsuitable"—or quantitatively by priority 1–9. The table can then be entered in the GIS and will serve as a reference for every record (ITU) in the attribute table that results from GIS analyses. The expert recommendations can thus be entered as a new attribute, or as a column in the attribute table, and be queried whenever reports are exported. It is clear that the answers provided by an expert system can be no more reliable than the professional expertise available when compiling the evaluations entered.

15.4.3 Linear combination

The linear combination model is used both in impact assessment and in suitability analysis. The starting point is a scale, as from 0 to 5, which is used to rank features with respect to a stated application. The attribute values of all attribute types are first coded by a knowledgeable expert. Rank on the scale indicates the mutual relationships of the attribute values. In the expert systems example above, for instance, the three categories of soil quality may be assigned 1 for low quality, 3 for medium quality, and 5 for high quality. Corresponding rankings may be assigned to the conservation and quaternary geology attributes. The result is that all attributes have rank.

In practice, the different attributes will have varying impacts on the end product of the analysis. Conservation might well weigh more heavily than geology in selecting an area for development. Consequently, the various types of attributes must be weighted to indicate their relative importance. All records (ITUs) in the attribute table are then weighted according to the attribute type involved. The result is a set of weighted features, which may then become new attribute values in the attribute table.

15.5 Practical Application of GIS Analytical Functions

15.5.1 Statement of the problem

In many towns, land use depends on natural resource data. However, urban area planning often makes poor use of such data. As a rule, extensive surveys are conducted, but further processing, including the superimposition of data and the assigning of priorities to categories of natural resources, is often neglected.

A digital map database could be established for the purpose of area planning, and natural resource data are often available on manuscript maps. The goal should be to produce maps of selected natural data and then superimpose them selectively to produce a map of priorities for wildlife, fishing, conservation, and outdoor activities. An additional goal should be to produce a composite ground resource map for use in town planning and land administration.

15.5.2 Procedure

Raw data

The raw data for the task comprised existing digital data and data digitized specifically for the following purpose:

- Digital basic data from 1:50,000 topographic maps: lake and watercourse contours, elevation contours, and roads
- Digital plan extract of township plan
- Digital data for precipitation area classification
- Manuscript maps for wildlife (reindeer, deer, moose, roe deer, black grouse, wood grouse, and ptarmigan), outdoor recreation, and conservation
- Extracts of computerized registers (attribute data)
  - Data having IDs in common with the manuscript maps for wildlife, outdoor recreation, and conservation
  - Fish registry data, including coordinates of fish monitoring stations

Relating geometry and attributes

All natural resource themes to be related to existing computerized registers were first digitized as line information and identified by codes. For example, the wildlife information on manuscript maps contained many overlapping symbols with consecutive numbers not associated with the species involved (Figure 9.3). Consequently, IDs were necessary to relate the information to attribute data and facilitate the subsequent sorting of the various species of wildlife. After digitization, the objects (points, lines, and area boundaries) were linked to the existing attribute data, which were also used to control the use of symbols in subsequent mapmaking.

Sorting themes

The attribute data for wildlife, fish, conservation, and outdoor recreation were sorted in layers (ARC/INFO coverage) so that no areas overlapped within a layer. Typical sorting criteria were species for wildlife data and conservation topic for conservation data. After sorting, the area data were edited to close all polygons. Then the attribute data originally associated with border lines were transferred to the areas themselves. This was possible because all lines contained attributes referenced to polygons on their right and left sides, and because the polygon numbers were available as numbers in the polygon topology table. The result of this neatening process formed the basis for both the thematic maps and the subsequent processing maps used to assess priorities.

Thematic maps

Data selected from the township plan were used as background data for the thematic maps. For instance, polygons representing industrial buildings or residential areas, respectively, were reclassified and joined to polygons representing built-up areas. Subjects intended to overlap in final map plotting were distinguished from each other by using differing pixel values.

One goal for the fish resources map was to represent fish population distributions accurately without direct reference to the discrete stock counts made in the field. This was accomplished in the following way. First, the digital precipitation data for partial and complete watercourses permitted computation of fish population averages for lakes and watercourses, which provided status data independent of species. Thereafter, a polygon layer representing the precipitation area borders was compiled from the watercourse register. Individual fish stock counts were then correlated to precipitation areas in a point-on-polygon routine. Finally, averages were computed for each area to produce the desired fish population distribution.

The first draft distributions of fish populations and those of other subjects often revealed inconsistencies or shortcomings in the original data. The professional staff in the relevant disciplines then supplemented or corrected the original data. With the new data, new thematic maps could be produced rapidly by replicating the processes used in producing the first drafts. This in turn required that all stages of the production process be well organized.

Use of overlays and buffer zones

Differing area categories were overlaid to identify areas with overlapping subjects, and union functions (NAND operations) were used in testing two layers at a time. Buffer zones were employed around such linelike information as wildlife tracks and hiking trails, to extrapolate them to areas.

Weighting

The three most prevalent wildlife species—moose, roe deer, and ptarmigan—were assigned priority weightings: respectively, 1, 2, and 3. These weightings, which were placed in a dedicated field in the attribute table, were retained through the overlaying process to produce aggregate weightings for all polygons. Wildlife tracks, represented by buffer zones of various widths, were also weighted.

Union functions were used in successive superimpositions of layers, so that the final result was a single layer containing the relevant

Moose

ID	Weight
1	1
2	1
3	0
4	1
5	3

+

Deer

ID	Weight
1	2
2	1
3	1
4	0
5	1

+

Ptarmigan

ID	Weight
1	0
2	2
3	1
4	3
5	2

=

ID	Moose Weight	Deer Weight	Ptarmigan Weight	SUM
1	1	2	0	3
2	0	1	2	3
3	1	2	3	6
4	2	0	1	3
5	1	0	2	3
6	0	0	0	0
7	1	1	1	3

Figure 15.13
Use of the union function to find the combined point sum for the various units in the figure (see Figure 15.14).

aggregate information (Figure 15.13). Attribute fields containing the weightings for the individual thematic layers were retained in the individual polygons, so total weighting could be computed for each. High total weightings dictated that polygons should have high priority; lower total weightings dictated lower priorities.

Unified priorities for wildlife, fish, conservation, and outdoor recreation
Weighting was also used to assign unified priorities for wildlife, fish, conservation, and outdoor recreation. Each subject was divided into two classes, assigned weightings of 1 and 2, respectively. Following the overlay process, the total weightings for the individual polygons provided the basis for plotting a priority map. Numerous statistical computations, including sum, average, and minimum/maximum, were performed in the attribute tables, both between various attributes for the same object and for like attributes for all objects in a thematic layer.

Adapting final results and plotting the priority map
Finally, an ordinal scale of priorities of the individual areas and polygons was compiled using the weighting sums resulting from the overlay processes. Again, the final assignments identified inconsistencies and omissions. These required revision of the foregoing weightings and computations, adjustments that were made relatively easily because the weightings of the individual subjects were retained throughout the various manipulations.

Once these adjustments had been completed, the final map was plotted (Figure 15.14). The ordinal priority ranks were used to control symbols assigned to the polygons, a project involving many themes that called for expert judgment in delineating borders. An important aspect of the entire "mechanics" of GIS processing was that human expertise was used continuously to evaluate intermediary and final results.

As is often the case in GIS applications, the converse of expert input in GIS processing is that GIS obliges experts to be more concise in their specifications. Another strength is that the ability of GIS to superimpose various alternatives provides rapid testing of hypotheses. GISs

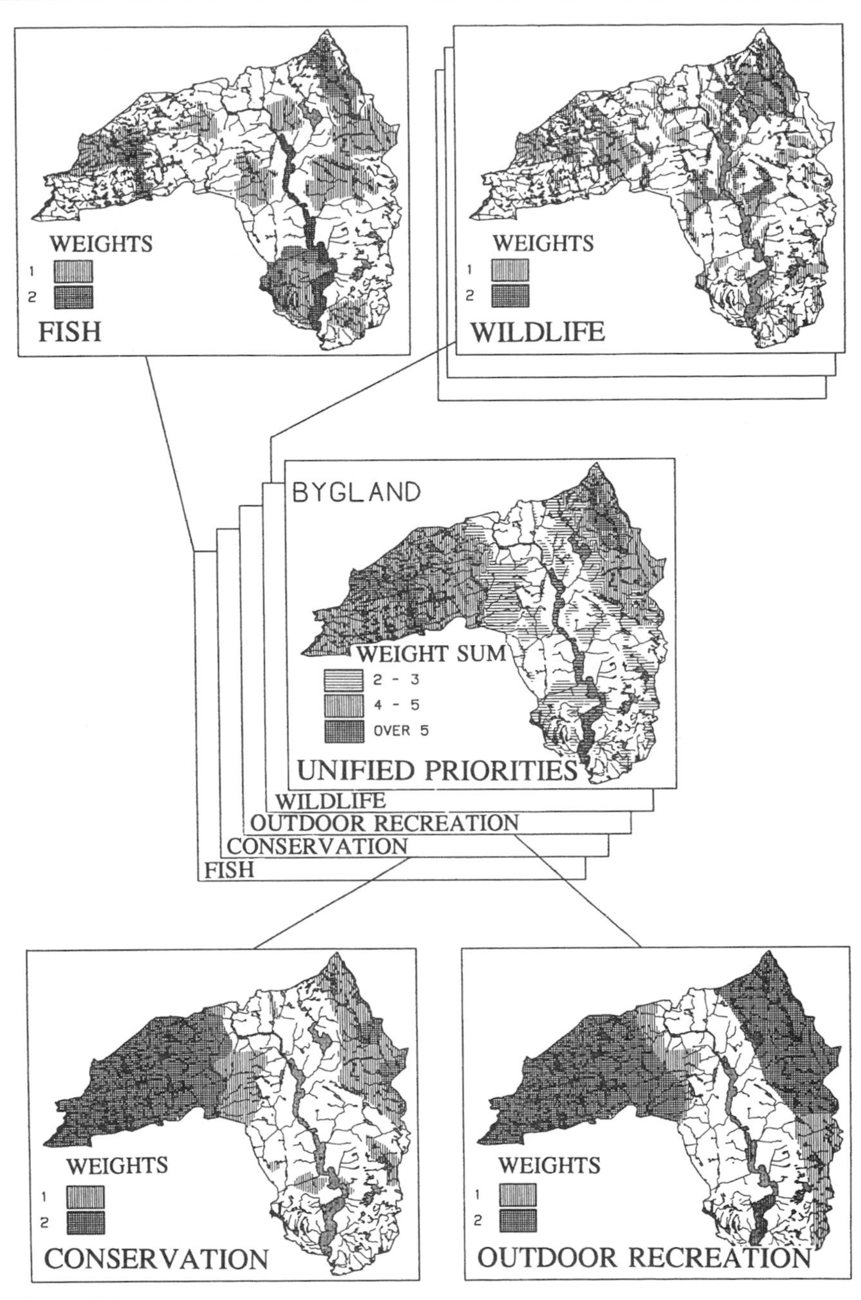

Figure 15.14
Unified priorities to natural resource data. (Courtesy of Asplan Viak.)

do not supplant professional expertise, but they do contribute to documenting it in the area planning process. The various weightings and the assignment of ordinal priorities have much in common with the processes involved in expert systems and linear combinations, as described in this Chapter.

15.6 Digital Terrain Models

Digital terrain models may be used in various operations which are partly included in the operations discussed above.

Elevations

Using elevation data stored as a point cloud, the elevation of a random point can be computed as the weighted median of the surrounding points, with the closest points having the greatest weight. Neighboring points are searched within a specified area (Figure 15.15).

Conflict and other mathematical functions more complex than the weighted median may also be used (kriging). The elevation of a new point in a grid model may be computed as the elevation of a plane passed through the closest grid points. Elevations may also be computed from the faces and corner elevations in a triangulated irregular network (TIN) model.

Slope and slope direction

A terrain's slope and direction may be computed relative to a plane through the elevation model points (Figure 15.16). Slope degrees and direction are useful in simulations of precipitation runoff and in defining drainage areas. The triangle terrain model, which comprises sloping triangular surfaces, is ideal for simulating terrain surface conditions. Slope and slope direction are computed directly from the coordinates of the corners of a triangle. Slope is usually expressed in degrees or percentage terms with respect to the horizontal. Slope direction is almost always expressed in degrees from north.

Contours

The contour lines of constant elevation may be interpolated directly from the digital elevation model. The program used identifies x and y coordinates for new points having the same interpolated elevation and then connects them together to form contour lines (Figure 15.17). Many methods are used. In the triangulated irregular network (TIN)

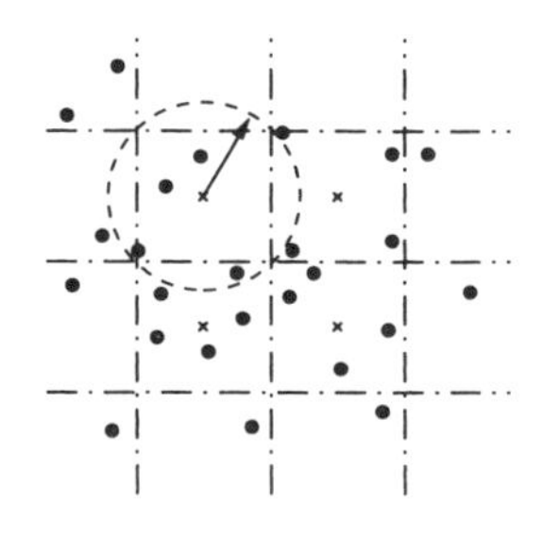

Figure 15.15
Area search.

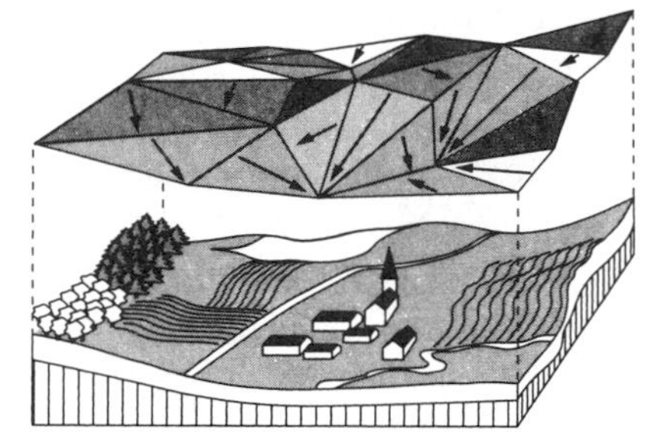

Figure 15.16
Slope calculation in a TIN model.

model, triangle faces intersecting a stipulated contour elevation are identified and the points of intersection determined by linear interpolation. In elevation data stored in the form of point clouds, triangles are first formed, then treated as are those of the triangular terrain model.

In grid models, the grid points on either side of a contour line are identified. A plane through the points or linear computations then locates the new point on the contour line. In all cases, the contour interval can be chosen to suit the accuracy required. The various approaches to generating contour lines can produce differing results on plotted maps.

Volumes

Accurate surveys of excavation volumes are crucial in planning road work, major industrial buildings, and so on. Soil borings are used to calculate cover quantities, such as down to bedrock, and replacement quantities, such as down to usable soil. From these quantities, haulage can be calculated down to the project level. Quantity surveys may be conducted in GIS by entering data in a terrain model program, which includes the extent of the excavations in elevation and ground plan, soil conditions, slopes for planned leveling, volume expansions due to blasting, and so on. The quantity survey computations are based on calculations of cross-sectional areas multiplied by the intervening distances between cross sections.

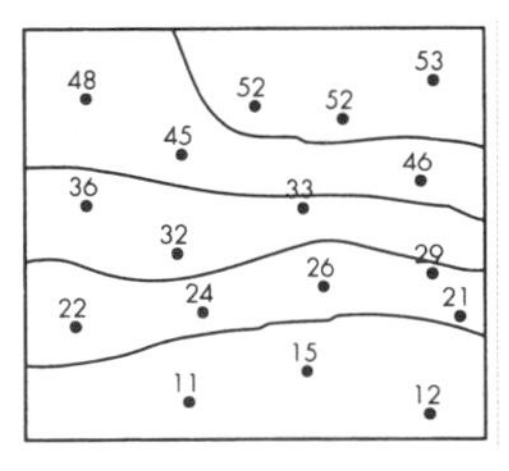

Figure 15.17
Calculation of isolines (contour lines) in a point cloud of spot heights.

Visualization techniques

The results of excavations may be visualized by excising a rectangular area in the terrain model and viewing it at an angle from above. Users are generally free to select the viewing angle that best displays the features sought. Profiles and grids in the area are plotted to suitable densities. Regular grids are best for the purpose, so several triangulated irregular network (TIN) models support regular grids for plotting. Drawing straight lines between successive grid points often produces a rough, serrated terrain surface. Consequently, some systems support splines, or curved-line segments, for plotting.

Perspective drawings employing profiles are usually based on isometric perspective images of terrain. That is, the units of measure are fixed along the axes, and the foreground of the image has the same scale as the background. Central perspective produces a more realistic image of terrain but does not permit measurements because scales vary in plots (Figure 15.18).

For the best replication of terrain, the program must delete profile lines and surfaces hidden by terrain rises as seen from a side view. The deletion process calls for the sequential sorting of lines, starting from the observation point and working backward. The lines are tested one by one for visibility. Computationally, the process is cumbersome and time consuming.

Figure 15.18
Perspective drawing. (Courtesy of Asplan Viak.)

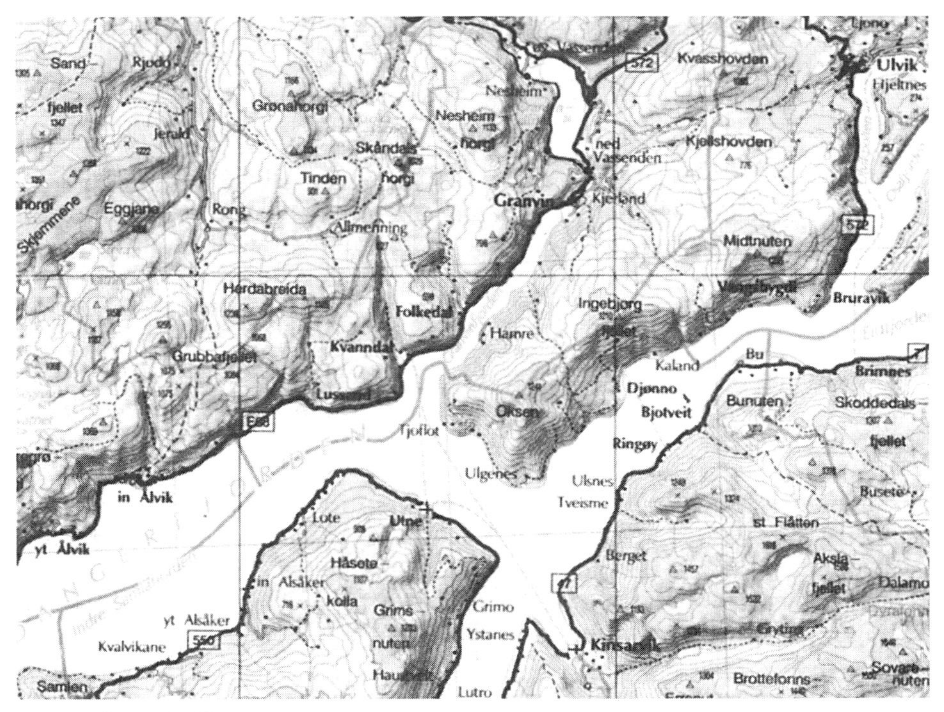

Figure 15.19
Shadowing of a topographic map. (Courtesy of the Norwegian Mapping Authority.)

Shaded and draping

A terrain model may be used to produce automatically relief maps which use shading to effect the appearance of the third dimension of height. These processes demand the computation of shaded areas from an assumed solar position. The simplest approach is to color the cells of a raster model. The colors can be varied with elevation, as on ordinary topographic maps. The color intensity can be varied with cell slope, to give the impression of sun and shadow (Figure 15.19).

Draping a digital aerial photo, satellite image, or thematic map over the surface of a terrain model produces a more realistic terrain image. However, draping two-dimensional images over three-dimensional surfaces poses problems because the two-dimensional image does not contain enough pixels to cover the terrain surfaces. The system must then compute pixel values for surfaces not covered, which deforms the two-dimensional image.

Miscellaneous applications of terrain models

The principles of interpolation and visualization in digital terrain models may be applied to other types of three-dimensional data. For example, the third dimension in georeferenced data may be time, population data, farm productivity, geophysical measurements, and so on. In visualizations, elevations may be replaced by any numerical values related to the observation points. Similarly, interpolation between contours may be performed or perspective drawings plotted.

15.7 Hydrologic Modeling

The use of a digital terrain model makes it possible to calculate a synthetic drainage system in the terrain, including the individual areas for drainage, drainage network, and nodes through which the water runs (Figure 15.20). This kind of hydrological model can be used to simulate different drainage situations, including the establishment of parameters for calculating the dimensions of the piping network in connection with the development of a housing estate.

A grid-based terrain model represents the continuous surface of the terrain. Drainage will always be in the direction with the lowest elevation. Each cell has eight adjoining cells and eight possible drainage directions (Figure 15.21). Comparison of the basis cell's elevation value with the values of the adjoining cells thereby determines the direction. The sum of the number of cells that run into each other gives the accumulated drainage and forms the basis for calculation of network. Cells with a high accumulated value show where the water collects and create a drainage pattern. Cells with the lowest accumulated value represent the limit for individual watersheds. The cells that represent the network (and the boundaries for the drainage areas) can be vectorized and attributes linked to the different nodes and links in the network. The links in the network can be numbered automatically (e.g., based on a ranking system where the lowest number is allocated to the links with no links upstream, and the highest number to the links with the highest number of links upstream).

When calculating the dimension of the water piping network, it is vital to know the probable water consumption, which could be represented, for example, by the number of persons who will be linked to

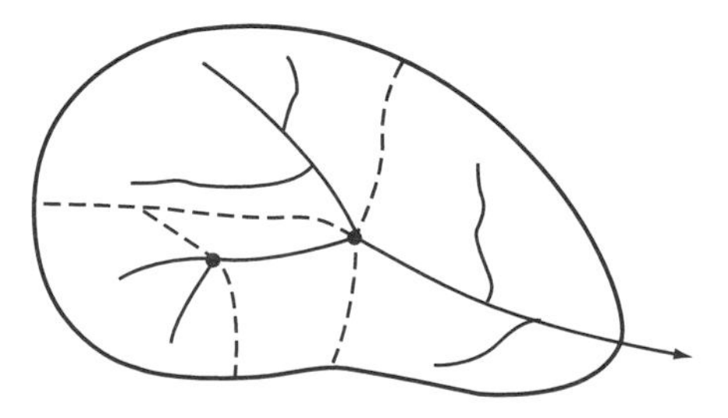

Figure 15.20
A watershed consists of watershed boundaries, subbasin, stream network, and outlets.

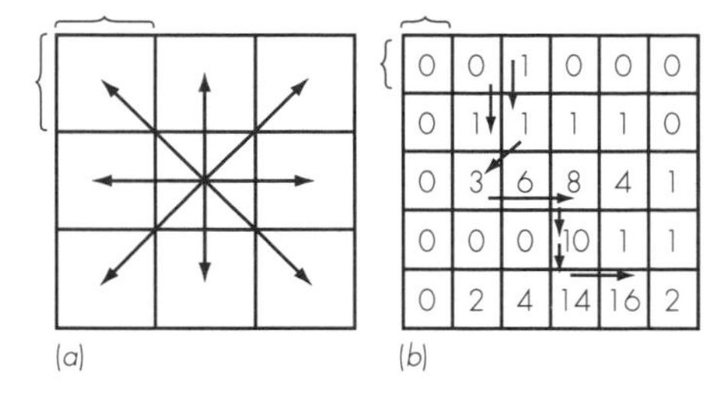

Figure 15.21
There are eight possible flow directions from a single cell (*a*). The number of cells that run into each other are accumulated (*b*). The cells with the highest numbers (values) represent the watershed network.

each node. The number of planned dwellings within each section of a rainfall area could also be a suitable parameter. When dimensioning the sewage network (surface water) it is also important to know the probable volume of water that will run through each node. The drainage volume is dependent on the water flow, among other factors, which is in turn dependent on the degree of slope and type of surface. The degree of slope can be calculated from the digital terrain model. As to type of surface, bituminous roads and corresponding surfaces will result in faster drainage of cultivated land. A digital map with surface structure, with use of overlay, can supply data on the area for sealed and permeable surfaces within each drainage field. The final calculation of piping dimensions is carried out in a special program for water and sewage networks.

15.8 Functions for Engineering GIS

Geographical data are often used in solving such engineering problems as locating basic points, surveying properties, and planning various phases of roads, water supplies, sewerage, residential areas, and so on. These tasks have previously been accomplished using standalone PC programs, but the trend is to integrate them into urban GISs (Figure 15.22).

Surveying and other measurements

Surveying programs usually include functions for computing coordinate geometry based on in-field measurements, with interfaces to various surveying instruments. For engineering applications, they may be extended to compute the intersections between lines, between lines and circles, and so on. They must also support computations of various geometric figures so that exact geometry may be computed.

Other operations may also be performed. The layout data of plans may be computed, and various map projections may be manipulated and transformations made between projections. Often, the systems

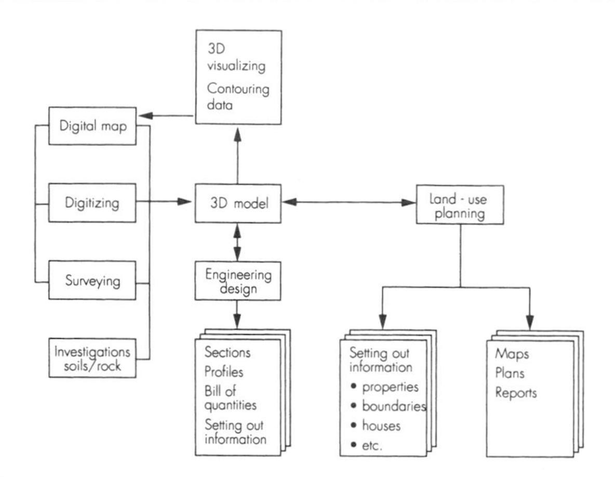

Figure 15.22
Flowchart illustrating the use of GIS in engineering.

can produce standardized deed or other maps in local or standard formats. Some systems support the method of least squares, as well as automatic quality designations in the form of error ellipses. Other systems can combine GPS vectors with conventional measurements.

Manipulating terrain
To aid engineering work, digital terrain model systems have been developed to manipulate simultaneously as many as 10 separate terrain layers, such as loose fill quantity, bedrock, and various geological strata. Volumes for cut-and-fill operations are computed and documented automatically in maps, perspective drawings, profiles, and cut specifications. For example, when a new road has been aligned, it may be designed with exact curve radii, transition curves, lengths, and cross sections. Thereafter, its plans may be plotted with center lines, driving lanes, shoulders, cuts, fills, building lines, and so on. These data are then used to set out lines in terrain. The documentation concerned may be standardized, for example to the standards of a roadwork commission.

As visualization prior to construction is desirable in many engineering applications, some GIS terrain models are integrated with three-dimensional computer-aided design (CAD) systems. Such a combination can present realistic images of the planned construction, perhaps showing a complete building, with facades, windows, and interiors surrounded by graphic symbols for trees, cars, and other extraneous objects.

CHAPTER 16

Visualization

As we have already seen, GISs are not merely automatic mapmakers but are capable of a wide range of tasks. Using GIS, the cartography of a map can be supplemented with textual information, digital images, sound, diagrams, and other miscellaneous graphical information. Today, GIS is linked with multimedia and virtual reality. Thus the concept of visualization is extended beyond the traditional view of cartography.

Nonetheless, whether displayed on screen or printed on paper, maps are the primary GIS product for end users. However, a map will generally only be an aid in a decision-making process or to identify a situation. The real end product could be a building permit, a road, or a new conservation area. By using GIS, the visual product can be quite closely related to the decision-making process. The end product is a map that has been created for a particular purpose, often with a short life span, based on data of varying origin and quality. This places new and greater demands on the map designer.

The cartographer's task, then, is to transform geographical data in digital form into a visual product, in an accurate and user-friendly way. Regardless of the production methods used, a map is a medium for visual communication, and as with other visual communications media, its appearance affects how the map is perceived and, consequently, how readily the information it contains is interpreted by the user. This is one reason that traditional cartography is a venerable craft in which artisans train for long periods to achieve the aesthetic appeal that is necessary to ensure that maps communicate information to users. Even the most skilled cartographers, though, are limited by the drawing implements at their disposal. Computerized GISs are not so limited and are therefore capable of producing a far greater variety of maps than can be drawn by cartographers, even when using the best hand tools of their craft.

This capability has the enormous potential of enabling less skilled mapmakers to produce truly professional maps; it also has the drawback of permitting the production of art-less maps that are at best unattractive and at worst misleading. Thus far, GIS facilities have no cartographic intelligence that can guide operators in the choice of map symbols and other graphic effects. Consequently, those who operate GIS facilities should be reasonably knowledgeable in all aspects of cartographic communications.

The use of EDB/GIS in map presentation makes it possible to divide maps into categories other than the traditional base maps and theme maps. Today, maps are often divided into the following categories (Balstrøm et al. 1994; Brande-Lavridsen 1994):

- Hard-copy maps, drawn or printed on paper
- "Soft" maps, visible on a computer screen
- Digital map data held in an electronic database
- Real-time maps, maps that show changes segmentially

16.1 Theoretical Foundation

16.1.1 From data to display

When visualizing geographical data, it is important to be aware that one often operates with a series of fundamentally different types of data, each of which need to be handled in a different way. We have already mentioned that geographical data are represented in the data model by points, lines, and areas (or raster). The same three geometric forms are essential to the presentation of geographical data in maps. In addition, geometric volumes such as terrain surface, buildings, and other three-dimensional objects can be represented.

Attribute data are divided into two categories. *Qualitative data* describe the type of data, such as different types of vegetation or different types of language. As an aid to visualization, it can be useful to subdivide qualitative data into those which are conceived as different but associative, and data that are conceived as clearly different (selective) (Bertin 1983). *Quantitative data* specify number; they are often split into three levels, according to precisely how they can be measured. Ordinal data are arranged in order; interval data can be placed in differently defined groups or classes; and relations data are indicated with exact numerical values measured on a scale in relation to a particular zero point and with a particular resolution (Figures 16.3 and 16.4). It can often be appropriate to place ordinal data in one group, with interval and relations data placed in another.

The temporal perspective is, in principle, a type of data that we wish to use to show changes over time. Finally, we need information as to the quality of data: its accuracy, for example. As GIS is used increasingly as a decision-making tool, we now need to be able to visualize the quality of this basis for decision making.

16.1.2 Cartographic instruments

Normally, the aim of cartography is to achieve an intuitive recognition and understanding of one or more messages in the map. Understanding maps is a mental process which to a great extent is based on the

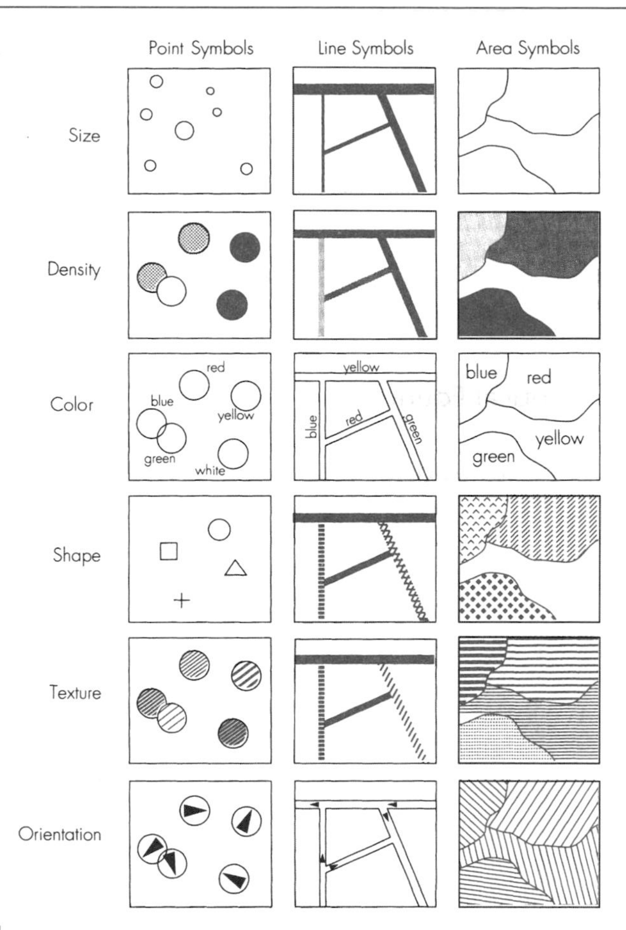

Figure 16.1
In principle, six different graphic parameters can be used to symbolize geographical phenomena: size, density, color, shape, texture, and orientation.

human ability to recognize patterns, contexts, and differences. This can be achieved by the use of certain cartographic instruments. Traditionally, cartographers command six graphic variables in symbolizing geographical phenomena: size, density (gray scale), color, shape, texture, and orientation (Bertin 1983) (see Figure 16.1). Of course, not every graphic symbol or effect is important in itself. A map with only one symbol, one color, or one gray tone has little value. It is the context in which the variables are presented, together with the geometric variables—points, lines—and areas—that are of importance. As far as graphic variables are concerned, it is their differences in size, color, and so on, which are associated with the differences in the measured values. The six variables differ in their application and can illustrate either the qualitative characteristics (two subgroups) of various object types, or the quantitative characteristics (two subgroups) of various attribute values.

Size

Variation in size is the most convenient way of illustrating variation in the quantity and, to a certain extent, sequence represented by a symbol, whose size can then be made a direct function of the magnitude measured. However, differences in size are not always readily perceived. As with most areas of human perception, the difference in magnitude that can be perceived is related to the initial magnitude involved. For example, the diameter of a circular symbol representing a given amount must be nearly tripled to convey the impression of being doubled.

Lightness and gray-scale value

Variation in density (gray scale) is used mainly to illustrate ranked quantitative data. Lightness and gray values are best suited for areas. As for all cartographic variables, there is a lower limit to the size of the differences that can be perceived by a user. Gray scales should be divided into no more than 10 sections, lest neighboring sections be mistaken for each other. However, if dot rasters are also employed, the number of classes represented may be increased to upward of 20. It is also important to be aware that the human eye cannot differentiate between more than seven to eight different classes on a map without closer consideration and comparison (Kraak and Ormeling 1997). In fact, the maximum number of different objects that can be observed by sight alone, without consciously connecting in the brain, is four. The properties of density are illustrated in Figure 16.2.

Maps in which symbols cover areas in a density proportional to the value of the phenomena being represented are often termed *choropleth maps,* from the Greek *khoros* (place) and *plethos* (value, number). Density implies variations of magnitude: darker means greater. Therefore the lightness of a single color usually conveys relative magnitude better than do variations of base color. Variations in the lightness of a single color are, in fact, often more obvious to users than are differences in color hue, and display an orderly sequence quite clearly.

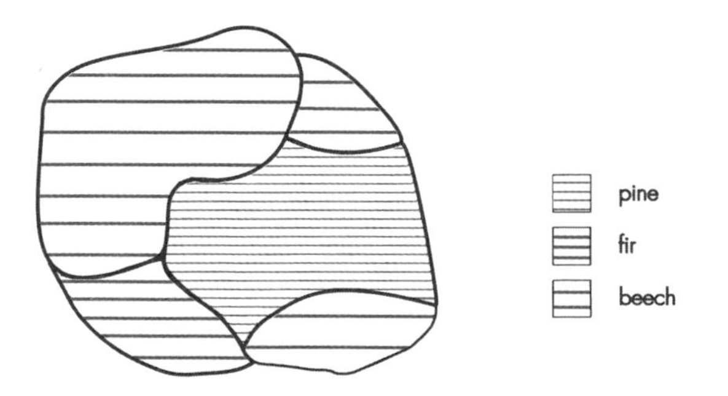

Figure 16.2
Incorrect use of density as a graphic parameter. The impression might easily be created that the map presents magnitudes, whereas in fact, the data shown are qualitative rather than quantitative. (From Bjørke 1988.)

Visual variables	Phenomena ability
size	quantitative
density	order
texture	quantitative
color	quality
orientation	quality
shape	quality

Figure 16.3
Relationship between visual graphic parameters and their abilities.

Color

In color theory it is usual to divide the concept of color into hue, lightness and saturation. *Hue* is associated with the name of the color (red, green, blue); *lightness* is changed by adding white or black to individual hues. Differences in hue can be both effective and pleasing and are particularly well suited to distinguishing various qualitative (selective) phenomena from each other. Unfortunately, colors are also the most frequently misused of all cartographic variables. Although colors are often coded to scales and used to indicate quantitative properties of areas, such as population densities, monthly temperature means, degrees of industrialization, and so on, colors themselves carry no subjective implication of rank or order. Red, blue, green, and orange might well represent the values 10, 20, 30, and 40 of some measurable quantity shown on a map, but users can only interpret the rank of a colored area correctly if the map has a color-code key that relates the colors used to the quantity involved. The use of colors also permits the exploitation of common associations. For example, red, which is commonly associated with danger, might be used to indicate dangerous areas on a map.

Shape

Varying geometric shapes are best suited to indicating qualitative differences (associative data). Shapes convey no overall impression, but may be used to convey details; they are usually noticed last of all the cartographic variables.

Texture

Very occasionally, variations in texture are used to show qualitative differences (associative data), usually by varying the printed dot density of a fixed percentage mix of two colors, such as 70% black and 30% white.

Orientation

The differences in orientation refer to different point or line patterns and are best suited to illustrating qualitative differences. Cartographic variables have varying ability both to perceive visibility and to distin-

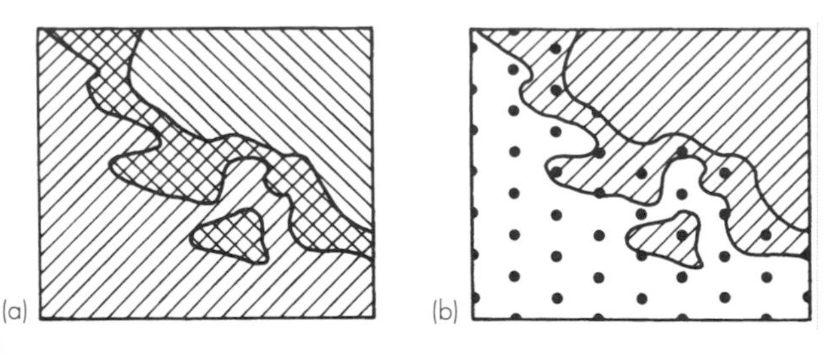

Figure 16.4
Representation of the diversification of language areas and overlapping of language groups: (*a*) the incorrect impression that the overlap area represents a new language group; (*b*) the correct visual impression. (From Baudouin and Anker 1984.)

guish between various phenomena. When selecting a variable it is not feasible to look at the individual variable, but one must also observe how the variables work together (interact; Figure 16.4).

Cartographic variables may be combined to display several phenomena simultaneously. For example, colors may be varied among differing dots or lines. Patterns that combine cartographic variables are frequently used to present data arranged in a particular order. In maps containing several cartographic variables, patterns differentiated by symbol size and scale of grays appear more visible than those differentiated by symbol shape and size.

In addition to the six graphic variables, several other means are also available. The geometric representations of point, line, and surface are cartographic instruments in themselves, represented on the map by dot, dash, and patch, respectively. All six graphic variables can be used in geometry. Map users are also aware of the different symbols that are used to make maps understandable: a cross represents a church, an airplane an airport, and so on. It is also possible to achieve a cartographic effect by use of more complex patterns, such as the positioning of symbols in certain groups.

Text and numbers

Text and numbers also play an important role in cartography (Kraak and also Ormeling 1997). Both text and numerals should be able to:

- Display nominal differences (e.g., differentiate between different types of objects and classes)
- Display hierarchy (e.g., differentiate between greater and less important types of object)
- Be associated with points, lines, and areas on the map (positioning on the map)

A hierarchy can be achieved by using variation in:

- Boldness

- Size
- Spacing between letters and words
- Color
- Style between upper- and lowercase letters

Nominal differences can be achieved by using variations in:

- Color
- Type

16.2 Selecting Map Symbols

The graphic display of information can be successful only if the variables displayed have the same mutual order, similarity, and magnitudes as the variables illustrated (Baudouin and Anker 1984). Map modeling consists to a large extent of choice, but the choices are seldom straightforward. The following guidelines to practice (Palm 1988) are useful:

1. Select point, line, or area symbols.
2. Consider the final application of the map (tourism, documentation, planning, etc.) and the probable capabilities of its users.
3. Consider the characteristics of the cartographic variables.
4. Use symbols of equal impact to represent equally important variables.
5. Use related symbols for related phenomena.
6. Consider visual phenomena and their characteristics. For example:
   a. Symmetrical symbols (roads, railways, etc.) are often visually dominant.
   b. Symbols appear smaller when they are surrounded by larger symbols, and vice versa.
   c. Whenever two areas adjoin without a border line, a light area appears lighter and a dark area appears darker.
7. Heed tradition (red means danger; the sea is blue).
8. Use existing standards.

The most common errors in cartographic presentations include:

- Incorrect use of cartographic variables
- Background color too strong in relation to the most important information presented
- Excessive number of dissimilar themes presented
- Inadequate legends or instructive material

Maps on which several phenomena are presented often contain areas where adjacent phenomena overlap. Graphic variables must then be chosen very carefully, so that users readily comprehend the over-

Figure 16.5
Typical symbol library. (© ESRI.)

laps and do not interpret them as separate phenomena (Figure 16.4). In principle, the cartographic variables are suitable only for showing statistical data. Should it be desirable to display dynamic data in addition, it will be necessary to use several maps, one for each data set. Arrow symbols can be used to indicate direction of movement.

16.3 Potentials and Limitations of GIS in Cartographic Communications

Practical solutions

The ability of a GIS facility to employ cartographic variables is determined primarily by the constraints imposed by the software systems on its screens and plotters. These programs usually contain predefined

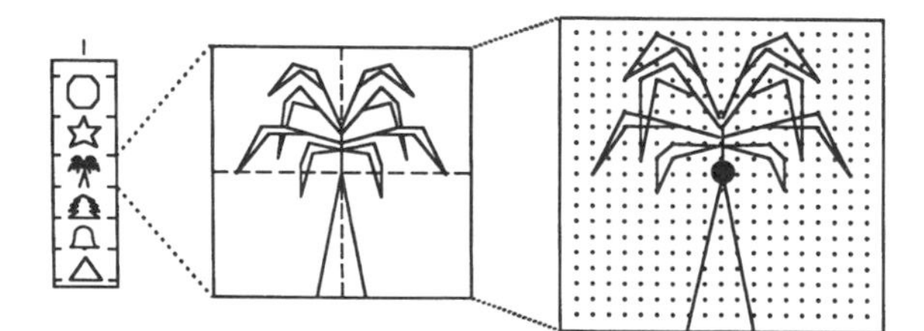

Figure 16.6
Symbol design. (© ESRI.)

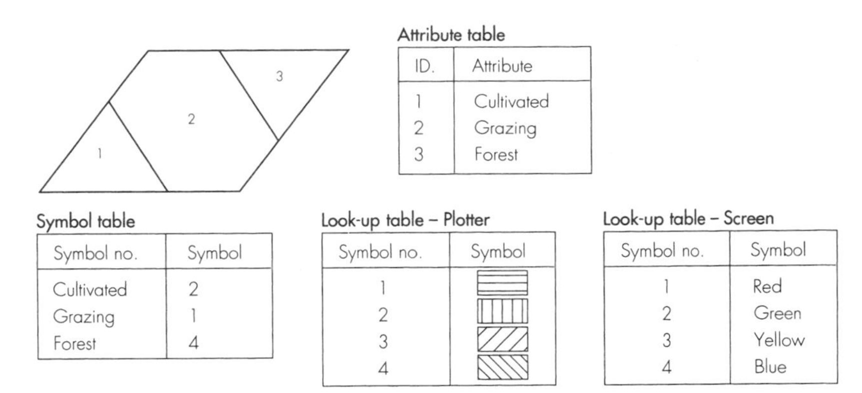

Figure 16.7
Look-up table.

symbol libraries (Figure 16.5). Users may often define their own symbols, either by digitizing symbols drawn or printed on paper or by using software tools to construct their own symbols (Figure 16.6). As shown in Figure 16.5, a large selection of different styles of lines, symbols, and text lettering is now available.

Normally, GIS will support printouts of defined attribute values, such as measured values and names of measurement stations, or permit attribute values to direct the plotting of selected symbols. The size and color of symbols may be related to the attribute value by checking with a reference table (look-up) table (Figure 16.7). Density of symbols or solid colors may be used to indicate phenomena covering areas on a map. For example, colors may be assigned numerical codes which are entered into the plotting codes for the map polygons, which in turn are activated when the map is plotted. Most GIS facilities support a range of letter fonts, either in basic system software or in plotter programs.

Raster displays support color assignments to each individual dot, so raster plots can include solid colors. However, raster resolution may limit the amount of detail in symbols. Similarly, variations in texture may be difficult to achieve. In general, screen resolution sets the upper limit to on-screen enlargement of a map: The greater the resolution, the more a displayed map can be enlarged.

Plotters, whether color raster plotters or film printers, seldom replicate on-screen colors exactly. This limitation may be vexing, particularly when color nuances are vital, as in the processing of remote sensing images. In such cases, extensive calibration may be required.

Raster plotters are best suited for producing graphic effects because their characteristics are similar to those of raster screens. The quality of the printed product depends on plotter resolution. Both raster screens and raster plotters may be limited in their ability to fill an area with solid color, for example by not being able to accept polygons larger than those that can be delineated by borders comprising a maximum number of points.

Geometric generalization is an important tool in map design. This includes the omission of objects, a reduction in number of points, changes in shape and size, fusions, and distortions. Automatic functions have been developed that carry out certain limited parts of the generalization. Some systems support overall changes, for instance by altering the size of symbols and the width of lines whenever the scale of a map is changed. There is no GIS available today that can carry out the entire generalization process automatically. To a great extent, therefore, the operator will still need to make decisions to the best of his or her ability.

In addition to geometric generalization, conceptual generalization is also used in map design. This is linked mainly to the attributes and assumes that the operator chooses to merge or remove some object types and enhance others. In this case, no machine treatment can be carried out before the operator has made a choice.

As we know, reality is three-dimensional whereas maps are two-dimensional. Normally, plotting of three-dimensional objects is carried out in GIS by using perspective drawing. Some systems can display several windows of the same segment, viewed from different angles, simultaneously on the screen. There are also examples of stereoscopic presentations on screen and virtual reality, where the viewer can freely wander through the terrain. Virtual reality requires unusually powerful computers and is still not available in normal commercial GIS form.

Multimedia

Multimedia and virtual reality are based on the simultaneous use of sound, text, animation, and pictures/video. For example, in tourist information systems and atlases, by pointing at special mapped objects on the screen, the viewer can release information in the form of speech or the playing of music (e.g., a national anthem). Through linking, electronic atlases and other specially adapted applications are able to supply supplementary text information by selecting map objects (e.g., property title owners, historical facts on a country, or similar). By linking pictures and video to objects on a map, it is possible to increase the information value. In its simplest version, animation is based on the presentation of fixed images on the screen in an edited sequence. To obtain a live effect, 25 to 30 pictures per second must be shown. Software is also available that can create new frames by interpolating automatically changing images between two key or main frames. The most advanced method for creating animation is based on algorithms that create new frames automatically based on the prespecification of objects, changes, and timing for changes (Kraak and Ormeling 1996). The basis for map-related multimedia is that the user can, to a greater or lesser extent, control the effects through predefined links to the map. The term *hyper maps* can be applied here because of the linking opportunities that have been developed.

16.4 Final Comments

Beautiful final products may, of course, hide poor-quality data. Data uncertainty is not easily shown in graphic form. Most experience relates to showing geometric accuracy, less to showing attribute accuracy, maintenance status, completeness, and logical consistency. Uncertain boundaries between polygons can be shown by dashed lines, double lines, a line darker in its middle and lighter toward its edges, or as a zone of a particular color or of no color at all. Uncertain points can be indicated by circles of differing size and color or as vectors. The network, deformed in proportion to the uncertainty in the point basis, is also one way of showing data quality. DTM can be used with very good effect by placing the size of the uncertainty on the *z* axis. It is also possible to decide to use a more powerful generalization of inaccurate data than for exact data. Attempts have been made to use blinking lights or sound signals to indicate uncertainty. Today there is no available GIS system that can automatically handle visualization of the quality of data in a satisfactory manner.

Maps are static presentations that are not particularly suited to showing temporal changes. But animation techniques can bring onscreen maps to life. For example, maps may be displayed in rapid sequence to illustrate a particular trend, such as the day-by-day spread of radioactive fallout from Chernobyl during the first two weeks following the disaster. The use of virtual reality will undoubtedly be able to open new opportunities for showing the time dimension.

In short, a GIS exploits its analytical prowess to generate a wide variety of maps rapidly from the same data. However, GIS maps do have cartographic limitations because few aesthetic capabilities are supported in GIS. The GIS user is unable to manipulate the overall aesthetic appeal of a map as can the conventional cartographer.

One reason for this disparity is that the art of mapmaking (and map printing) has developed over the centuries; the art, like the artisan, is mature. GIS is still young; in consequence, its cartographic aesthetics are relatively crude and the disciplines involved are less mature than those of conventional cartography. The skills of a GIS user are related to the analytical *manipulation* of spatial relationships, whereas those of a conventional cartographer are related to the *representation* of spatial relationships. The GIS user relies on the technological end products of the sciences: computational methods, mathematical operations, and the like. The conventional cartographer relies more on the arts, which involve perception and are therefore integral in both aesthetics and communications.

CHAPTER 17

Choosing a GIS: Organizational Issues

17.1 Historical Background

In many ways, the evolution of the applications of geographic information mirrors the course of history. From the earliest seagoing charts to the most recent star atlases, maps have reflected not only a contemporary knowledge of the world but also the contemporary technologies employed to gather it (Figure 17.1). From the earliest Roman census compiled for taxation purposes, to the Domesday Book, to modern demographic studies, overviews of populations and their distribution have recorded stages in the development of commerce, trade, and government. Conversely, many of the quests for improved georeferenced information have spawned changes that have affected history. The search for a reliable method of measuring longitude during the sixteenth, seventeenth, and early eighteenth centuries triggered the first government sponsorship of scientific research projects. Similarly, research into nuclear weaponry during the 1930s and 1940s led to development of the digital computer, one of the earlier examples of civilian benefits resulting from military R&D spending.

Initially, individual mapmakers expended considerable time for a limited gain in knowledge of their immediate areas, even though maps were established as useful sources of geographic information. The desire for innovation spurred progress: the need for accurate longitudinal measurement, the development of the theodolite, machine printing, photogrammetry, aerial photography, satellite remote sensing, and a host of lesser technological breakthroughs all prompted progress. As a result, the curve of qualitative cartographic improvement versus the effort expended to gain it soared—although the curve subsequently leveled off as the effort given to development produced fewer further gains.

Map sheets produced by traditional manual means in the late twentieth century may be both more attractive and more accurate than those of the late nineteenth century, but updating them still requires a considerable investment in worker-hours. The details of a public util-

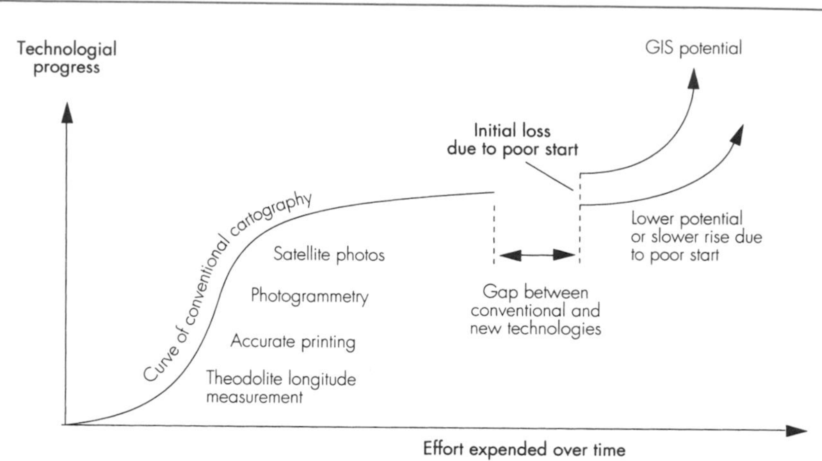

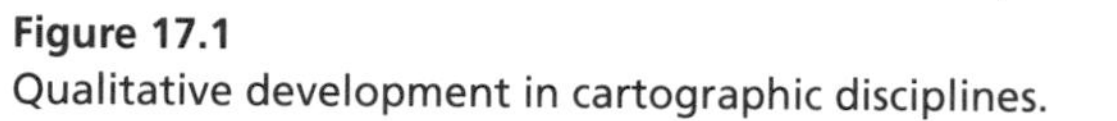

Figure 17.1
Qualitative development in cartographic disciplines.

ity network of the 1990s will probably be more involved and far more extensive than those of the same network of the 1890s, but the filing of information, on paper in folders in file drawers may have remained unchanged. So the human element—the speed and accuracy with which a cartographer can draw, the time required for a clerk to find and fetch a file—is clearly the ultimate constraint. Computerized GIS can change this overall picture quite dramatically because it dispenses with much of the human intervention required for the conventional acquisition, manipulation, and storage of geographic information.

An organization dealing with georeferenced information may initiate GIS to realize a gain, but as implied in Figure 17.1, effort (in time and money) is required to switch from conventional means to GIS. Bridging the effort gap successfully is therefore the first step toward realizing the benefits afforded by GIS.

17.1.1 Bridging the gap

Bridging the gap invariably entails switching to a new technology and changing procedures (Figure 17.2). In modern business language, the

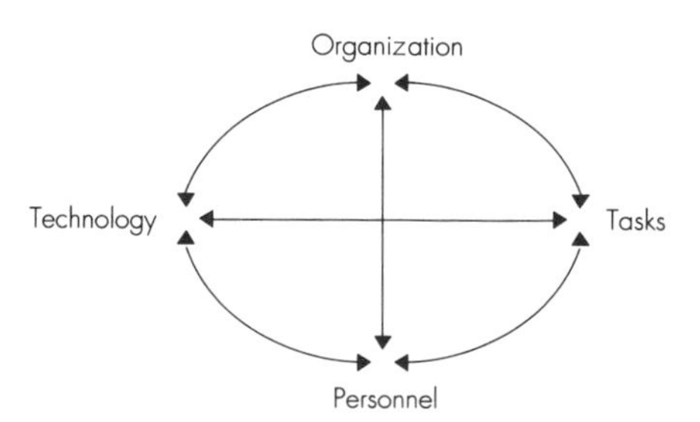

Figure 17.2
Successful initiation of GIS depends on a prudent balance of technology, organization, personnel, and tasks.

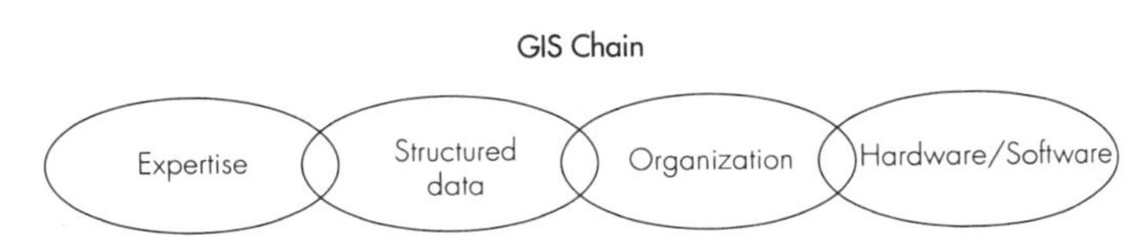

Figure 17.3
Interlinked chain of a GIS facility.

first steps to initiating GIS in an organization call for a number of strategic choices. For many high-tech approaches, the benefits accrued are related to these initial choices. A GIS facility cannot be bought off the shelf; rather, it is an assemblage of items of equipment that can become a useful tool only when properly placed in an organization and supported by expertise, structured data, and organizational routines (Figure 17.3).

The principal choices concern four aspects of GIS:

1. Data
2. Hardware and software
3. Expertise
4. Structuring

Definitions of these four aspects are usually interdependent and may change over time in ways that are difficult to predict. So a systematic approach is essential.

The tasks and issues involved in initiating GIS may be classified either as organizational or technological. This chapter concentrates on the organizational questions. Technical questions are dealt with in Chapter 18.

In practice, GIS projects tend to overfocus on technology and underestimate the organizational tasks. This has also been the case in other areas of IT. Studies in the United Kingdom show that 30% of large IT projects have budget overruns, owing primarily to organizational problems. In the United States corresponding studies show that 10% of the problems in IT projects are technology related and 90% are related to organizational questions. Nordic studies (Tveitdal 1987; Kylen and Hekland 1990) show that many organizations with GIS experience see in retrospect that their GIS activities could have been started more profitably in other ways.

Even though introduction of a GIS is very similar to other IT, it would appear that GIS is particularly difficult to handle from an organizational viewpoint. There are several reasons for this. With the introduction of GIS, many different users will normally be supplied with data from a variety of sources. This makes it necessary to develop new and complex organizational relations to obtain a good flow of data. In addition, it can be problematic for an organization to handle the long data establishment period; this often leads to frustration because of the long period between the investment time and the realization of the utilitarian value. In IT projects it is usual to allow a payback period of from one to three years, while with larger GIS projects one should normally allow for six to eight years. In that sense, there is a large difference between CAD (Computer Aided Design) and GIS. It is possible to start using CAD as soon as the software has been installed (and training given). It is not possible to obtain any benefit from GIS until at least some of the database has been established.

As for other major facilities, the process leading to the implementation of a GIS usually comprises:

- Planning
- Analysis
- Design
- Implementation
- Operation and maintenance

The budget for the planning phases should, of course, reflect the cost and complexity of the system envisioned. An investment in the preparatory phase of between 5 and 10% of an overall project budget is usually adequate to minimize project risks.

17.1.2 Impact of GIS

The organizations that make choices relevant to GIS may start from widely varying points. For example, many may already have access to a computer system that includes a network, as well as to other systems that need to communicate with GIS. Some may have digital georeferenced data available and possess some expertise in the use of software, hardware, and GIS in general. Others may have a form of GIS already in operation. In certain countries, organizations tend to choose final solutions in preference to systems. In such cases system integrators have become an important element in composing major systems using software, hardware, and data from different sources. The extreme solution is to out source the entire operation for a longer or shorter period.

The scope of a GIS implementation project is related to the final impact of GIS on an organization. Two categories can be established according to impact:

1. *Minimum impact.* Within the organization, GIS is seen as a useful new tool to be acquired, much as a new computer or telephone exchange. Typically, the GIS users may be within a division of a larger company or a bureau of a governmental agency.
2. *Major impact.* Initiating GIS changes the way in which the organization operates. Consequently, the problems addressed affect virtually all phases of the organizational hierarchy and operational paradigms. Typical organizations include municipal authorities, public utilities, and cartographic agencies.

The demarcation between these two categories is a blurred zone rather than a sharp line. GIS implemented in one small bureau of a large governmental agency may radically change that bureau. For the bureau, the impact is major; for the parent agency, GIS is merely a tool that enables one of its many bureaus to accomplish its task. Conversely, a large agency in a small country might need to be reorganized completely in its conversion to GIS. In terms of equivalent efforts in larger countries, however, its efforts may be small. Hence the qualifier "from the users' viewpoint" should be applied to all assessments of the impact of implementing GIS. The bulk of the following discussion is devoted to the implementation of GIS in organizations for which it effects a major impact. A third category of organization spawned by GIS, such as a national geographic database, is also described later in this chapter.

17.2 Phases in Organizational Issues

As we have seen, GIS differs considerably from its conventional predecessors. Consequently, the effect that the introduction of GIS has on an organization may result in the need for restructuring. The prudent, realistic approach usually requires a balance between remodeling the organization to accommodate GIS and shaping GIS to suit the needs of the organization.

Evolution of the organizational changes normally includes:

- Development of a business concept and identification of goals
- Appraisal of current:
  - Tasks, work functions, and routines
  - Basic data
  - Information products
  - Information and data flow
  - Data processing infrastructures
  - Potential users
- Review of the experience of others with GIS
- Identification of user requirements
- Cost–benefit analyses, including delineation of:
  - Assumptions for realizing yields
  - Measures to be enacted to realize benefits
- Choice of strategies related to:
  - What is to be automated
  - Start date
  - Level of investment
  - Geographical coverage
  - Future data flow
  - Organizational matters
  - Budget and financing

- Development of a plan for future conduct, including project reorganization
- Data structuring

Current trends are to regard information as a widely available and independent resource available in numerous ways through the integration of data processing schemes. Consequently, planning should not be limited to the organization itself, but should include external users, data suppliers, and other cooperating organizations. Hence several organizations may participate in a single GIS project. Preparatory work should also aim to instill motivation, build competence, and promote maturity within the organization. Indeed, the organizational evolution should itself draw on the expertise of:

- Professionals who know the tasks
- Executives who know the organization
- Staff
- System specialists and/or external consultants

Evaluations of the various aspects of data processing and computer hardware and software should be aligned with activities relating to the evolution of organizational changes.

17.3 Development of a Business Concept and the Identification of Goals

A thoroughly prepared business concept should form the basis for initiating GIS in any organization. Many different motives may lie behind a decision to introduce GIS. Management may wish to increase internal productivity or external competitive ability. There may be a professional interest in new technologies, or the organization may be obliged to respond to internal or external pressures to approach tasks in a new or more vigorous way.

The motives for choosing GIS must be thoroughly understood, for only then can goals be identified with sufficient clarity as to become attainable. This obvious requirement often means limiting the overall scope of the GIS project. The business concept forms the basis for defining tasks to be implemented by the organization, and the user requirements are a consequence of these tasks.

The aims should be formulated in such a way as to give as precise a description as possible of what results one hopes to attain through the project, and they must be realistic for the business. It would be natural to specify one short-term and one long-term aim for the introduction of GIS. In principle, this means setting objectives, then reviewing the results compared to those objectives. For this reason, the aims should be concrete and, as far as possible, testable. For example, they can be expressed in percentage form as an improvement in efficiency or as an increase in income, a reduction in costs, in new services and products, or in increased service.

Those parts of the organization concerned with the introduction of these new technologies must also be identified to permit assignment of internal and external markets. The marketable entities consist of the GIS goods and services generated. These must, in turn, be defined and test marketed. Initially, GISs must often be "sold" within an organization and GIS salespeople enthusiasts capable of promoting GISs—are sometimes needed to trigger the processes leading to the implementation of a GIS facility.

17.4 Appraisal of Current Tasks, Users, Data, and Data Flow

An assumption fundamental to the implementation of any new technology is that it performs old tasks in new ways, performs tasks that were previously extremely difficult or impossible, or offers a combination of these two advantages. One of the fundamental starting points for appraising a new technology, therefore, is clear identification and description of the current tasks. The appraisal should, of course, include *all* the users, regardless of whether the system to be implemented is limited to a single bureau or agency, or is to be a larger, multiuser system.

The first step of an appraisal, then, is to compile an overview of actual and potential users: the next is to elicit the following relevant information from these users:

- All production processes
- All data sources
- All files and filing systems
- Descriptions of all data (including content, accuracy, completeness, and timeliness)
- All data users
- Data flow, documents, and communications between processes, staff, and internal and external files
- How data are used
  - Tasks and products
  - Decision making
- How data are managed and maintained
- Descriptions of current methods, including any current computer systems
- Human factors, including the working environment and individual expertise

The depth of detail should, of course, suit the goals of the assessment. For example, a detailed overview of a production process is essential only if one of the purposes of the transfer to GIS is to rationalize production. Many organizations are already computerized and thus have data processing infrastructures in place. As these infrastructures may affect the introduction of GIS facilities, they should therefore be evaluated with respect to system approach and software systems, user interfaces, databases, operating systems, programming languages, hardware, and communications facilities.

Product information is vital. In some cases, production processes may depend on the information and organizational structures required to produce specific products. In other cases, product improvement or process optimization may be important factors. A clear diagram of information flow provides an overview of the various processes and their interrelationships. However, because data flow may be complex and therefore difficult to describe, special techniques, such as structured systems analysis, may be necessary.

Assessments of current tasks must also contain evaluation of factors having financial impact, including:

- Problems related to the conduct of such current tasks as:
  - Proliferating delays
  - A decline in quality
- The ideal situation sought by individual users of geographic information, and which can be resolved by use of new technology, such as:
  - The demand for new products and services
  - Ways of changing current products and services
  - Necessary organizational changes
- Projected time or cost savings resulting from realization of the ideal situation
- Side effects of the new technology
- Number of present and potential future users

This information can form the basis for later cost–benefit analyses.

17.5 Review of Others' Experience with GIS

All organizations and activities have their own particular requirements. Nonetheless, a review of the experience of others is useful whenever new technology is implemented. As is the case with all technologies, operating GIS facilities fall into three broad classes:

1. *State of the art* or *leading edge.* Usually only a few organizations are at the forefront of the field because keeping abreast of the latest developments is expensive and requires expertise to deal with unproved approaches.
2. *Proven.* In the burgeoning field of GIS, the bulk of facilities now in operation comprise off-the-shelf hardware and software adapted

to specific tasks and proven in operation. Directly or indirectly, the cost–benefit ratio of such a facility is usually near optimum.

3. *Outmoded.* Aging facilities in the proven category are often reliable producers. However, as constituent parts or whole facilities become outmoded by newer designs, cost–benefit ratios are less favorable because greater benefits may be realized through upgrading.

All organizations have their own requirements, but in most cases the experience of organizations with proven facilities are the most valuable in defining user needs. The bulk of facilities now in operation comprise commercial off-the-shelf (COTS) hardware and software, but realization of user needs should be balanced against the availability of new technology. Usually, only a handful of organizations are able to keep at the forefront of the field because keeping abreast of the latest developments means that cost-prohibitive and special expertise is needed to deal with an unproven approach. It is not surprising, therefore, that no one ever got fired for buying from IBM or Microsoft. It is, however, important to balance this conservative tendency. Whenever GIS is to be implemented, therefore, a review should be carried out based on a variety of sources, including user experience of state-of-the-art facilities and information from system suppliers and others familiar with the market and technological development.

In most cases, the experience of organizations with proven facilities are the most valuable in designing a new GIS facility. State-of-the-art facilities are usually either too expensive or too exotic for organizations not involved directly in GIS R&D. Outmoded facilities are seldom of interest to an organization contemplating a new technology.

Whenever there are few or no relevant users, as when GIS is to be employed in a totally new sector, studies must be conducted in two phases. First, aspects of the new application, which may be similar to those of existing applications, should be singled out. Second, those aspects should be evaluated, one by one, to form an impression of the mix of proven and new approaches in the new system.

17.6 Identification of User Requirements

The business concept forms the basis for defining tasks to be implemented by the organization, and the user requirements are a consequence of these tasks. It is only possible to identify the requirements once user tasks have been identified and are clearly understood. Definition of user requirements is closely related to the development of the business concept and could form an integral part in building the concept.

As discussed above, defining user requirements is a process where a great deal of work could take place during and even after system procurement. Traditionally, corporate GIS acquisitions for large organizations with operational responsibilities involved detailed specification of the user needs up front, often with the aid of a feasibility project. Increasingly, however, the greater part of the work to define user needs takes place during and even after system procurement has been specified, as the first stage in system integration. This exposition is based primarily on an increasing partnership between users and vendors; but whether identification of user needs is done pre- or postprocurement, it is a basic activity that should be treated in a structured way. Each organization has to find the best balance between pre- and postprocurement in their definition of user requirements.

Preprocurement definition of user requirements

Identification of user requirements in the preprocurement phase should be based on the following activities:

- User surveys
- Workshops
- Review of documented experience

Thus user requirements for most corporate implementation of GIS in both government and utilities are normally identified from user surveys, often as part of a feasibility study to establish the overall potential for GIS in the business. A standard questionnaire should be used in the user surveys to compile homogeneous information from varying sources, and care should be taken in selecting representative respondents if the information gleaned is to be meaningful. Because questionnaires are always subject to misinterpretation and users are not always aware of their specific needs, user studies are best conducted through personal interviews. Unfortunately, these are both time consuming and costly. In most cases the survey should be limited to a representative sample including, of course, the major users who will benefit most from the new GIS technology. The sample should include a broad spectrum of other users to give the best overall picture of what is required.

To sum up, interviews with hands-on users are often more productive than those with an organization's executive. The interviews should be carried out by someone with broad GIS experience who can guide the respondents through the interviews. One difficulty with such surveys is that even the most experienced users seldom fully exploit the potential of a GIS facility. The potential benefits of a system are evident only after its goals have been defined. Consequently, user surveys can provide an overview that is valid only at the time they are conducted, and the results should be assessed and analyzed by GIS experts, who can more readily see the potential and the limitations.

Since GIS awareness is a process in the organization, workshops could serve as a forum for speeding up this process. Both workshops and user surveys could be carried out as integrated activities for identification of user requirements. Workshops could also be used as an approach to building a total business model for GIS, but business modeling for the sole purpose of GIS implementation is now relatively rare. Many organizations have already built them as part of a wider program, and generic models are now also available for local government and the utilities.

Information compiled from user surveys, workshops, and reviews will normally lead to a long list of different user needs, which will have to be given a structured and neutral evaluation. The business model should form the basis for all evaluations—only requirements that are within the business model should be evaluated. (This is one important reason for business modeling.) Ideally, the ranking of needs should be based on a solid cost–benefit analysis, or at least on an evaluation where both costs and benefits have been taken into consideration.

Postprocurement definition of user requirements

The basic assumption for the postprocurement activities of GIS is that the organization has selected a supplier of GIS software to develop applications that meet the business concept of the organization. This selection will be made on the basis of vendors' responses to some relatively simple previous research. The requirements common to all such cases, irrespective of size or sector of the organization, will constitute some form of design specification. Although the terminology may vary, such design specifications normally comprise two parts: a specification of applications and a functional specification. The application specification sets out the organization or user needs for GIS in terms of the applications required to be met, a logical data model, and the required service levels, such as user numbers, user locations, and requirements for uptime and other performance levels. As in the prerequirements activity, user requirement may be identified through user surveys, workshops, and a review of experience from other users. The functional specification defines how the chosen GIS software will be implemented to meet these needs. It is usual for the suppliers to produce the functional specification, based on their own knowledge of the software. The nature of the postprocurement activity varies according to the scale of implementation and the nature of the application, but user involvement should be relatively high since GIS technology is mainly end-user oriented.

17.7 Cost–Benefit Analyses

A prerequisite for obtaining funding for GIS should be that the proposals for introduction be based on solid plans and financial evaluations. It is important to know what can be achieved by GIS so that the right decisions can be reached as to what should be carried out and how the profit can be achieved. Cost–benefit analyses ensure that projects can be selected which can show profitability. User surveys and (possibly) pilot projects provide the bases for analyses.

The purpose of financial evaluations is primarily to:

- Rank prospective projects
- Decide if projects shall be carried out

Such analyses can also delineate the need for further studies or planning and rework the project for greater profitability. Other motives include building awareness of financial impact, developing motivation, and enhancing planning.

17.7.1 Theoretical basis

In business, costing tends to focus on the improved efficiency and increased productivity resulting from a new technology. This is usually expressed in terms of profitability. Government agencies evaluating the change to new technologies often take socioeconomic factors into consideration. This mandates a broad view that includes both internal efficiency and external effectiveness. Socioeconomic costing, therefore, is based on the inclusion of all parties likely to be affected by the change under consideration.

In a company or a municipal department, inner efficiency (benefit) can be measured either in terms of increased productivity or as savings in time and money. However, time savings are relevant only if they can be used to equal advantage elsewhere. From a socioeconomic viewpoint, such savings seldom pose problems when there is a shortage of qualified workers. During high unemployment, however, they may cause redundancies, thus increasing the number of unemployment benefit claims. Considerations such as these dictate that for analytical purposes, a minimum tangible benefit level should be set.

The external effectiveness of an organization is determined by how well it serves its users, both public and private. For users, increased effectiveness is manifested in lower costs, higher productivity, or both.

Cost–benefit analyses customarily weigh various options against each other with respect to their applicability over a prescribed time period. The zero option, which retains the status quo with no change of technology, is almost always included. This requires a prediction of future events without the new technology. Other options may incorporate varying degrees of technological change, but they must be pure; that is, each individual cost–benefit assessment must be independent of the costs and benefits that would accrue without technological change.

In principle, socioeconomic evaluations and business economic evaluations comprise the same phases:

1. Define different projects/alternative actions.
2. Establish assumptions for calculation: time perspective, rates of interest, etc.
3. Chart the positive and negative effects of the various project alternatives.
4. Quantify the effects (i.e., cost and benefit).
5. Calculate the cost–benefit ratio.
6. Consider the nonquantifiable effects.
7. Carry out sensitivity analyses, if necessary.
8. Recommend or choose solutions.
9. Establish prerequisites (preconditions) for the realization of gains.
10. Work out strategies or plans for realization of gains (including budget proposals).

17.7.2 Negative effects: costs

Commercial cost accounting is usually straightforward. Yet when socioeconomic factors are involved, cost accounting becomes tenuous, because the complete effects of investments elude concise definition. For example, a national map service may accurately estimate the initial cost of a new map series, as the costs of geodetic data, mapmaking, field surveys, reproduction, and the like are easily ascertained. However, the ultimate distribution of costs among the various end products, as ordinary map sheets, digital map data, and so on, is seldom known as accurately.

Cost accounts usually include the following:

1. Planning costs
   - Internal personnel costs
   - External consultants
   - Equipment and material in the planning stage
   - Travel and meetings
2. Establishment costs
   - Data conversion
   - Applications development
   - Equipment and software
     - PCs
     - Peripherals
     - Software
     - Installation costs
     - Software development
     - Communications or data network

  - Computer room
  - Security systems
3. Operations and maintenance
  - Internal personnel costs (training, operation, user support, and administration)
  - External consultants
  - Equipment maintenance
  - Software maintenance
  - Database creation
  - Updating databases
  - Conversions
  - System operation
  - Supplies
  - Administration
  - Cost of capital
  - Insurance
  - Rent

In addition, there may be intangible liabilities, including:

- Increased vulnerability due to reliance on computers, which may fail due to hardware or software malfunctions or failures
- Poorer working environment, due to equipment-generated noise, tedious digitalization tasks, etc.
- Increased entry-level competence, which can bar some staff members

17.7.3 Positive effects: benefits

The tangible benefits that may accrue can be divided into three categories:

1. Benefits in resources
   - Reduction in permanent staff required
   - Reduction in temporary staff required
   - Less overtime
   - Savings in consultant fees
   - Reduced costs
2. Benefits in products and services
   - Greater task turnover
   - More rapid production
   - Faster processing
   - Few production errors
   - Sale of new products
   - More efficient invoicing of fees, etc.
3. Benefits in effect
   - Greater turnover
   - Simpler access to information

- Simpler treatment
- Fewer persons involved
- Improved internal and external communication
- Lower costs and less work for others

The intangible benefits may include:

- More rapidly available information
- Higher-quality goods and services
- More conclusive decisions
- More rapid decisions
- More and better used information and improved service
- Finer analyses
- Superior plans; strategic positioning and management
- Greater understanding of problems
- Enhanced expertise and more challenging work
- Stronger competitive ability
- Upgraded capability of organization
- More career options for staff
- Work that is less routine
- Keener financial management

All vital intangible effects, positive and negative, should be described and incorporated in the decision-making process.

17.7.4 Quantification and analyses

Cost–benefit analyses appraise costs and benefits in measurable terms, according to the prime rule of the relationship

$$\text{benefits (or costs)} = \text{quantity} \times \text{unit price} \tag{17.1}$$

Normally, quantification of the positive effects will be based on a reduction in working time, reduction in costs, and a possible increase in income. A typical computation would be of labor savings expressed in worker-years, based on a productivity increase of 10% per staff member. The unit price would then be the average staff labor cost per year. It is generally recommended that relatively conservative estimates form the basis of all quantification, to avoid presenting overly optimistic figures, or, alternatively, use sensitivity analyses (more on this later).

Costs and benefits often occur at different times, with costs usually being incurred before benefits accrue (Figure 17.4). Both costs and benefits may be divided into non-recurring and annual values. These differences must be resolved so that the various values may be compared directly. Future values must be stated in terms of present values, uncorrected for price-level fluctuations. For example, an income

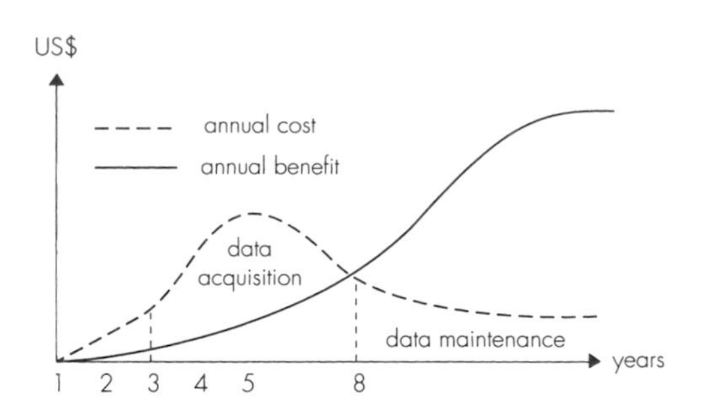

Figure 17.4
The benefit will come after a period of investment. Cost and benefit curves indicate that data acquisition and maintenance normally account for 60 to 80% of aggregate costs.

of 1000 monetary units received five years hence will, due to loss of interest, be worth less than a present income of 1000 of the same monetary units. Consequently, future values must be discounted to permit assessment in terms of present values.

The usual procedure involves computing the present value (PV) of a potential investment or project. The PV is the aggregate discounted value of the expected benefits, and the aggregate discounted value of the expected costs. Net present value (NPV) is computed by the relationship

$$\text{NPV} = \sum_{t=0}^{T} \frac{B_t - C_t}{(l + r)^t} \qquad (17.2)$$

where B is the benefit, or return, and C the cost at time t in years, r is the discount rate, and T is the time horizon (the point in time defined as the end of the economic life of the project or the time at which the assessment no longer evaluates costs and benefits). A typical computation is shown in Figure 17.5.

The final computational result depends strongly on the interest rate chosen. In general, the interest rate should be high for high-risk projects, short-term projects, or combinations thereof. Only effects that occur within the project(s) life span will be dealt with here. This means that the lifetime has to be long enough to include the main effects of the project(s). One should bear in mind that effects that occur late in the life span have relatively little influence on the current value estimates.

The time horizon is determined by the projected economic life of the investment or project involved. The time horizon of most computerization projects is five years. However, the time horizon of GIS projects is longer, usually about 10 to 20 years. This is because the data employed in GIS are valid over longer periods of time.

Year	Annual costs 1.000 US$		Annual benefit 1.000 US$	
	1999 US$	Present value (1999)	1999 US$	Present value (1999)
1999	200	200	50	50
2000	300	280	100	93
2001	300	262	200	175
2002	200	163	400	327
2003	200	153	700	534
2004	200	143	800	571
2005	200	134	800	535
SUM:	1.600	1.335	3.050	2.285

Cost	Benefit	Benefit/cost	Benefit ÷ cost
1.335	2.285	1.71 : 1	US$ 950

Figure 17.5
Typical computation of NPV, based on a discount rate of 7% and a time horizon of five years.

Alternatives are often ranked in terms of their respective benefit–cost ratios, which are the ratios of the present values of their aggregate benefits to aggregate costs:

$$\text{benefit–cost ratio} = \frac{\text{present value of benefits}}{\text{present value of costs}} \tag{17.3}$$

The benefit–cost ratio usually provides the best index for ranking options. However, in some cases, where options involving considerably different levels of costs are involved, the aggregate net benefit is also useful. This represents the difference between benefits and costs:

$$\text{net benefit} = \text{present value of benefits} - \text{present value of costs} \tag{17.4}$$

Methods that do not include considerations of time and interest rates cannot be used as profitability criteria.

17.7.5 Prerequisites for harvesting benefits

The cost–benefit analyses involved in decision problems are invariably based on assumptions. The prerequisites that form the basis of the project will have great significance for the recommendations that the cost–benefit analysis will result in.

The main prerequisites will often be:

- Which decisions have to be made
- Which measures have to be initiated

The following aspects can form the basis for clarification of the prerequisites:

- When can the gains be taken out?
- Is there any need for new developments?
- Do we need new competence?
- Is reorganization necessary?
- Will this mean adaptation for the employees?
- What is the relationship to other projects?
- What is the relationship to other internal and external activities?
- What is the timetable for completion of the project?
- In how many budget periods will there be need for investments?
- Are there any particular cost and resource restraints?
- Who is to approve the startup of the project?

17.7.6 Sensitivity analysis

The role of the assumptions, however, is not always clear at the outset; therefore an analysis may be performed to identify the assumptions to which the outcome of the cost–benefit analysis is sensitive. This involves changing the assumptions (parameters) of the decision problem to reveal how the changes affect the outcome. The analysis involves individual computations, which may be based on:

- Pessimistic or optimistic assumptions for costs and benefits
- Postponement of various activities or investments
- Change in the order of activities
- Changes in the rate of investment

17.7.7 Review of the cost–benefit relationship

There are no exact measures for assessing cost–benefit alternatives. However, rule-of-thumb experience indicates the following:

1. Measures that result in a cost–benefit ratio of 1:2 or more should be enacted, when assessed both in commercial terms and in socioeconomic terms.
2. Measures that result in a cost–benefit ratio between 1:0.8 and 1:2 should be analyzed further with an eye to cutting costs or increasing benefits.
3. Measures that result in a cost–benefit ratio of less than 1:0.8 should be abandoned.

The cost–benefit relationship is an insufficient basis for a decision as to which projects should be carried out. We should also take into account the nonquantifiable effects.

17.7.8 Evaluation of the nonquantifiable effects

The nonquantifiable effects should be handled equally for all the projects to be ranked. To ensure that this is done, a precise description has to be given. An evaluation should also be made as to whether each single effect has a great or small influence on the project and what the chances are that this effect will occur. Thus there can be four categories, which will be weighted differently in evaluation of the projects:

- Great effect and good chances of realization
- Great effect but slight chance of realization
- Little effect and slight chance of realization
- Little effect and good chance of realization

17.8 Developing a Strategic Plan

The term *strategy* originates from the Greek *strategos* (*stratos,* army + *agein,* to lead). Today, *strategy* is normally used to describe a plan as to how different situations should be handled to reach goals that have been defined. In the planning process of a strategy, the prerequisites should be clarified for the choices to be made. Ideally, the process should result in clear directions as to what decisions and steps should be taken, how they should be carried out, and which resources should be used and by whom. As stated above, the cost–benefit analysis will form a solid basis for this type of evaluation.

The technologies involved in GIS are developing rapidly and continuously, so GIS users may improvise new applications at any time. Consequently, the strategic plan should be sufficiently flexible to allow its conduct to be influenced by prevailing conditions. The ranking of alternatives in the plan is based on the results of the cost–benefit analyses, in which investments in GIS are tested against imminent or current investments elsewhere in the organization. In addition, public agencies must consider their official duties. With a suitable strategy, the optimum initiation of GIS activities might be viewed as shown in Figure 17.6.

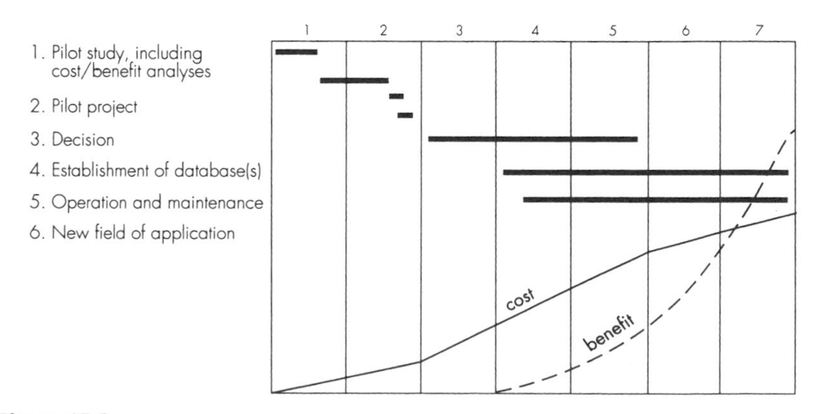

Figure 17.6
Idealized initiation of GIS, where costs and benefits accrue during a project.

A strategy plan should identify and illuminate:

- What is to be automated
- Start date(s)
- Mode of introduction
- Level of investment
- Geographical coverage
- Future data flow
- Organizational matters
- Project organization
- Budget and financing

17.8.1 What is to be automated?

Current data flow, production lines, products, and cost–benefit analyses form the basis for the macro view of what is to be automated. The macro view may include:

- Which work processes are to be replaced by GIS
- Which files are to be replaced by GIS databases
- Which lines of communication are to be replaced by electronic communications via data networks

17.8.2 Start date

One of the prime decisions to be made in all changes of technology is when to start. Studies conducted in Scandinavia (Bernhardsen 1985) indicate that no users with operational GIS facilities felt that they should have postponed their changes to GIS. In fact, some felt they should have started earlier. The only regret voiced by some was that they might have started in a different way. In short, the study indicated that an organization contemplating GIS should initiate the relevant activities as soon as possible.

In general, the later the startup, the longer the time before benefits are realized. As illustrated in Figure 17.7, a late start may cost no more than an early one, but may result in fewer aggregate benefits. The benefit–cost ratio is then less.

One of the more common justifications for postponing an investment in new technology is that newer and better technology will soon be available. The experience of GIS users thus far indicates that such postponements seldom prove profitable if the system is implemented in a proper way. The continuous evolution of all the technologies involved in GIS implies that the basic decision of whether to invest now or later remains, no matter at what time it is made.

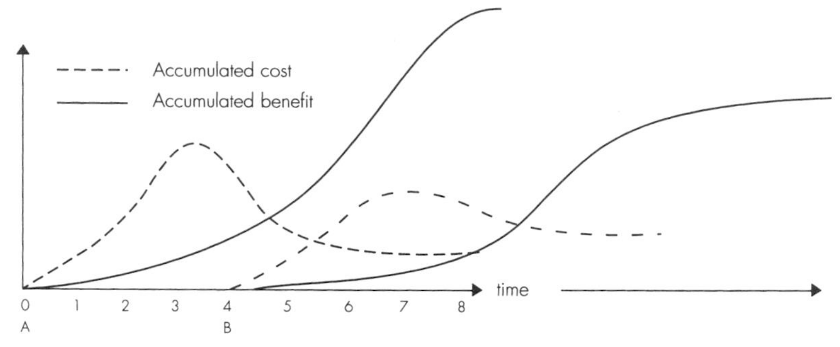

Figure 17.7
Early versus late startup of a GIS facility. An early start date (point A) results in a higher benefit–cost ratio than does a later start date (point B).

To summarize, the factors that should influence the decision of exactly when to start include:

- The relevant technologies are available, so they no longer comprise a bottleneck.
- Postponing a start postpones benefits and thereby lessens project profitability.
- All technologies now evolve continuously, so there is no single best time to start.

17.8.3 Method of introduction

As emphasized earlier, there will almost always be a relatively long period for data establishment before a GIS can be fully operative. This has a strong influence on how new technology can be introduced into an organization. An introductory period with system training, adjustment to new routines, will also be long drawn out, so that an overnight change in technology is not a suitable method for introducing GIS.

A lengthy introduction period will normally lead to both old and new routines being run in parallel for a certain time. This is unfortunate from a financial standpoint because it requires extra resources (see below), but it gives increased security in case the new technology should fail. The maintenance of parallel routines can also be very frustrating for employees, even though they would normally benefit from a relatively long introduction period, which gives opportunities for thorough training.

System training, changes in working routines, and other organizational changes which often accompany GIS can be implemented in several different ways. Many larger organizations choose to carry out a pilot project to:

- Assess organizational problems
- Assess the staffing requirements for and time consumed in transferring data from conventional to digital form

- Help develop the organization
  - Increase appreciation of the GIS concept
  - Mediate organizational problems
  - Contribute to verification of cost–benefit analyses
  - Better time estimates for data acquisition, etc.
  - Better assessments of time gains, etc.

Other organizations choose a gradual approach where different parts of the system are introduced in a particular order so that employees are never confronted with major changes. There is no particular method of introduction that will always be the best. What is certain is that the method of introduction should be *planned and accepted by the employees.*

17.8.4 Level of investment

The crucial choice to be made in investment is usually between constraining investments in new technologies within conventional budgets and considering more extensive investments. As is the case for almost all changes from traditional to electronic technologies, the initial investments are relatively large, yet short term. This trend should be exploited, as in establishing GIS databases quickly to realize the benefits they generate. Prolonging the establishment of a database by stretching it over phases involving both conventional analog and computer digital data only escalates costs without correspondingly increasing benefits. As shown in Figure 17.8, cautious investment often yields less than aggressive investment.

In conclusion, the guidelines for levels of investment may be summarized as follows:

- Think big but start prudently.
- Make major investments after a pilot project phase.

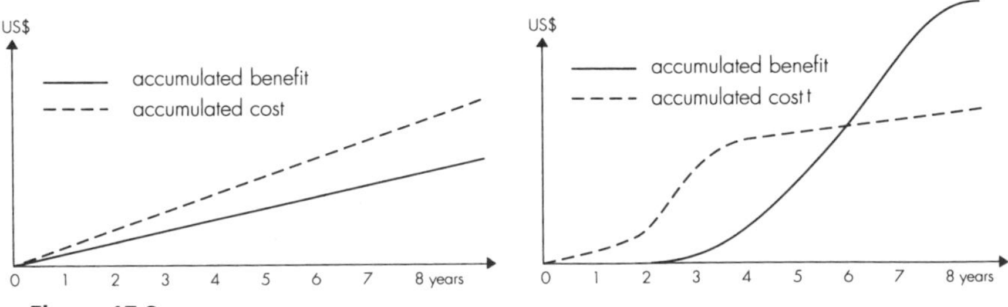

Figure 17.8
Aggressive investment often yields more than cautious investment. (*a*) Linear investment within the constraints of a conventional budget produces linear benefits, which may prove unprofitable. (*b*) A major early investment, such as in the conversion of an entire database, can yield benefits that rapidly exceed costs, for greater profitability.

- In switching from conventional means to GIS, major investment should be committed in the initial implementation phase in order that the project be profitable.
- Data should be converted rapidly to digital form.

17.8.5 Geographic coverage

As a rule, the basic decision to be made concerning geographical coverage within the scope of a GIS is whether to provide full digital coverage of all regions and areas involved or, rather, to digitize information fully only for areas of greater activity and more rapid change. Again, the economics of the relevant benefit–cost ratios are the best criteria.

Different user organizations have differing needs for the geographical coverage of digital information. The distinctions involved often are dictated by functional or administrative divisions (townships, regions, counties, or entire countries). In general, a need for digital geographic coverage must be fulfilled completely before the user can realize the benefits. This is because staged shifts invariably include periods when both conventional and digital data are in use, which usually increases the work involved in a given final product.

As a rule, therefore, the most cost-effective approach is for an organization to start with full geographic coverage, even though such coverage requires a greater initial investment in data acquisition, operation, and maintenance. The distinction between the coverage needs of various organizations is then not one of coverage itself, but rather of the regions or areas of responsibility involved. A managerial organization such as a major governmental agency must usually maintain geographic coverage of its entire area of responsibility. An organization dealing with projects such as roadwork need only cover areas involved in its projects.

Once a GIS facility is planned, the question arises as to which part should be installed first. From an investment viewpoint, the part that yields the greatest proven benefit should be first. But for most users, a part comprising several georeferenced themes concerning a single area or region provides a more comprehensive initial benefit. This is because the various themes are often synergistic. For example, a digitized property register enables GIS to provide information on properties, and a digitized road data register enables it to provide information concerning roads. However, if both digitized registers are available, the GIS can provide additional information. It can then provide answers to such queries as: "Which properties front onto which roads?" and "Which properties are affected by the proposed rerouting of this road?"

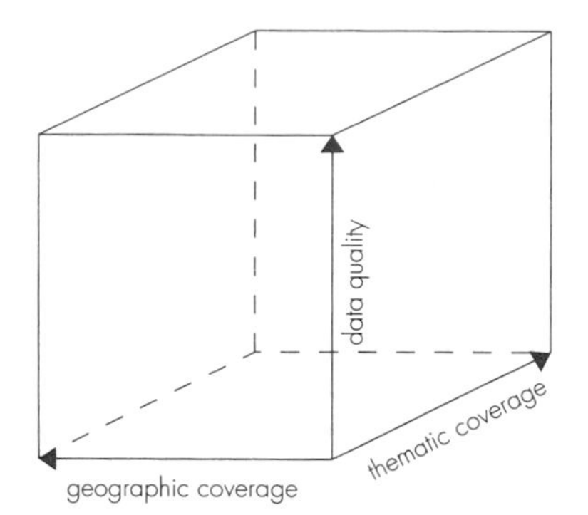

Figure 17.9
Usefulness depends on geographic coverage, thematic coverage, and data quality. These three parameters may be viewed as defining a *usefulness cube.* (Courtesy of Ken Jones.)

In conclusion, the guidelines for geographic coverage may be summarized as follows:

- At least one theme covering an entire region or area of responsibility must be converted before benefits may be realized.
- Staged conversion to GIS, in which conventional and digital data may be used simultaneously, should be avoided.
- Managerial organizations have the greatest need for coverage of entire regions or areas; organizations operating on a project basis have less need.
- Several themes pertaining to a region or area can synergistically increase benefits to users.

The potential benefit accrued (usefulness) depends to a great degree on geographic coverage, thematic coverage, and data quality. These three parameters may be regarded as a *usefulness cube,* as shown in Figure 17.9.

17.8.6 Future data flow

The description of future data flow must relate to the functions to be performed. Comparing future to current data flow helps identify which work functions will become superfluous, which must be added, and which should be modified. In a public agency, for example, the change of overall data flow caused by a change from conventional means to GIS may be regarded as being a change from series to parallel processing of work, as illustrated in Figure 17.10.

17.8.7 Organizational matters

The efficient exploitation of a new technology in an organization often mandates alterations in work routines and chains of command which,

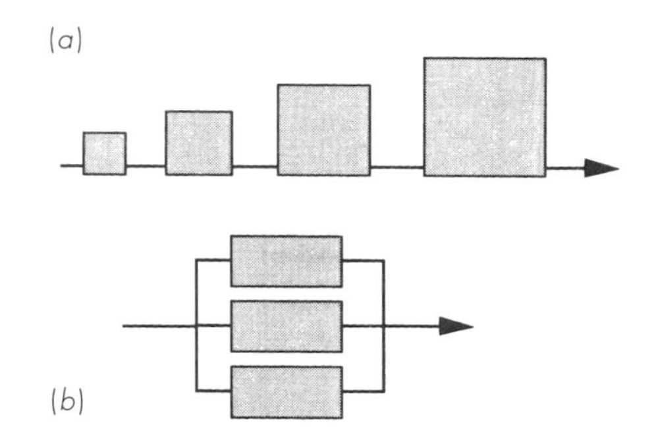

Figure 17.10
In a public agency, implementing GIS may change work processing from series (*a*) to parallel (*b*) form.

in turn, affect the overall organization (Figure 17.11). In practice, altering an organization may prove difficult because of the formal and informal positions that always exist in chains of command. Changing the organization changes relationships between personnel. Staff changes always bring in human factors that are difficult to predict or control.

Consequently, organizational matters are vital in all initial implementations of GIS facilities; the problems are often more complex and more crucial to success than are the technical problems involved. As a rule, the latter can be solved in a straightforward manner, by purchasing and installing new equipment, new software modules, and so on. Purchasing incurs costs, of course, but in a well-planned project these are anticipated. Changing and replacing staff members are less straightforward and may trigger unanticipated problems. Hence organizational matters usually require more continuous management attention than do technical problems.

Consequently, one goal of any inception study should be to delineate alternative future organization models and to recommend ways

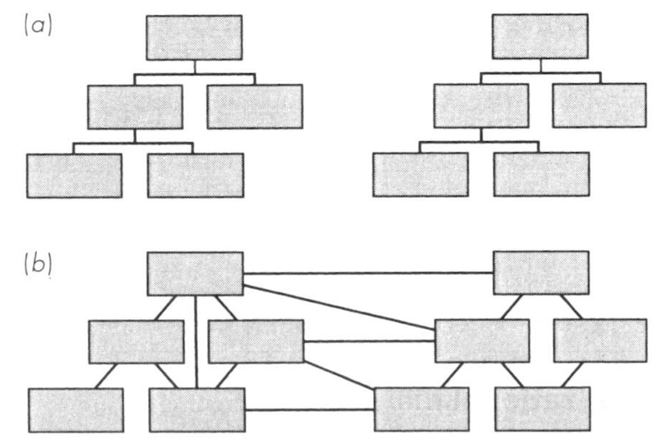

Figure 17.11
Transition from a hierarchical organization (*a*) to a network organization (*b*).

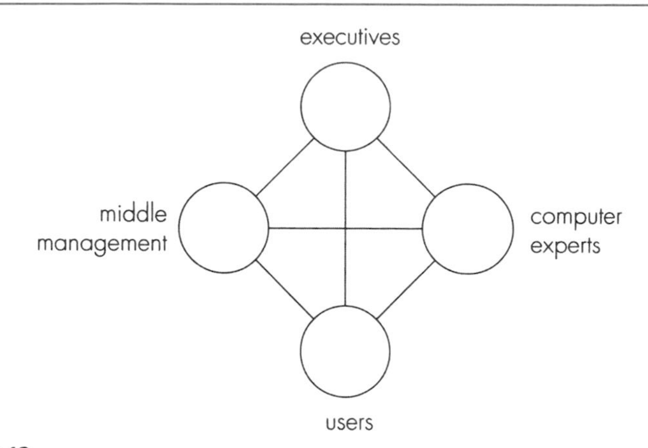

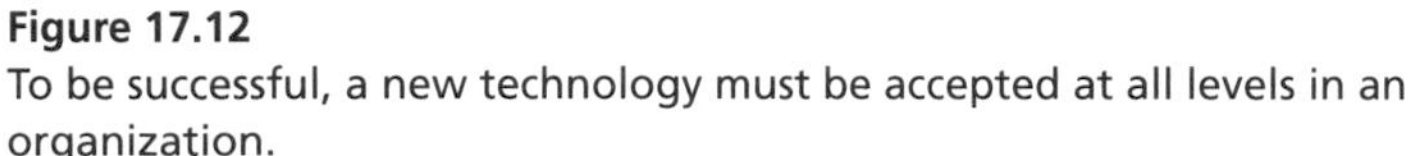
Figure 17.12
To be successful, a new technology must be accepted at all levels in an organization.

of testing their validity. New staff positions must be described in detail. In principle, new tasks and new data flows may be described independently of the persons, groups, or departments ultimately responsible. So new staff structures must be defined and management modified accordingly.

Availability of information is vital to a healthy GIS facility (Figure 17.12). Whenever information availability is restricted, GIS utility suffers. So the common tendency of an agency or department to monopolize its own information is one of the major foes of successful GIS. Consequently, one of the problems encountered in implementing a GIS facility may involve combating the bureaucracy that blocks information flow.

Other subjective factors can further complicate the initiation of a GIS facility. Human nature apparently dictates that about a quarter of the personnel in any organization always prefer the status quo and will oppose any change whatsoever. Except in high-tech firms, executives are often indifferent to newer information technologies, partly out of ignorance and partly from being overly concerned with cutting costs rather than increasing benefits. However, as executive support is often crucial to the success of a GIS project, initial presentations may be received more readily if they contain clear executive summaries and concise examples of the utility of GIS in sectors known to be of interest to the executives concerned. Successful projects are often conducted enthusiastically by middle management, who often seek the benefits of new technologies while resisting extensive organizational changes.

Specialists are essential in the execution of a GIS project but are seldom capable of addressing the myriad detail involved. For example, computer experts are usually so engrossed in their disciplines that they lack the broader view needed for dealing with organizational

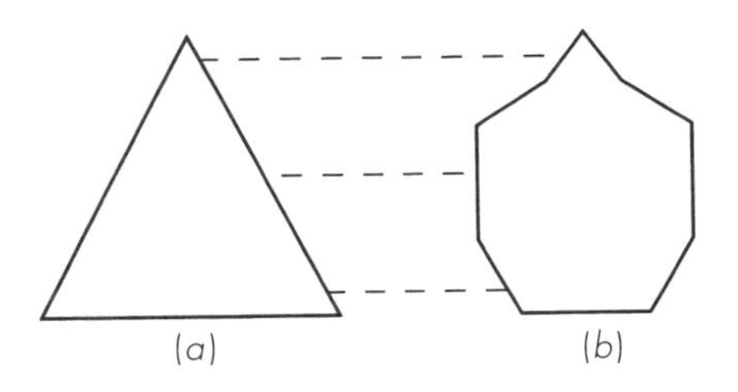

Figure 17.13
The introduction of a new technology in an organization often changes its power pyramid from the traditional triangle (*a*) to a barrel profile (*b*), in which fewer employees are involved in routine tasks and middle managers assume more comprehensive roles, both upward and downward from their location in the traditional pyramid.

problems. Operators and other direct users are similarly constrained within the framework of their jobs. As shown in Figure 17.13, the introduction of a new technology in an organization often changes its power pyramid.

Project organization

Strong, independent project organization is essential to the success of any major GIS implementation (Figure 17.14). The project need not be permanent, but can be disbanded when the GIS facility becomes operational. GIS, computer, and other experts are usually more readily hired for a project than trained for the tasks involved. All projects need committed staff and problem solvers, who need not be specialists and who can advantageously be recruited from among the users of the information products involved.

Organization and staffing of the project will depend on its complexity and size. In larger projects, the project manager should answer to both a control group and a technical working group. The control

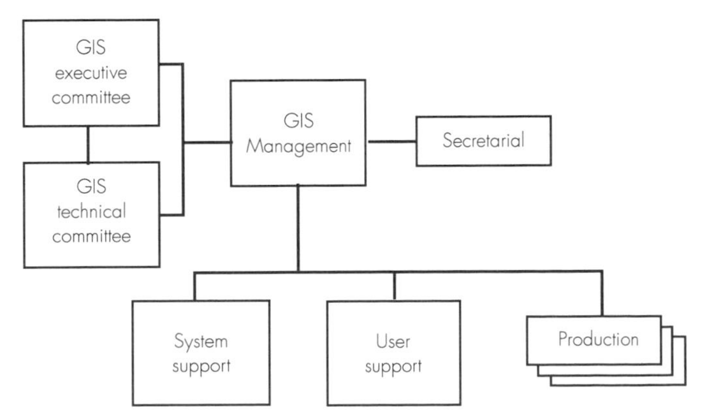

Figure 17.14
Organization of a GIS project.

group should lay down political guidelines and other management support (e.g., by solving disagreements that cannot be solved in other parts of the project organization). It should be staffed by managers from the various departments that are involved in the GIS. The technical working group should make the important technical decisions and be the driving force behind design and development. This group can be staffed by middle managers, who might be the same people who are leaders for the various groups carrying out the practical work. The project manager should ensure that clear aims for the work are established at all times. He or she should also be responsible for the planning (technical and financial), performance, and follow-up of these aims. The practical performance, such as data structure and data conversion, can be organized in individual departments. Once the project has reached the operational phase, a separate group could be established for end users, to exchange experiences and define new needs.

Positioning the project in the organizational structure

There can often be several alternatives for the formal positioning of a GIS project in an existing organizational structure. It can be placed directly in the line organization (e.g., in the relevant department, such as planning or water and sewage). One choice for a municipal project is to locate it either in the city manager's office or in a support department, such as IT. Many multiuser GIS projects in the United States have, in practice, been located within a department in the line organization (Somers 1994b). Even though this can create problems because of a perceived lack of neutrality, this can be counterbalanced by the advantages of clear GIS needs and good accessibility to technical support and budget, thereby allowing speedy establishment within the main organization.

A more radical solution can be to establish an independent consortium for operating a joint GIS. The consortium can include participants from both the public and private sectors. Such a solution will often have a high-risk political and financial profile, but it has been carried out with success in such places as San Diego, California (Dangermond 1997).

17.8.8 Personnel and training

There will be a need for a series of specialist functions in a larger GIS project. These can include project management, system support, database and network support, user support, and production. In practice, GIS implementation projects are often understaffed. Therefore staff requirements should be carefully assessed at the outset and the staff identified should be made available to the project. This usually requires either that staff members be relieved of their customary tasks and are assigned to the project, or that personnel be hired specifically for it.

A personnel management program for transition to a new technology should include:

- Identification of factors affecting personnel
- A plan for training, retraining, and relocation
- Concise job descriptions, including responsibilities
- Employee involvement in planning
- Clear definitions of positions and salaries
- A plan for job rotation
- Delineation of the simplest possible organizational structure, including:
  - Management authority
  - Delegation of responsibility
  - Division of labor
- Clear communication to and from executives between and within individual organizations
- Retention of the organization's advantages and strengthening of:
  - Teamwork
  - Goal orientation and adaptability
  - Identification with the project
- Adapting staff proficiencies and size to the tasks involved
- Creation of an environment of challenging work and career advancement

Many GIS projects, particularly in the public sector, suffer from lack of training of both operators and users. Personnel management for the transition to a new technology should thus be given top priority. What type and level of expertise are to be attained? This depends on the level of the project at the outset and on how it will be staffed, as well as on whether external services are to be involved. Ideally, all staff members who will be in contact with GIS and/or its products, either directly or indirectly, need suitable training. This also applies to executives, who should have a general understanding of both the potential and the limitations of GIS as a tool in decision making. Middle managers should have an adequate technical background to enable them to coordinate the implementation of GIS. Field staff and others who acquire data should be trained in its conversion and updating.

Most organizations elect to train their own personnel in technology conversion instead of hiring outside expertise. But who trains the trainer? The benefits of a system can be fully realized only if all functionaries, planners, sales, and other personnel are adequately prepared. Training should comprise on-the-job experience as well as formal education. External consultants are often most beneficial in the strategic sectors of a project, as they are not involved in the organization and can frequently resolve difficulties and disputes more easily than can members of the organization's staff. The need for expertise is frequently underestimated. Without the requisite skills, progress can be frustratingly slow; consequently, maintaining and building staff expertise should be a top-priority task.

We may summarize the most important organizational aspects of implementing GIS as follows:

- Executive involvement is crucial to success.
- Organizational problems are usually greater than technical problems.
- Organizing or reorganizing should hinder monopolizing of information.
- The introduction of GIS effects changes in existing routines for information interchange between and within individual units of an organization.
- Altered work routines mandate organizational changes.
- At least one-fourth of the personnel of an organization can be expected to oppose change.
- There should be both a technical troubleshooter and an enthusiast who can market the project.
- Operator cooperation must be enlisted.
- The initial stages of implementing a GIS facility may be project oriented; organizational alterations should be tested before being finally enacted.
- Long-term organizational changes may need to be made after the initial operational phase of a new GIS facility.
- Successful GIS projects have a plan.

17.8.9 Budgeting and financing

Any GIS implementation project should have a budget that allows for distribution of activities over the duration of the project and also permits monitoring. Projects based on meticulous cost–benefit analyses are usually the healthiest. In many cases, there is not a single project, as several organizations may be involved in establishing a joint GIS facility. In such cases, costs are usually apportioned among the organizations according to the benefits accrued. However, the cost of facilities used by many organizations may be met in other ways, such as charging annual fees, assessing an initial outlay plus an annual fee, paying according to ability, and so on.

Government agencies must often decide whether to finance a new facility through their normal fiscal allocations or by charging user fees at market rates. As discussed later in the chapter, the problem is central to the organization of a national database.

17.9 Developing a Logical Data Model

In the computer sciences, data modeling is considered to be part of the development of a data processing system. This view is common but not always correct. The development of a logical data model should not be linked to any particular technology, but should be devised by professionals who are expert in the field of application involved. It is therefore most natural to place this activity under organizational development.

Data models were discussed in detail in Chapters 3, 4, and 5. To summarize: Data models comprise objects, specified by five parameters: object type, geometry, attributes, relations, and quality. These parameters should be described in a data dictionary. The procedure for data structuring includes several topics discussed in Chapter 12:

1. Defining applications
2. Stipulating constraints, tasks, and requirements
3. Ranking according to benefit and cost
4. Descriptions of attributes
5. Coordination of conditions and definitions
6. Choice of geometric representation: raster or vector, points, lines, and surfaces
7. Descriptions of relations between objects
8. Quality requirements based on cost–benefit analysis: geometric accuracy, updating frequency, completeness and resolution, geographic coverage, logical consistency
9. Coding of data with respect to identity, object types, and attribute values

A good data model is part of the specification of a data processing system, not the result of it.

17.10 Alternative Introduction Strategy

As we have seen in Section 17.1, GIS facilities may be divided into two groups, according to whether or not they affect an organization overall. There are, of course, many borderline cases, but in general, many smaller organizations, such as small towns, utilities, or agencies, will implement smaller GIS facilities for special and often internal purposes. From the organizational viewpoint, such facilities may be regarded as ancillaries.

Nonetheless, the guidelines for assessing, planning, and implementing such an ancillary GIS facility are similar to those discussed for major facilities. The differences involved are those of scale; the tasks involved are usually more easily defined; the economic risk involved is usually more modest; and data flow is more readily delineated. Finally, since the GIS facility is ancillary to the activities of the organization, the organization is able to adapt to it more rapidly.

Many ancillary GIS facilities are project oriented. Typical projects include digital terrain models used in planning residential developments, automation of township property registers, and so on. In most cases, the applications are clearly defined and the design challenge reduced to finding the most suitable system for the tasks involved. In practice, the system choices are often between the various PC systems available.

The introduction of GIS into this kind of activity will often produce fewer and simpler approaches to the problem, thereby allowing for a more homogeneous and schematic strategy. This has also proved to be the case in more complex GIS projects. Today it is often the case that a number of organizations acquire GIS first and define user needs afterward.

Simplified GIS strategy

A simplified strategy can popularly be described by an elementary "four S" guideline:

- Small
- Sure
- Seen
- Success

The initial system should be *small* to minimize financial risk and to speed familiarization with the new technology.

The organization should be *sure* that the tasks identified are clearly defined, necessary, and sufficiently limited that they may easily be realized.

The results should be *seen* soon, both to justify investment and to encourage the staff involved.

When these three requirements have been met, a measure of *success* will be at hand, relatively quickly and for a minimum investment.

17.11 Creating National Geographic Databases and Developing New Business Sectors

Creating a national database raises numerous strategic questions, such as public versus private financing, product organization, data content and quality, and so on. Many countries have addressed these questions and have created national databases of a variety of data, including:

- Property registers
- Road data

- Environmental data
- Elevation data
- Statistical data
- Addresses
- Building data

The creation of a national database may be compared to the development of a new market—the guidelines are equally valid in the public and private sectors.

Principles of evolving strategies

All strategies start with the initial decision of whether to maintain the status quo (the *zero alternative*) or to select among alternatives to create a national database. Of course, the status quo decision requires *no* strategy. Strategies afford the means of selecting among alternative courses of action.

Strategies are best based on cost–benefit analyses, which rely on both costs and benefits being expressed in monetary units. Quantifying the costs and benefits of various courses of action is perhaps the greatest challenge of the entire strategic decision process. Strategies must also include considerations of organizational factors, such as production, storage, distribution, and updating.

All planning presupposes some experience with the subject. However, as a national geographic database is often the first of its kind in a country, the relevant experience is usually sparse. Consequently, the evolution of a strategy usually includes activities dedicated to defining the issues at hand. In general, a strategic plan may evolve in five phases:

1. Identification of current status, or zero alternative: the first user study:
   - Among users
   - Among providers (including own organization)
2. Elaboration of a product concept defining:
   - Offerings that data users consider vital
   - Characteristics of data offered: accuracy, timeliness, etc.
   - Initial matrix of products offered, as in three categories of increasingly finer service
3. Formation of a cost model:
   - That simulates costs of the various product alternatives
   - That ascertains product cost chains
4. Delineation and conduct of market analyses: the second user study, which aims to identify:
   - User ranking of products and inclinations to pay for products in the various application groups
   - Size and potential of the sundry market segments, including factors that may detract from market development

5. Devising a model, based on the cost and market analyses, that promotes identification of an optimum product concept

The first step in the market analysis is to detail how users currently accomplish tasks without the benefit of the proposed product. Ideally, the users surveyed should have some experience in using their own digital data at the national level. Business criteria should be used to analyze the material compiled in the first user study. That is, cost–benefit analyses are conducted using net present values of incomes and costs. Knowledge of the behavior of current markets permits an estimate of future activities. In other words, products may be specified in terms of data content, accuracy, completeness, timeliness, distribution, costs, operation, financing, pricing, and so on.

A structured attack is used to specify products suited to identified market needs. The aspects of various products may be conveniently ordered in a matrix (Figure 17.15). A cost–benefit analysis presupposes that a cost is associated with each item in a product matrix. A realistic cost model should relate to a basic product. The model may then be used to compute the cost of the basic alternative of the product and to compute the marginal costs of the other alternatives.

The goal of the second user study is to measure the reactions of potential users to the various products proposed. The users involved in the first user study should be interviewed again. The user study and the supplementary treatment of the material should be carried out in a structured way. For example, a potential user might range the differ-

	LOW SERVICE	MEDIUM SERVICE	HIGH SERVICE
A. GEOMETRY	**Inter-regional** All public roads No private roads	**Intra-regional** All public roads All forest roads	**All-to-all** All public roads All private road
B. ATTRIBUTES	**Road identity** Road class Road number Public ferries	**Cruising advice** Speed limits Critical slopes Weight/height etc.	**Advanced advice** Addresses Traffic lights Zip code etc.
C. GEOMETRIC ACCURACY	**Low precision** +/- 50 m x,y	**Medium precision** +/- 10 m x,y	**High precision** +/- 2 m x,y,z
D. CURRENTNESS	**Revised every** 5 years	**Yearly** Revised annually	**Continuous** 1 - 3 months
E. CORRECTNESS * OF ATTRIBUTES	**>95% of all data**	**>99% within 1 yr**	**>99% right away**
F. USER ACCESS	**Magnetic medium** national format	**On-line access** without GIS-supp.	**Full GIS support** on-line access
G. DELIVERY TIME TABLE	**Normal cycle** within 8 years	**Speeded** within 3 years	**Crash program** within 1 year
H. PRICE	Low	Medium	High

Figure 17.15
Typical product matrix for a national road database.

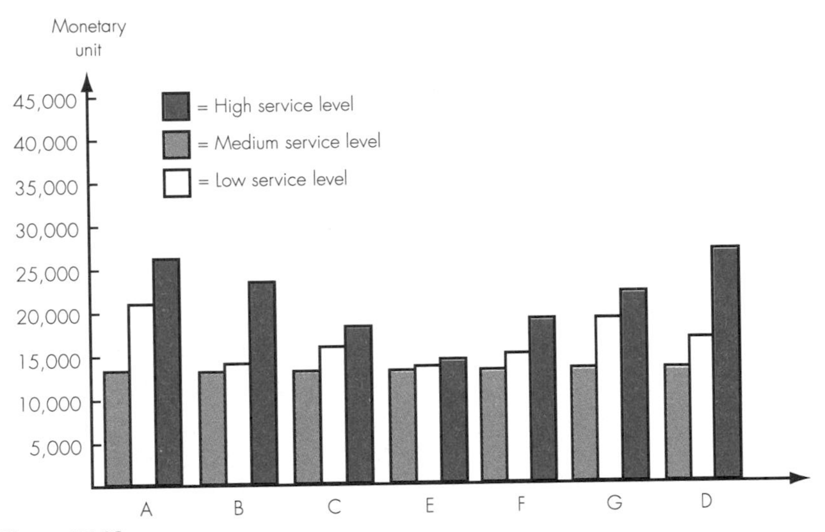

Figure 17.16
Effects on cost of a partial increase in attribute support level (see Figure 17.15).

ent alternatives according to a specific points table. This awarding of points can also be used in grouping the material.

Delineating strategies

Depending on production organization and the market, public agencies, which often operate national databases, may choose among various business approaches, such as:

1. Supplement basic data with other standard data and distribute the combination as a standard package to all users (extended map production).
2. Supplement basic data with other standard data to compile tailormade user products.
3. Supplement basic data with user-specified data to compile tailormade user products.
4. Coordinate data interchange without being involved in producing but with responsibility for standardization.

17.12 Strategy for a National Map Service

Recommended strategies for national road databases in Scandinavia include the following necessary conditions:

1. *General strategy.* The national map services should adopt an offensive strategy to adapt to rapidly changing technologies and markets.

2. *Creating databases.* National road databases created with data from various sources should be aimed toward a small nucleus group of users willing to pay for the services provided. This will make the project feasible. This product cannot be classified as a public service, as many financially weaker groups will be excluded.
3. *Currentness.* Currentness is paramount, that is, updating every three months for important items and at least once a year for all other items.
4. *Financing.* A multitheme road database should be user financed. As the market demand is so varied, the product is not a characteristic public service. Semipublic financing may gradually increase during the updating phase as products become standardized and more users access the data. The portion of the road database dedicated to rationalization of national map production can be financed by normal public funding.
5. *Pricing.* Market pricing should involve a high entry price to cover updating costs. Otherwise, the road data will be inadequately financed, which will create problems in the updating phase and pricing problems for both the initial group and for subsequent users. In the long run, the most called for data will undoubtedly become a public responsibility, a fact that may influence marginal costing in pricing.
6. *Organization.* The national map services operate both as administrators of public map policies and as producers of products for the commercial market. Consequently, the map services should be divided into two units. The basic, public service unit will administer the geometrical information (infrastructure). A more commercial unit will compete in the private market and should focus on value-adding activities.
7. *Access to basic data.* All private and public organizations should have equal access to basic data that are produced in connection with national map production. In this way, provider and user interests balance, and various interests may compete on an equal footing.

17.13 Introduction of GIS in Developing Countries

Even though GIS technologies are well suited to solving the information problems of developing countries, the high-tech approach may involve particular, nonquantifiable problems, such as:

- Political circumstances
- Cultural traditions
- Management conditions
- Lack of expertise

- Inadequate infrastructure
- Extreme climates
- Insufficient financial resources

This means that special considerations often have to be made in connection with the introduction of GIS in developing countries. In such countries, the primary users of geographic information are frequently unfamiliar with GIS and its benefits. Therefore GIS implementations should start with pilot projects that aim both to provide experience and to rapidly meet short-term, high-benefit information needs.

Without an implementation strategy, projects in developing countries often become technology driven, and technology takes the upper hand, ahead of the tasks it is to perform. The result may be an unprofitable overinvestment in hardware and software at the expense of primary tasks, which remain undone.

In developing countries, the challenges involved in introducing GIS are often institutional and organizational rather than purely technical. Orchestrating cooperation between providers and users so that the GIS facility fulfills its purpose may be more demanding than commissioning the system itself.

17.13.1 Building capabilities in developing countries

The International Institute for Aerospace Survey and Earth Sciences, formerly the International Training Centre (ITC) of the Netherlands, which is experienced in training programs in developing countries, recommends that for GIS technologies to succeed, minimum personnel training should comprise the following groups (van Genderen 1991):

- Decision makers and planners, including officials and administrators, who need a general understanding of the practical possibilities and limitations of GIS as a decision-making tool
- Leading personnel in institutions, public management agencies, and private companies, who need sufficient technical knowledge to coordinate the introduction of GIS
- Personnel involved in the practical surveying and acquisition of digital data used in GIS
- Technicians responsible for the operation and maintenance of equipment and programs
- Research workers knowledgeable in GIS and with expertise in applications development and in GIS as an analytical tool
- Instructors knowledgeable in GIS technologies and their practical applications who are responsible for training and teaching the various categories of personnel
- School and university students

As a rule of thumb, when GIS is introduced in an established organization, one-tenth of the staff should go through primary training. When new organizations are established, about 30% of the staff should be trained in GIS per year.

Training may be conducted either by sending personnel to an industrialized country or by sending experts to a developing country. Training in an industrialized country is often expensive and therefore limited in the number of personnel it can accommodate. However, it is often the most effective, as relevant equipment and well-qualified instructors are often available.

Training in the uses of GIS in natural resource management is now offered by several UN agencies and programs, including the Global Environment Monitoring System (GEMS), United Nations Institute for Training and Research (UNITAR), the United Nations Environment Programme (UNEP), the Global Resource Information Database (GRID), and the Food and Agricultural Organization of the United Nations (FAO).

The International Institute for Aerospace Survey and Earth Sciences and many universities also offer courses, as do regional and national training centers such as the Asian Institute of Technology (AIT) in Bangkok.

CHAPTER 18

Choosing a GIS: Technical Issues

We have already seen in Chapter 17 that the issues involved in initiating GIS may be classified as either organizational or technological. In this chapter we deal with the technical issues.

The development of a GIS system pertains to all aspects of the computerization and data processing involved, and may include:

- Pilot project
- Design requirements
- Choice of hardware and software
- System implementation
- Database design
- Creating databases
- Operation and maintenance
- Evolving new applications

A GIS system is usually developed partly in parallel with the relevant organizational activities. For example, design requirements may be compiled when the organizational strategy and future data flow have been defined. The choice of a system or supplier may be made after the benefit/cost ratios have been finalized. Work routines may be delineated as the system nears completion.

18.1 Pilot Project

A pilot project entails an experimental small-scale implementation in which techniques, equipment, and subsystems for use in a full-scale GIS are tested in advance. In respect of system development, pilot projects should:

- Provide a better basis for choice of system(s)
- Test various production methods
- Identify system faults

The scope of a pilot project depends on the complexity of the organization(s) implementing the GIS facility. From a practical viewpoint, a pilot project should be budgeted in both time and money. The equipment and subsystems used should be as simple as possible, both to minimize loss should the chosen approach prove unsuitable and to ease the requirements for staff training and maintenance. As a rule, pilot projects should not take longer than one to two years. Staff involvement in a pilot project should be restricted to those persons most concerned with the GIS facility. Although widespread familiarity with GIS technologies and with the pilot project is advantageous, project progress may be complicated if an excessive number of conflicting interests are involved.

Normally, one or more limited geographic test areas should be chosen. Ideally, the test areas should be dissimilar: for example, one of high complexity and one of relatively simple content. The basic goals of a pilot project—to test approaches and prove viability—dictate maximum use of the experimental system(s) involved. Staff should be made available accordingly, lest the pilot project be delayed by lack of personnel.

Pilot project outcome and the decision to change technologies

If the outcome of a pilot project countermands GIS implementation, either the pilot phase may be extended, different approaches may be taken, or the GIS project may be abandoned completely. Before any of these steps are taken, the reasons underlying the negative outcome should be examined. As is often the case in the assessment of new technologies, a GIS pilot project can flounder if initial expectations are unduly high. Whenever suppositions of technical, financial, or organizational benefit outstrip realistic goals, disappointment invariably results. These effects can scuttle even the best of projects. Hence, workable goals at the outset of a pilot project are to be recommended.

If the outcome of a pilot project supports GIS implementation, work should start as soon as possible, particularly in choosing a system and starting data conversion activities. With a rapid decision and no dead time, the wave of enthusiasm from the pilot project can carry over into the implementation phase.

18.2 Choosing Hardware and Software for GIS

The right choice of computer hardware and software is essential. In practice, this may prove both easy and difficult. The choice is easy because it deals with comprehensible details, such as technical characteristics and prices which may be compared. Yet it is also difficult, because future applications are unknown and computer technologies change continuously. The impact of implementing GIS in an organization is uncertain and pressure from aggressively competing system vendors also complicates choice.

Part of GIS facility	Average useful lifetime (years)	Relative cost (%)
Computer equipment	2 to 5	10
Software programs	3 to 6	20
Data	15 to 50 or more	70

Figure 18.1
Longevity of a typical GIS facility.

As in other disciplines, computerization efforts in the geographical disciplines often overemphasize equipment and software. Regrettably, data, the greatest resource and value of a GIS facility, often seems of lesser interest. The fallacy of focusing system concern on hardware and software to the exclusion of data is illustrated clearly by the longevity of a typical GIS facility, shown in Figure 18.1. The shorter expected useful lifetime of computer hardware and software combined with the lesser investment involved—from 10 to 30% of the aggregate cost of a GIS implementation project—indicate that the major concern is best directed elsewhere, to the data, which are the backbone of an operating GIS.

18.2.1 Purchasing rules

Major purchasing of hardware and software may be subject to company rules or guidelines, as in larger corporations or in the public sector. Today, international economic agreements* also govern public-sector purchasing. Rules are often linked to stipulated cost levels, so that a rule comes into effect when total cost exceeds a set amount.

Specific cost limits apply for different product sectors in EEA, such as goods and service, building and labor supplies and communication, and so on. At present the limit for public-sector procurement at the municipal level for IT goods and services is ECU 200,000. The corresponding limit for central government procurement is lower, ECU 140,000. Procurement notices have to be published throughout the EEA for purchases in excess of these limits. It is worth noting that these limits apply to the total contract value and to a certain extent, to total purchases within one product sector. Regulations are quite strict in regard to splitting the procurement into several contracts. In the case of a contract that extends over a number of years and involves several subcontracts, individual procurement notices may be required for each supply. Familiarity with these rules ensures that purchasing proceeds without undue delay.

* Examples include the World Trade Organization's (WTO, former GATT) Public Government Procurement Agreement of April 15, 1994 and the regulations agreed within the European Economic Area (EEA) by Directives of Public Procurement, no. 93–96 of June 14, 1993.

18.2.2 Organization of the selection and implementation process

At this stage, responsibility for the implementation phase of the project may be defined. An organization should clearly define its own role before requests for proposals (RFPs) are sent to vendors. The assignment of responsibility for the selection and implementation process normally varies considerably, depending on the competence of the organization involved and the nature of the market supplying the goods and services. At this stage, an organization elects one of the following strategies:

1. *Retains system responsibility.* The organization, or its outside consultant, designs the system in detail and issues specifications for equipment, subsystems, and the like. The organization itself is then ultimately responsible for the system(s) being able to fulfill the design requirements.
2. *Assigns system responsibility.* The organization issues its design requirements and invites suppliers to respond with bids that include the detailed specifications of how the requirements are to be met. The supplier(s) selected carry the ultimate responsibility for the system(s) being able to fulfill the design requirements.

There are many variations of these two main approaches. An overall system may be split up into subsystems, some with supplier subsystem responsibility and some without. The basic decision as to who performs detail design and compiles specifications and who assumes system or subsystem responsibility invariably involves trade-offs between in-house and external activities.

18.2.3 Design requirements and system specifications

The preprocurement requirements (see Section 17.6) pertaining to the various tasks, routines, sources, models, components, and subsystems comprising a GIS facility are usually incorporated in the specifications when invitations to bid are issued to selected suppliers. In most cases, these design requirements reflect the results of user surveys, cost–benefit analyses, and strategic planning, and involve compromises in respect of what is currently available on the market. Design requirements may be compiled for:

- Organization(s) evolved for the new facility
- Tasks automated, including possible adjustment to existing systems
- Operations conducted
- Products produced
- Data employed
  - Type
  - Amount
  - Quality

- Data sources
- Data model(s)
- Time horizon
- Training
- Standards and their validity
- Documentation
- Data processing infrastructures already in place

Hence the design requirements of a system stipulate how it is to function, *not* how it is to be implemented. In turn, specifications are compiled for the equipment, subsystems, routines, and other specifics of implementation that provide the various functions. Today, most organizations have software, equipment and digital data that have to be adapted to GIS. This can form the framework for the type of technological solutions to be selected and should be clearly outlined in the specification.

The final requirement specifications are usually included in an RFP, which aims to provide potential vendors with adequate information for bidding. A comprehensive RFP is the basis for realistic and reliable bids. As we have already noted, the design requirements of a system stipulate how it should function, not how it should be implemented. In most cases, the design requirements are based on an appraisal of the current situation, the results of user surveys, cost–benefit analyses, strategic planning, a pilot project, existing technologies, and compromises based on market choices. Requirement specifications fulfill several functions. While a requirement specification tells bidders what kind of applications a customer wants, it also acts to enhance in-house understanding of the system to be purchased. It serves as the gauge against which compliance to contract is measured.

Normally, requirements vary in their importance, so they can be grouped according to:

- Requirements that must be fulfilled
- Requirements that may be fulfilled
- Future requirements that must or may be fulfilled when the technology becomes available
- Requirements that are secondary or optional, on which the vendor may bid

Cost–benefit analyses can be used to rank the importance of requirements (see Section 17.7). Design requirements can conveniently be divided into main requirements and special requirements. Both the reasons and goals for implementing GIS should be stated. The tasks to be addressed in the organization and the future functionality envisioned for GIS should be identified, such as in stipulating whether the system should process vector data, raster data, or both. Data security may also be a factor to be included.

The functional requirements vary from organization to organization. Typical requirements include:

- Descriptions of different applications
- Data conversion that depends on the types of data processed, the technologies employed in data conversion (scanner, digitizer), and the volume of data to be converted
- Functions for importing and exporting data, including formats
- Database designs and storage functions, which depend on factors such as the data models to be preserved and the data volumes to be stored
- Printout and presentation of data, such as to meet specified standards

Another example is that of search functions, their search criteria, and search speeds. Analytical functions are often needed for overlays and in network analyses. Many organizations already have systems (client lists, property registers, pipe network descriptions, operation and maintenance systems, etc.) which should be integrated with GIS. In such cases, the degree of integration must be described. However, in practice, experience indicates that overly ambitious integration plans may well degrade a GIS project that is otherwise feasible. The functions of standard operations, such as customer counter services and printouts, may also be described. The data used by all subsystems will need to be updated, so updating functions must be described. A facility may entail special functions, such as for geometrical design and for the computations of lengths, areas, volumes, and other parameters. In all cases, there should be a clear statement as to how vendors should respond to requirements that their systems do not meet but which they wish to develop.

Of course, most organizations considering implementing or updating GIS already have more or less advanced computer system(s). Therefore, the existing computer systems must be described to ensure compatibility of the new equipment and software. The descriptive parameters should include type and capacity of equipment (storage, processing speed, number of workstations, etc.), type of operating system, and other relevant software.

Many organizations have a general information technology strategy that will govern a new GIS installation. The factors involved include operating system, hardware platform, network, and so on. In addition to the general information technology strategy, special requirements may be stated for a GIS facility, such as for user interfaces and response times. The number of workstations and numbers and types of peripherals incorporated in the GIS facility must be stated. The chief parameters of the system configuration and attendant network design must be listed. Normally, the details of a new system are not described in great detail, as technology changes so rapidly. Vendors should be encouraged to offer new technologies and to describe them in terms of the benefits they have for the applications involved.

The system specifications, which are based on the design requirements, include stipulations of:

- System solution
  - Computer (capacity and speed)
  - Peripheral devices (number and types)
  - Communication
- Software
  - Functionalism
  - User friendliness
  - Accessory modules
  - Customizing
- Maintenance
- Training
- Costs

There is no single ideal set of specifications. Organizations combine specifications in varying degrees to suit their individual needs. For example, organizations with ongoing products and administrative tasks, such as utilities, national map services, and municipal authorities, usually emphasize stability and service, and choose systems accordingly. Agencies responsible for planning and resource management may favor greater flexibility and diversified applications. Schools and universities usually favor inexpensive elementary systems.

18.2.4 Request for proposal

A well-prepared RFP, based on design requirements, is a prerequisite for receiving meaningful proposals. Evaluation of responses can be simplified by compiling part of the invitation to bid as a list of questions to which bidders respond. This is a form of compliance list, in which the response to a question is "comply" or "not comply."

An RFP, or invitation to bid (ITB), should include a description of how a bid should be structured and which criteria will be used in assessing the bid. For example, the importance of fulfilling the functional and technical requirements in relation to price, the size and economic stability of the bidder, the support and training, and the documentation offered may be described.

The manner in which queries from vendors and other relevant information are handled during bidding should be described. For example, responses may be written to all queries, with copies to all bidders. An ITB should be sent to no more than three or four selected vendors, who are expected to be able to meet the requirements of the system as well as the need for future support.

18.2.5 System evaluation

System proposals should be evaluated in terms of how closely the bid meets the requirements stated in the RFP and on the quality of the vendor's bidding. Often, requirements may be met not by a single system but by a combination of several systems. Choosing a new system that will provide integration of the new system(s) into an existing system can complicate the situation. Criteria for determining the final choice of vendors usually includes service capability, market position, company stability, price policy, references, and so on. At the end of the day, it is both the system and its vendor which are chosen.

Several objective parameters may be drawn upon to simplify choices and aid decisions. Requirements should reflect needs, not only system parameters. Systems may be ranked numerically by assigning weighted values to all selection criteria (Figure 18.2). The weighted sums for all systems can then be ranked.

Functions not available but offered as developable must be assessed separately. The fulfillment of functional requirements may conveniently be assessed separately, before the bid prices are compared. This prevents unconscious favoring of the system bid at the lowest price. Normally, no one system bid is superior in all aspects. Consequently (and as stated in Aronoff 1989), the final choice often involves compromises.

Benchmark testing

The performance of a system *as installed* is crucial to its payoff potential; its performance at some future date is of less immediate interest. Performance may be assessed in terms of how a system executes specific functions, how rapidly it works, how simple it is for users, and

	weight	rank				weighted rank			
Selection criterion		a	b	c	d	a	b	c	d
System cost	5	7	7	6	9	35	35	30	45
Database design	4	6	7	8	7	24	28	32	28
Human-machine interface	5	8	6	7	6	40	30	35	30
Equipment functionality	3	7	8	7	7	21	24	21	21
Own programming	3	7	5	5	6	21	15	15	18
Follow-up	4	7	6	6	8	28	24	24	32
Data interchange	5	8	7	6	5	40	35	30	35
Expansion possibilities	4	7	8	8	8	28	32	32	32
Documentation	4	7	6	6	5	28	24	24	20
Maintenance costs	4	9	5	6	6	36	20	24	24
Drawing functions	2	7	9	9	6	14	18	18	12
Unweighted sum		80	74	74	73				
Weighted sum		315	225	255	287				
Rank		1	4	3	2				

Figure 18.2
Typical comparison of four GISs (a, b, c, d), using an initial ranking of 1 to 10 and weightings of 1 to 5.

how flexible it is in current applications. Such questions are best answered by testing.

Different systems are best compared by running specific test programs using uniform test data. This benchmark approach, or functional test, is valuable in assessing various systems and approaches. Each system should be tested intensively over several days.

Parameters to be considered

The parameters to be considered in selecting and installing equipment include:

1. Installation details
   - Floor and other structural loads
   - Furnishings
   - Main power supply
   - Ambient temperature and humidity
   - Noise environment
2. Computer characteristics
   - Computational capacity
   - Computational accuracy
   - Communication
     - Local area network
     - Data communications protocols
     - Speeds
   - Number of peripherals supported
     - Graphic terminals
     - Alphanumeric terminals
     - Printers
     - Plotters
     - Digitizers
   - Expansions accommodated
   - Compatibility

In practice, equipment use is often impeded initially by computing capacity and the number of input/output interfaces available. The more important characteristics of the peripheral units include:

1. Graphic displays
   - Screen size
   - Image quality
   - Colors
   - National character sets
   - Ergonomics
2. Plotters
   - Accuracy
   - Resolution (raster plotters)
   - Output format
   - Plotting speed

- Plotting medium
- Plotting tools (pens, etc.)

3. Digitizers
   - Accuracy
   - Resolution
   - Size

The characteristics of tape stations, internal storage facilities, disk drives, alphanumeric terminals, printers, local area networks supported, data communications protocols available, and so on, should be listed and evaluated in a similar manner.

The software requirements should include:

- Structure of basic system and application programs
- System openness to support user-defined application programs, communications programs, etc.
- Macroprogramming capability
- User instructions and system support documentation
- Freedom to change databases
- Facilities for restricted access
- Operational records and user accounting
- Tools for restructuring databases
- Selection of languages available for entry of error messages and commands
- Programming language
- Support of data interchange with other systems
- Human–machine interface (response time, error messages, commands)
- Support and follow-up

The list above may, of course, be much longer, depending on the software application involved.

In general, the criteria involved in choosing a system depend on the nature and goals of the organization involved. A typical case in GIS applications is the implementation of a joint-use facility shown to have a benefit/cost ratio of 4:1. The guidelines for its specifications are:

- Start with a general-purpose system.
- Note that general systems generally entail more initial effort.
- Modular systems are easily expanded.
- Tailor-made systems constrain further development and limit the utility of databases.
- Tailor-made systems are rapidly obsolete and often expensive to maintain.
- General-purpose systems ease information interchange and interorganizational communication, which enhances benefit/cost ratios.

- Specialized systems can limit benefit/cost ratios when information is digitized.
- Evaluation should be as structured and as objective as possible.

From the early days of computer systems up to the explosive expansion of computerization triggered by the introduction of personal computers (PCs) in the early 1980s, systems were inevitably proprietary. Hence the choice of a system was tantamount to choosing a supplier. Vestiges of this trend remain, as many systems remain proprietary. However, internationalization and intercompany tie-ups in the computer industry, the trend toward open computer architectures, increasing interprogram compatibility, and, of course, cloning and competition have all contributed to increasing the selection of suppliers capable of delivering a suitable system or subsystem. The choice among competing suppliers usually includes considerations of:

- Capabilities and service
- Market position
- Company stability
- Price policy
- References

18.3 Contracts

A mutually acceptable, clear, and complete contract is essential and benefits both purchaser and vendor. For a complete system, two contracts should be entered into, one for purchase and one for maintenance. The purchase agreement should contain a short description of the equipment scope and the mode of payment. Normally, the delivery will have a guarantee period, so the goods and services supplied under guarantee must be specified. Joint activities and responsibilities may be described as needed. The contract should contain guidelines for confidentiality between the contracting parties and how ownership (hardware) and use rights (software) are to be transferred to the purchaser. The treatment of delivery delays must be specified and of how any breach of contract is to be handled.

A maintenance agreement should contain a description of its scope and limitations and of the services to be supplied for its duration. It should also describe the financial and payment conditions and of how any breach of contract is to be handled.

The final decision

The final decision to procure a GIS should be made only after contract negotiations have been completed and all financial and legal aspects have been clarified. Careful specification, bid evaluation, and contracting procedures, as discussed above, ensure the best decision. But only future use can determine if the right choices have been made.

When a system has been chosen

The vendor should fulfill the contract concerning:

- Installation/integration
- Customizing, database design, etc.
- Acceptance testing

Acceptance testing may be necessary to determine whether a system functions per specifications and in conformance to the contract.

18.4 Technical Database Design

A thorough database design is essential, both for realizing the functions required and for determining:

- Scope, so that only relevant data are entered
- Ease of access to data
- Efficiency with which functions are executed
- Facility with which data are updated
- Receptiveness to new types of data
- Ease of restructuring data in the database

As discussed in Chapter 12, a data model provides the basis for a database design. The detailed design of a database depends on the system software and hardware chosen. In joint-use facilities, databases must be further designed for individual organization data or for data common to all the organizations involved.

The best approach is for each organization to design databases that best suit its specific needs, and to incorporate direct or indirect interorganizational communication via a joint database (Figure 18.3). Such a distributed approach requires that distinctions be made between joint data common to all organizations and individual data belonging to the individual organizations. The various individual and joint databases are then connected by a data communications network. The physical separation between the databases and the communication among them are not obvious to users, who access the entire network. The distributed database approach also requires good updating routines and extensive standardization.

As the number of users having on-line access to a database or interconnected databases increases, the need to restrict access to certain parts of the data may also increase, for security, strategic, or maintenance reasons. Such considerations must also be included in database design.

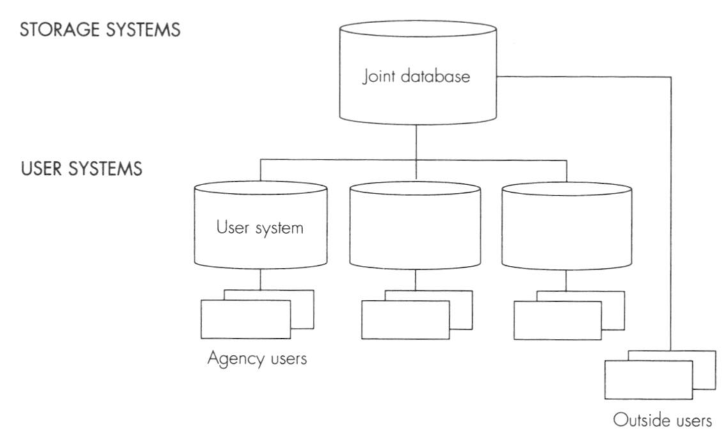

Figure 18.3
In a joint database, some information is available to all users, while parts of the total information set are stored in user databases. The joint database is usually separate and may be networked. Different types of user databases may be used. Outside users may be permitted to retrieve and present data.

Logical (software-dependent) database design

Logical database design is software dependent. Different GISs will therefore be able to offer different solutions. This is discussed in detail in Chapter 12. The principal aim is to define the most suitable storage form, such as map sheets, as well as the division of types of objects into various subject layers. This implies that this design computes only geometric data. The choice of storage form and division into thematic layers determine how effectively the data can be used. Once the final choice has been made, it can be difficult to restructure the data.

Physical (hardware-dependent) database design

The physical database design is dependent on the choice of hardware. That means that it is dependent on relationships such as storage medium, accessibility, and speed. This is discussed in detail in Chapter 12. In many cases, however, the constraints imposed by data storage facilities are more administrative than technical, as organizations implementing new systems frequently wish to employ an existing data processing infrastructure.

18.5 Creating a Database

Creating a database is costly, time consuming, and a drain on facilities and staff. Therefore, the databases employed in a GIS facility should be compiled as rapidly as possible, in order that the data they contain be accessible as soon as possible. Consequently, both private firms and public agencies often arrange data conversion and entry work in round-the-clock shifts. As they know their own data and professional fields, most organizations elect to convert and enter data in-house. As the workload involved may be enormous, outside temporary staff are often hired for the routine tasks.

Identified user needs, preferably ranked according to benefit-cost ratios, should determine the data entered in a database. Clearly, data accessed by many users have the greatest utility. Therefore, database creation should start with joint data. At the bottom of the data ranking list could be information entered on the premise that it may be useful at some time. In many cases, if its use is uncertain, it need not be entered.

The practical side of the establishment will generally comprise the following activities:

- Development of templates if needed
- Thematic map manuscript preparation
- Digitizing and editing of objects
- Attribute coding and verification

18.6 System Operation and Maintenance

Data must be maintained to retain their value. Normally, data maintenance in a major facility is considered only after the facility is fully operational, some three to seven years after the first data have been entered. However, data maintenance is vital from the very start.

Persons active in data acquisition are usually those most qualified to maintain the same data. Surveyors can both acquire and then maintain utility network data, planners can compile and then maintain zoning plans, and so on. The most direct approach is for the responsible professionals to maintain their own data by means of workstations accessing a joint database.

Maintenance routines should include deletion of duplicate entries, which are common in conventional manual systems. This implies that the various organizations accessing a joint database are obliged to depend more on each other. Consequently, all organizations involved must agree upon the maintenance routines to be used.

18.7 Safekeeping and Security Routines

In many ways, computerized information is more vulnerable than conventional information. Errors in equipment, storage media, and computer programs, as well as user mistakes, can have serious consequences. Security routines are therefore necessary to guard against errors. As the collection of geographic information is very costly, damage to storage media (tapes, disks, etc.) can result in considerable financial loss. Moreover, as updating, access and new data entry become more frequent, security needs to be increased. On-line users can hardly be expected to wait several days while data are reconstructed following damage or an accident. Electrical outages or losses of data are expensive and time consuming for both users and operating organizations. Security routines are therefore vital.

There are many causes of data loss. The storage medium might be damaged through misuse; equipment might fail; fire or other damage might occur. Data in daily use in a primary memory may be lost due to user errors, operative system errors, equipment failures, or electric power failures. Without prior warning, disk sectors or entire disks may become unreadable, or data stored on tapes may deteriorate.

Backup copies of disk files should be made periodically to minimize reconstruction work should file data be lost. A customary rule for systems in daily use is that all database changes are automatically backed up daily from working disks onto tapes or tape cassettes. The entire database may be backed up less frequently, perhaps once a week or every other week. Some backup systems back up all transactions continually.

Working databases independent of original databases are often used in GIS work, for example, for tasks concerning a limited geographic area. A working database should never be stored on the same disk as an original database, lest a total disk failure, *disk crash* in computer jargon, destroy both databases. Data on tapes should also be backed up frequently. Although tapes store data well, data can deteriorate on all magnetic media in time. Therefore, tapes should be recopied periodically; once a year is recommended. Backup copies should not be stored with original data or databases, lest fire destroy both, but should be stored in a separate fireproof enclosure.

Some GISs contain data that must be protected against unauthorized use, such as personal data, details of telecommunications networks, and the like. The operative system can be configured to permit only authorized access to data or parts of data stored. Complete safeguarding and security of data would be extremely expensive and time consuming. For example, meticulous storage and handling routines can reduce the risk of disk damage, at a cost proportional to the measures enacted and the disk storage boxes and cabinets acquired for the purpose. Uninterruptible power supplies (UPSs) and filters may be installed to guard against power outages and large fluctuations in power voltage. These are often a necessity in developing countries. However, these add to the cost of equipment and may be expensive, not least because each UPS unit consists of a rectifier that converts the main ac into dc for float-charging batteries, and an inverter for conversion from dc to main voltage and frequency to power the equipment.

Therefore, safekeeping and security routines must be evaluated in terms of the costs of data loss. As in other sectors and disciplines, calculated risk must be taken into account. Insurance policies may be taken out against financial loss. Normally, it is possible to insure equipment against physical damage due to fire, flood, vandalism, and theft. Some insurance is available to cover the costs of reconstructing data and the losses incurred during operational interruptions. Finally, security measures may be required to guard against the potential losses of deliberate break-in and data destruction, *data crime* as it is now commonly called.

18.8 Evaluating New Applications

With time, most GIS users discover new needs and therefore new applications for systems and data. In many major GIS facilities that have been operational for some time, users report as many as 40 new applications implemented after the system became operational. The new applications often differ considerably from those originally anticipated for the facility, and are therefore among the factors that increase the benefits GIS provides.

The procedure for assessing new applications is straightforward. After a GIS facility has been in operation for a few years, a survey of user needs may be made to chart new needs. Identified new needs may be evaluated in terms of benefit/cost ratios, just as the initial needs for which the system was originally designed should be reevaluated. Similarly, the needs for organizational alterations should be evaluated, and alterations made using the same criteria as used initially. Most new applications require programming work, and the ease with which they may be realized depends on the flexibility of the system implemented.

CHAPTER 19

Standards and Access to Data

19.1 Introduction

Maps and other forms of geographical information constitute one of the world's oldest scientific fields and an area that has been the subject of standardization for many centuries. Long before digital methods came into use, it was generally accepted that the efficient interpretation and reading of maps was dependent on a uniform and congruous presentation. This applied not least to sea charts, where a safe sea voyage was (and still is) based on being able to change to a new map sheet independent of regional and national borders.

The introduction of data-based methods over the past 20 to 30 years has made it even more apparent that the efficient use of geographical information is dependent on a uniform description of the data. A large number of countries have therefore started work during the past 10 to 15 years on the development of national standards, especially for the exchange of digital geographical information between different computer systems. The basis is that the producer of data and the end user can interpret and understand the data in the same way. A standard will provide definition of data structures, data content, and rules that will:

- Increase mutual understanding of the geographic data among users
- Eliminate the technical problems of exchanging geographic data between different (geographical) information systems
- Increase integration and combination of geographical data and related information

Standardization can be carried out at three levels:

- *Level 1:* Generic standards (mainly IT-based; e.g., data description language, query language, transfer syntax)
- *Level 2:* GIS application independent standards (e.g., geometry, topology, quality, metadata)
- *Level 3:* GIS application specific standards (e.g., cadaster, utilities, roads, base maps, urban planning, etc.)

Levels 1 and 2 are normally covered by international standards, while level 3 is a typical national task.

The uniform transfer and understanding of data is dependent on a series of factors. Transfer of spatial data involves modeling spatial data, data structures, and logical and physical file structure. To be useful, the data must also be meaningful in terms of content and quality. In addition, a common terminology has to be developed at each level. Comprehensive international work has been carried out in this field. In 1992 the decision was made to establish a committee under the European Standardization Committee—Comité Européen de Normalisation (CEN).* This work has won broad support throughout Europe and most countries participate in it. In 1994 a corresponding initiative was taken by the International Standardization Organization (ISO).† Whereas CEN is most concerned with describing the more traditional aspects of geographical information such as topology, geometry, metadata, reference systems, and exchange/coding, ISO's work has, in addition, a description of the service relationship linked to geographical information (such as interface, etc.). The relationship between ISO and CEN is regulated by an agreement aimed at ensuring the establishment of a joint international standard in 1999.

NATO is also carrying out international standardization work and is the first organization to develop its own standard, Digital Geographical Information Exchange Standard (DIGEST) with object catalog FACC. Some years ago, a global data set was established, based on DIGEST, called the Digital Chart of the World. This data set corresponds in detail and accuracy to a scale of 1:1,000,000. The International Hydrographic Organization (IHO) has developed a standard known as DX90 with object catalog S57. Key elements in this development are the definition of the Electronic Navigational Chart (ENC) and the Electronic Chart Display and Information System (ECDIS). Work is being carried out to harmonize all these standards with ISO.

In recent years, information technology companies, database system vendors, GIS vendors, and others have established the Open GIS Consortium (OGC). This organization works on the development of open solutions linked to the processing of geographical information. The purpose is to develop the necessary interface to enable total mobility between the systems, as well as the integration of data and processing into mainstream computing. The prominent market position of these companies gives them considerable influence on the work that is carried out within ISO.

European system suppliers have developed a culture whereby they adapt to customers' and public bodies' requirements in matters concerning standardization. This differs somewhat from the United

* This committee has been named CEN/TC-287—geographic information.

† This committee is known as ISO/TC-211.

States, where powerful suppliers often control the market through *de facto* standards. But the Federal Geographic Data Committee (FGDC) and the U.S. Geological Survey have had the main responsibility for developing the Spatial Data Transfer Standard (SDTS). SDTS will serve as a national spatial data transfer mechanism for the United States, and provides a solution to the problem of spatial data transfer from conceptual level to the details of physical file encoding.

In this standardization work, great emphasis is placed on the efficient transfer of GIS data from one computer platform to another, as well as on the possibility that a GIS can use data that are dispersed through a variety of systems. These standards will probably result in a much stronger integration of data and systems in the future across the bodies of different platforms, based on standards such as Open System Environment (OSE).

It is unlikely that the entire industry will move to a single standard in the near future. It is probable that there will be a few important international standards in addition to all the proprietary commercial GIS formats. In Canada, Open Geographic Data Store Interface (OGDI) has been developed to overcome these problems. OGDI is an application programming interface that provides a solution for the most usual geographical data integration problems, including:

- Conversion of various formats into uniform transient data structure
- Transformation of coordinate systems and projection systems
- Searching and retrieving geometric and attribute data
- Accessing different geodatabases and formats
- Using the Internet for distribution and search of geographical data

19.2 Elements for Standardization

A series of factors have to be taken into consideration when developing standards for a geographical area (town, region, country) and miscellaneous subjects.

Fundamental principles

Usually, there are a number of fundamental and general elements that are suitable for standardization, such as spatial referencing and spatial referencing systems, where, for example, only the use of geographical coordinates (latitude and longitude) and UTM can be permitted. Should other systems be used, transformation parameters have to be stated. All measurements and quantities (geometry and attribute values) can be defined as valid only by use of the International System of Units (SI units), for example. Specification of date and time can be standardized, as in the sequence year/day/hour/minute, and hours given in 24-hour format.

A statement of data quality must be given. Such a statement should consist of:

- Lineage
- Positional accuracy
- Attribute accuracy
- Timeliness
- Completeness
- Logical consistency

The positional accuracy must be given in units related to nominal ground, not to units on a map scale. Data should be given at the lowest possible unit and not aggregated. Data should also be free of bias and must be unambiguous.

Standard transfer formats

It is still commonplace for different GISs and DBMSs to store data in their own internal format. The exchange of data often requires the transformation of formats and is therefore no trivial task when we know that a wide range of different formats is available. Were a common standard transfer format available, all system suppliers would be able to relate to this and develop regular export and import routines (Figure 19.1). Naturally, both users and producers of data have realized this, and that is why many countries have initiated the process of developing such standard formats.

The transfer format should provide transfer of geographical datasets that include metadata, spatial data, and attribute data. In general, a minimum additional programming should be necessary for creating a standard dataset (from GIS-A) or reading a standard dataset (to GIS-B). Data exchange between the most common GIS and CAD

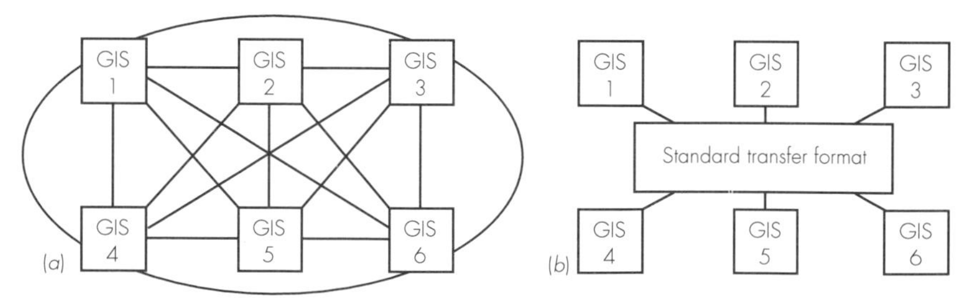

Figure 19.1
By means of standardized exchange formats, digital data can be exchanged between different GISs. A standard exchange format considerably reduces the need for translators—in this example from 15 (*a*) to 6 (*b*).

could also be included. Standard transfer format for files should include rules for:

- Transfer of geometry
- Transfer of links between geometry and attribute data
- Attribute data transfer
- Transfer of metadata

Standard transfer formats can also be used as storage format if necessary.

A special language (syntax) is often used (e.g., <. .> is used to enclose the name of different format elements, and ::= can mean "is defined as"). Special signs can be used to collect data in groups at different levels. The permissible number of signs or digits has to be specified for each element. For example, one person can be described as follows, where the group level is indicated by 1 to 4 dots:

• Person		
•• National identity number	H11	11—figure number
••• Name	T30	Text with 30 digits
••• Address		
•••• Street	T30	Text with 30 digits
•••• Street number	H4	4—figure number
•••• Postal code	H5	5—figure number

The description of a standard format should also include a code list for all the elements so that the same code will always represent the same type of element or object regardless of where it is produced (see Figure 19.2).

In the United States the U.S. Geological Survey has been responsible for the development of Spatial Data Transfer Standard (SDTS), which also specifies the implementation of SDTS in terms of the ISO for a Data Descriptive File for Information Interchange. The U.S. Geological Survey has also developed a topologically structured format, Digital Line Graph (DLG). Germany has an ambitious program in ATKIS (Amtlicher Topographisch–Kartographisches Informasionssystem), which includes both a topological data model and a cartographic data model.

Numerical code series	Object group
1 000	Survey control stations
2 000	Terrain formation
3 000	Hydrography
4 000	Boundaries
5 000	Built-up areas
6 000	Buildings and facilities
7 000	Communications
8 000	Technical facilities

Numerical code	Object type
4 001	National border
4 002	County boundary
4 003	Township boundary
4 011	Property boundary
4 022	National park border
	etc.

Figure 19.2
Many countries have created national standards for coding geographical data.

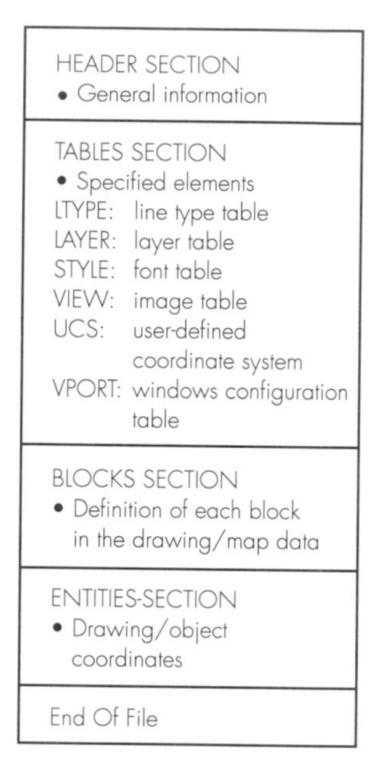

Figure 19.3
DXFs handle vector data and are in reality adapted CAD (AutoCAD format) data. The structure is based on a *header,* with general information; a *table,* which contains definitions of line table, font table, and so on; a *block,* which contains definitions for each block in the drawing; as well as *entities,* which contain the coordinates. If we only need to transfer the object's geometry, it will only be necessary to fill out the entities part.

AutoCAD's DXF format (Figure 19.3), Intergraph's SIF format, and Hewlett-Packard's HPGL format are typical industrial standards. PostScript, which is a graphic language, also has potential in the plotting, printout, and transfer of geographic data. In countries that have not established a national transfer format, industrial standards such as DXF (AutoCAD) and ASCII tables are used.

Computer Graphic Metafile (CGM) is an international standard (ISO 8632) for digital imagery. CGM files are system independent and are used in GIS to unite graphics from different systems. Tagged Image File Format (TIFF; Figure 19.4) is an industrial standard developed in the United States, used primarily for transferring image data between different desktop publishing systems. TIFF is hardware independent. Continuous Tone Image Format (COT) is also an industrial standard developed in the United States, used in image treatment programs. These formats are also appropriate for the transfer of raster data between different GISs.

Data volume often depends on how data are formatted and designed. General formats must provide leeway for all possible situations. Consequently, in practice the data often include basic adminis-

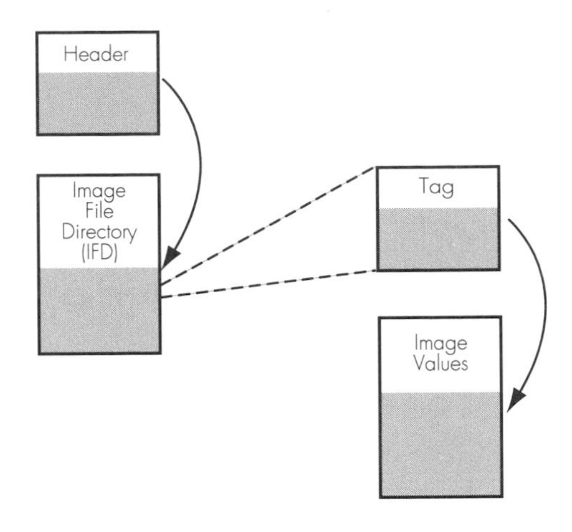

Figure 19.4
TIFF handles raster and image data. At the beginning of the file are a header and a pointer, which points at the address of the image file catalog. In the catalog are stored information on the relevant data and pointers to the data itself.

trative data which are not used, or space for attribute data which is not used. A series of software modules is available for packing data for transfer.

All standardization work is time consuming since it requires a high degree of unity and many parties are involved, all with different interests, thus using different systems. It is now possible to find solutions that can handle simultaneously many types of data warehouses with different formats.

Standardization on application level

On the application level, it is possible to standardize buildings, transport, utilities, land use, administration units, water, heights, control points, annotation, and text. For each application it is possible to define the following:

- Objective and limitations for the standard
- Data model, including digital representation of objects and relationship between objects
- Object attributes
- Permissible attribute values
- Geometry
- Conversion table
- Presentation

The definition of a geographical data standard at the application level could be done using data modeling techniques. A graphic schema [entity-relationship (ER) model] covers the definition of main object and relationships between object types. An example for buildings is shown in Figure 19.5.

A building will normally be described by its roof outline. A small building can be represented by a center point that will represent the spatial position of the building. Built-up areas are used for representing groups of buildings where the density at the actual scale does not make it possible to represent each building. A built-up area has its own geometry but can also be related to a set of buildings. A building describing a part of a building complex can be related to a building "parent" describing the outline of the entire complex. This model covers the essential requirement for defining the fundamentals for georeferencing building information and for mapping buildings and built-up areas at different scales.

The relevant object attributes can be presented as a list, as shown for buildings in Figure 19.6. Permissible attribute values can be set up in a corresponding list (Figure 19.7). (This figure and the following examples are taken from the National Standard for Zimbabwe–Digital Geographical Information.) The rules for geometry can also be set up in a list (Figure 19.8). Rules for presentation can also be made, in tabular form, as shown in Figure 19.9. It will often be necessary to make conversion tables between a standard's code system and codes used previously.

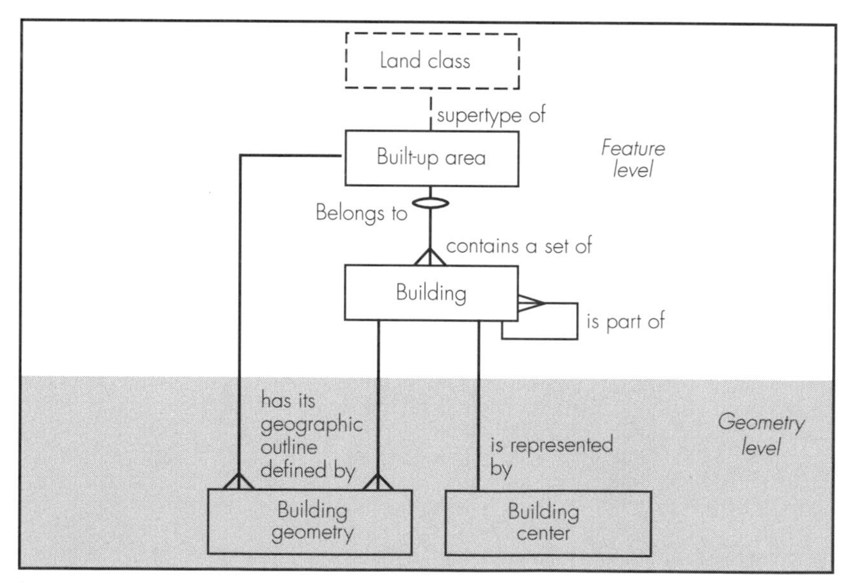

Figure 19.5
Data model for built-up areas and buildings.

Object	Attribute	Explanation
Built-up area	Area_Name	Offical name of the built-up area
	Area_Type	Main classification of built-up area
Building	Building_Name	Name of the building
	Building_ID	?

Figure 19.6
Standard for object attributes.

Attribute	Legal value	Comments
Area_Type	Commercial	
	Industrial heavy	
	Industrial light	
	Residential high	
	Staff quarters	
	Unclassified	Default
Building_Type	Apartment block	
	Bus terminal	
	Car parking	
	Etc.	

Figure 19.7
Standard for attribute domain.

Geometry entity	Element	Legal values
Building center	Geometric primitives	Point
	Layer	Building center
	Feature-key	Optional
Building geometry	Geometric primitive	Line, polyline
	Layer	Roof outline
		Wall outline
		Area outline
	Feature-key	Mandatory

Figure 19.8
Standard for geometry.

Map Product	Legend element	Derived from:		
		Layer	Attribute	Attribute value
1:50,000				

Figure 19.9
Standard for rules of presentation.

19.3 Metadata

Metadata are "data about data." Metadata are stored in a database and intended mainly as a tool to enable users to find out which data are available, if they are suitable for the purpose intended, where the data are stored, whether there are any limitations linked to their access and use, and possibly, how the data can be transferred to a suitable system (Figure 19.10). Metadatabases are intended to be an effective link between producer and user but can also be useful for internal use by producers to enable them to maintain a view of their own data over time (to avoid work duplication, among other things). Metadatabases have to be established and maintained by data producers. A decisive factor for users is that data sets should be described in a comprehensive and standard way. The main purpose of a metadata standard is to define those data structures that should be used to describe a geographical dataset. The standard for metadata does not specify how original data are organized in the computer or how data transfer shall be effected in practice.

Metadata attributes

A series of elements should be included in a metadatabase for describing a data set, of which the following are the most important:

- *Dataset identification:* sufficient data proving an ambiguous identification of the dataset (e.g., the owner organization and the unique name of the dataset within that organization).

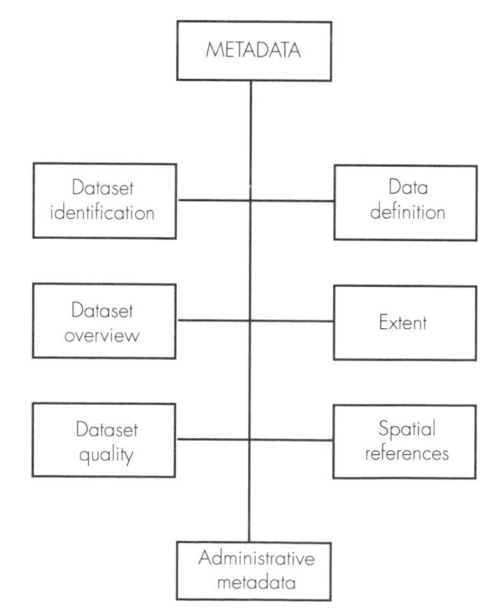

Figure 19.10
Data model for a metadatabase.

- *Dataset overview:* an overall description of the dataset (e.g., a brief summary of the content, the intention of the production, language and character set used) and the mechanism used to represent spatial information in the data set (e.g., direct georeferencing with use of vector and raster, or indirectly by use of postal codes or similar). The total number of objects can also be useful information. References to additional documents about the data set could also be given.
- *Data quality:* how well the data set overall meets its nominal ground location (relative to the specification). The elements are lineage, positional accuracy, attribute accuracy, timeliness, completeness, and logical consistency.
- *Spatial references:* description of the spatial reference and projection system. For direct reference systems, the following should be specified: datum, ellipsoid, map projection, height reference system, and coordinate system resolution. For indirect spatial systems, the structure has to be explained (e.g., address system, map sheet, postal code administration units).
- *Extent:* space covered by the dataset with respect to planes, vertical and temporal dimensions; name of the area covered, bounding area and minimum and maximum (x, y) coordinates; minimum and maximum vertical coordinates; the date on which the status and description of the extent are valid.
- *Data definition:* information on the content of the data set; object types and their attributes; description of the object type or the reference to an existing standard coding list; the domains from which attribute values may be assigned.
- *Administrative information:* the name and address of the organization that has produced the data; the responsibility of the organization in relation to the data set (e.g., the creator or the owner, copyright, etc.).
- *Distribution:* how to acquire a data set; where and how the data set is held and how it may be transferred; constraints regulating access to and use of the data set, other than copyright; copyright owner; and information about charges for the data set, including price per unit and discount. Data about the geographic and/or thematic partitioning of the data set; material in or on which the data set can be recorded and from which it can be retrieved (CD-ROM, etc.); information about how to access the data set eventually on-line, information about how to order the data; and future support service to users.

Not all information is equally important for storage in a metadatabase. Priorities should be set, for example, by giving information that is mandatory or optional. For each metadata attribute, therefore, the cardinality should be presented as one or many and type (string/text,

Metadata attribute	Description	Con-straint	Cardin-ality	Type
Dataset identification		M		
Dataset title	The unique name of the dataset within the owning organization(s)	M	1	string
Alternative title	Another name of the dataset, which is either in the same language or another language	O	1	string
Dataset overview		M		
Abstract	A brief summary of the content of the dataset	M	1	string
Purpose of the production	A summary of the intentions with which the dataset was developed	M	1	string
Spatial reference system used	Data about whether geographic features in the dataset are positioned in space explicitly by a co-ordinate system or implicitly by an indirect spatial reference system: for example, a postal address	M	N	string
Dataset language	The language(s) and the character set used within the dataset	M	1	string
Document reference	Reference(s) to additional documentation about the dataset, including documentation language(s)	O	1	string
Dataset quality		M		
Source process	A summary of the processes that the dataset has undergone, including time	M	1	string
Source method	A summary of the methods that the dataset has undergone	M	1	string
Source material	A summary of the original material from where the dataset is produced	M	1	string

Figure 19.11
Some standards for object attributes for metadata. Constraints: mandatory (M) or optional (O). Cardinality: one (1) or many (N). Type: string (text), numeric, date, address, point, area.

numeric, date, point, era). Figure 19.11 shows an example of some data which could be used to describe geographic data sets.

19.4 Data Access

19.4.1 Digital libraries

Today, the Internet is the most commonly used technology for realizing digital libraries. Several major suppliers of GIS software have developed systems that allow a certain GIS functionality on the Internet. This has made it possible to develop efficient user interfaces. A possible solution for a digital library is shown in Figure 19.12.

Normally, a digital library will cover the following basic functions:

- User interface for search and selection of data
- Catalog service/metadatabase

In addition, there will often be automatic functions for ordering and invoicing.

19.4.2 Organizational infrastructure

A number of countries have now initiated a process to improve the infrastructure for the distribution and use of digital geographical data. Technical progress such as the Internet and GPS have made this type

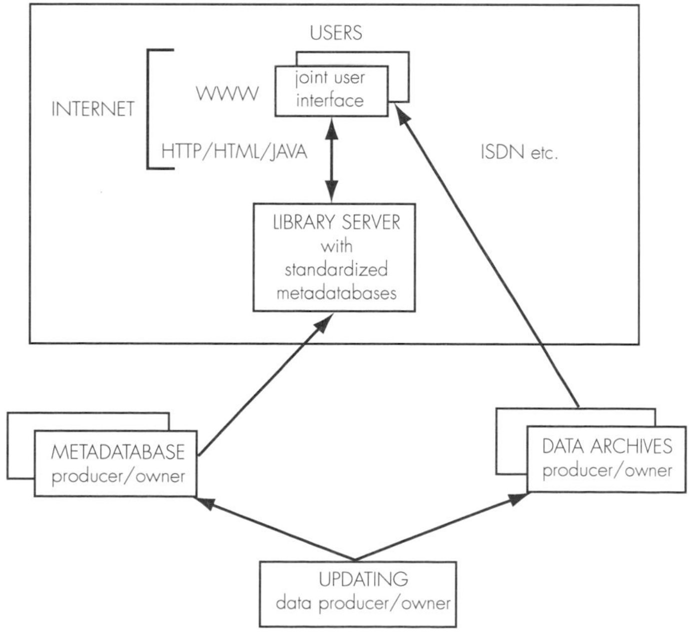

Figure 19.12
A digital library for geographical data is being offered to users on the Internet by clearinghouses.

of work very relevant. The *infrastructure* can be defined as "a data series and electronically based services which satisfy the common needs of different user groups for accurate positioning and georeferenced data." The broadcasting of GPS corrections over the radio could be one possible service. Standards and metadatabases are important elements in a national infrastructure. There should also be participants to cover the storage and distribution functions. The maintenance of an infrastructure presupposes the existence of organizations that can cooperate in the various functions.

Traditionally, the infrastructure in most countries has been taken care of by the national mapping authorities, but today we see the need to create new alliances to promote the sale of geographical data. In the United States, for example, the National Spatial Data Infrastructure (NSDI) has been established as an umbrella organization to handle the search, query, finding, access, and use of geographical data. NSDI is composed of organizations and individuals who generate or use geographic information, of the technologies that facilitate use and transfer of data, and the actual data. A joint organization has been established, the European Geographic Information Infrastructure (EGII), under the control of the European Space Agency (ESA). ESA and the European Commission (EC) have agreed to work toward a more efficient and cost-effective use of satellite data, by combining their expertise to establish a coordinated, decentralized earth observation network, the European Earth Observation System (EEOS). The EC contribution to EEOS is the Centre for Earth Observation (CEO), where a standard user interface is being developed for searching in metadatabases and archives. In addition, a common program is being developed for the search, ordering, and visualization of data, as well as the choice of standardized transfer formats.

Other organizations, such as brokers and clearinghouses, have also entered the market. Data brokers ensure the efficient turnover of data being offered on the market. Clearinghouses represent the technical solution, to allow producers to describe what data sets they have available and to allow users to find the data sets they need. Digital libraries based on the Internet and metadatabases are the most important technical factors.

CHAPTER 20

Formal Problems in Establishing GIS

20.1 Ownership and Copyright

The use of geographical data often raises the question of ownership. A country's laws and statutes usually regulate who may claim to own intellectual property. *Intellectual productions* is a useful term for devices whose value is not based on the ownership of a physical article. Those who can claim ownership may also regulate its use through pricing mechanisms and general licensing terms.

The legal status of geographical information is not always clear. If it fulfills the requirements of national law, it can be protected by copyright law. It can also be protected under national legislation on such issues as database protection and unfair competition. Where the data are procured in full by private firms, companies, or individuals, there would normally be no doubt that ownership belongs to those who procured them. In most cases, however, computerized maps and registers have been partially or completely generated in the public sector. The question of ownership then depends on whether public officials regard geographical data as belonging in the public domain or as a product protected under copyright law. The ownership situation becomes even more complex in view of the fact that a number of private companies now process data from public sources with the addition of their own data with the object of increasing the commercial resale value.

The Bern Convention of 1951 specified common international regulations for the protection of copyright and can be used as the basis for copyright considerations linked to mapping and geographical data for countries signatories to the convention. The new (not yet adopted) database directive from the EU gives protection to so-called "sweat of the brow" databases. This means that all data have to be registered/cataloged by one means or another in order to be protected.

Objective "facts" are not normally covered by copyright law. Certain geographical information products as well as certain elements in maps or charts, however, may not have the necessary level of achievement to be copyright protected. This type of data can therefore be copied freely. For example, elevation contours will normally be based on a certain amount of adaptation and are therefore protected by copyright, whereas place names alone are not protected. A database of geographical names may thus not be copyright protected, but can be catalog protected (Aslesen 1996).

Basically, the same rules apply whether the case concerns data that are reproduced in digital or in analog format. Copying from one electronic storage device to another is a reproduction under copyright law. Copying of screen display is more of a problem since the display is only temporary.

Public practice with regard to data ownership claims varies considerably from country to country, and sometimes within a country, as from state to state in the United States. In addition, the United States operates with different policies at the federal and state levels. In the United Kingdom, the public sector claims copyright. Hence permission must be granted by the Ordnance Survey (the national map agency) for *all* digitizing of existing maps. On termination of the digitizing, a copy of the resultant digital data must be provided free of charge to the Ordnance Survey. In addition, the Ordnance Survey levies a charge proportional to the amount of digitizing work carried out. In Canada, the public sector also claims copyright over map data but not as rigorously as in the United Kingdom. Usage fees are charged whenever the value of the digitized data exceeds the costs incurred in acquisition and digitizing.

National geographic databases and GISs may be regarded as parts of future infrastructures, as with road networks and electricity supply networks. Consequently, in the United States, governmental maps and digital map data are regarded as being in the public domain, and all copying is free. That view is based on the belief that society as a whole profits from the enhanced value of products and data that initially were financed by the taxpayers.

Increasing electronic access to public geographical data through the telephone network and the Internet appears to make fixed regulations a necessity in this field. GRID—the United Nations Global Resources Information Database—offers free access to public environmental data. Many countries are now in the process of formulating this kind of regulation, but there still appears to be a certain need to obtain practical experience before approving such regulations. Since the telephone network is independent of national boundaries, an initiative has been taken at the international level to coordinate the regulations for different countries.

20.2 Cost Recovery and Pricing

Cost-recovery methods are normally used to determine commercial prices for public data. Prices are thus expected to cover such costs as data establishment, storage, and distribution. The price is usually dependent on the quality of the data with regard to detail, timeliness, completeness, topological continuity, and so on. In many countries, price is also dependent on usage (application). Thus some National Mapping Authorities set lower usage fees for the public sector's use of data and higher fees for commercial uses in the private sector. Correct pricing, however, is difficult. High prices encourage users to compile their own digital map data, which then reduces the importance of the public sector. Low prices can result in data not being kept up to date because of failure to cover costs, and the value of quality data is not appreciated.

We have already seen that the United Kingdom and Canada enforce copyright regulations with corresponding fees/pricing based on the cost-recovery principle. In 1996 the United Kingdom had a total cost-recovering rate of 70 to 80% for the Ordnance Survey. A number of other European national mapping authorities (Norway, Sweden) have a cost recovery rate of 30 to 40%. In the United States, national data are available only against payment of copying and distribution costs, and cost recovery is thus quite low. The general impression today is that there is little point in a cost recovery rate of 100% for public data. There are a number of situations where geographical data/maps are of vital importance in special situations, such as accidents in wilderness areas. Data of this type are usually of little commercial interest and these costs cannot therefore normally be covered. Such data would therefore not be produced without public support. The need for public co-financing, however, becomes most apparent when seen from a military perspective. Historically, military needs have initiated the majority of national map series. Military requirements are often enough to justify total public financing of national data coverage at the smaller scales. However, it is probably an accepted principle in most countries that the taxpayers' burden be eased by allowing those who gain special benefit from the data to cover some of the costs.

In addition, product price should depend on risk. Should the public sector or a private company take a high investment risk in connection with the establishment of a database, it makes economic sense to charge a higher price than if one could be sure of a certain degree of cost recovery. Joint market ventures can lower prices since the risk is spread between several partners.

The business aspects of GIS have yet to be fully developed, primarily because so far, most GIS facilities have been established to meet the needs of particular public agencies. However, commercial uses will undoubtedly become more important in the future. The development of electronic marine charts and systems for car navigation require equipment suppliers to deliver digital map data together with the equipment sold. Databases may then become commercial wares with considerable profit potential. One particular development in Europe today is that several private consortia are in the process of establishing road databases for vehicle navigation and other transport purposes. Pricing of this type of data will probably follow normal market principles, based on competition from several suppliers.

20.3 Public or Private Organization of Geodata

In the discussion as to public or private organization of geodata, strategies must suit prevailing market conditions. But as market conditions change, prudent strategies must encompass alternative approaches with respect to organization of production and the market situation. Organization of production can be based on a decentralized, centralized, or semi-decentralized approach in which the producing organization is responsible for reasonable standardized data interchange between providers and users.

The market situation is likely to be comprised of a variety of user groups with common interests only in limited (local) geographical areas and in cost sharing of centralized geodata. Market demand from a single major user would normally be inadequate fully to finance such data provision. Only a homogenization of demand would enable any uniformity of supply to be implemented.

Depending on the organization of the production side and the market, public agencies, which often operate national databases, may choose among various business approaches:

1. Supplement basic data with other standard data and distribute the combination as a standard package to all users (extended map production).
2. Supplement basic data with other standard data to compile tailormade user products.
3. Supplement basic data with user-specified data to compile tailormade user products.
4. Coordinate data interchange without being involved in production but with responsibility for standardization.

The question of public versus private involvement is part of the evaluation picture, as are questions of regulation and accessibility. Economies of scale and market homogeneity also influence users' choice of strategies. Financing through national budget appropriations may increase the availability of products to users, but may also hold prices artificially low so that providers cannot meet the maintenance costs. If private financing is to be involved, the market must tolerate competition and price variation, which may exclude some potential users who cannot afford the products offered.

Socioeconomic considerations dictate that public databases should be accessible to all, with few limitations. Since the acquisition, processing, and distribution of geodata often involve both national and local authorities as well as commercial interests, several countries have taken the initiative to establish a partnership to solve such tasks. The National Performance Review (1993) has established this as U.S. government policy in the field. In this way the burden on individual institutions' budgets will be reduced.

20.4 Data Security

Many countries have laws pertaining to the establishment and use of registers of persons. Official permission is almost always required for such registers, a restriction imposed as a means of safeguarding individual rights. Geographical data that contain information about individuals may similarly be restricted. Sensitive material such as maps of military areas or data concerning endangered species or rare plants may be considered of national interest, and restricted accordingly.

In some cases, combining data from open registers may be illegal. Limited information may be openly available, but when many data sets are combined to cover larger areas or effect undesirable combinations, the result may constitute a security risk. For example, for national security reasons telecommunication administrations prohibit unauthorized persons from handling the network information contained on several map sheets. In general, the more detailed the information, the stricter the constraints on its use.

Access to computerized information may also be restricted because those who compiled it may not be confident of its quality and therefore prefer that it should not be widely used. Examples of this include data gathered ad hoc for special projects and not based on sound methods or quality controlled. As public geodatabases become increasingly accessible over the telenetwork, access limitations will probably have to be introduced. Perhaps different users will be granted differing access by way of, say, the coding of certain data, so that only those with access to the code will be able to interpret the data.

CHAPTER 21

A Vision for the Future

This book

In this book we have attempted to describe the most important aspects covered by the term *GIS* to give the reader a total picture of the subject. Our idea was to produce a book that would be suitable for the reader with little or no knowledge of GIS. Since this is an introductory book, many aspects have been handled relatively superficially and others are mentioned only in passing. A number of subject areas are not discussed at all. For example, there is little discussion of the different mathematical methods and formulas for handling the geometry. However, a number of textbooks are available which describe these parts of the GIS theory in greater detail.

Other sources

Those who wish to learn more about the subject should refer to other textbooks which give more detailed coverage of GIS. Other opportunities to study GIS are available in the form of an interactive GIS tutorial for self-study and organized courses. Courses at various academic levels are available in many countries and these are the best way of obtaining a thorough knowledge. A number of GIS publications are available containing both popular and academic news. The reader should also consult the References section in the present book.

Status of GIS today

Just as maps have been used in many different areas of society, GIS is now gradually taking over these functions and, in addition, exploiting totally new areas. In broad terms, we can say that GIS can be used for documentation (how things used to be and how they are today) and for analyses and simulations (how things could be in the future). Although GIS has proved to be a great help in many fields, we must not ignore the fact that it is only an aid and that the results from GIS will have to be used in combination with good judgment and professional skill. When used correctly, however, GIS has proved to be very useful in improving many aspects of daily life both directly and indirectly, and this should be the primary reason for using this technology.

The future

The results of market research indicate that GIS will increase in volume and will be applied in new fields of activity.

Technological developments

Technological developments in the IT sector appear to continue unabated. The increase in processor speed and transfer speed in networks, storage capacity, improved databases, and user interfaces are all factors that will make it easier to work with GIS in the future. Even though the GIS market in many ways can be described as technology driven (although it should, in principle, be user driven), this is a development that users welcome. We have established that today's technology is good enough to solve the main problems linked to the use of GIS, but in the future we are sure to see GIS-related solutions that are far ahead of today's (in many ways) boring monitor-oriented solutions.

We can see the outlines of this already in the form of multimedia (hypertext, sound, video, virtual reality) now available on the Internet. Impressive developments can be expected in this field, not least in respect to interaction between the user and the system. While such development will also benefit GIS users, it will also set new requirements for data accessibility and updating.

New applications

Geographical data and GIS are continuously finding new areas of application. To date, the GIS market has been dominated by products aimed at the professional user. New and untraditional user groups have appeared in recent years. *GIS in business,* for example, has become a common term. Looking at the relatively low number of potential users, we can see that professional users are relatively few compared to normal consumers. As GIS and GIS-related solutions become available on the Internet, we will be better able to reach a broader viewer group (Figure 21.1). There are users who are only interested in

Figure 21.1
Possible future scenario: politicians grouped around a graphic display board while discussing matters pertaining to geographical data.

obtaining a quick view of the data without having to pay anything for it. The financing of geographical data on the Internet will probably be based on advertising income (e.g., from hotels and other companies willing to pay to appear on the maps).

As to the professional use of GIS, the greatest social significance in the future will be found within environmental issues. This is a relationship that is difficult to quantify, but without systematic surveillance and documentation of the status of natural resources and analysis of future developments, large amounts will have to be invested to correct prior mistakes. The future success of GIS will also depend on how well we manage to communicate with decision makers. Once again we have to remember that it is not the operators who are the real users, but those who make decisions on the basis of maps and tables produced in GIS.

Standardization

International standardization (ISO and CEN) and cooperation between the system suppliers (Open GIS Consortium, etc.) will make it much easier in the future for GIS operators to bring up data from different sources. We can also expect that requirements for the standardization of data in the form of descriptions in metadatabases, etc., will also increase in the future. The efficient marketing of data will be dependent on suppliers being able to describe the data in an appropriate and correct way.

Data access

In the future, data will be available to a much greater extent over the Internet. In many countries today, clearinghouses and similar organizations have appeared on the scene that disseminate data via the Internet. We will probably have national and regional digital libraries for geographical data and images (aerial photos and satellite images). We will experience increased international competition in the field. For example, national mapping authorities will have to compare with both international and national companies for the supply of national data. Data on London could be bought over the Internet just as easily in Hong Kong as in Southampton.

Organization

Technological developments will always affect the way that work is organized. Some might view this as a threat, others as an advantage. Thus, organization is concerned with human factors, which have remained constant for many millennia (human genes have remained unchanged for at least 35,000 years). It is not likely, therefore, that it will be easier to obtain optimal organizational solutions in the future than it has been in the past.

New players in the market will change the traditional structures for the sale of data. Discussions are going on in many countries as to how much the public sector, through grants to national mapping authorities, should become involved in the production and sale of data. In most countries there would appear to be agreement that the public sector should be responsible for production of certain basic geographical data. This means that in the future most countries will have national mapping authorities which are responsible for carrying out national policy in this field. However, there will be wide variations in the extent to which production and any appreciation in value will be organized as a monopoly.

Research and development

Most of the research and development in the field of GIS has been linked to technological improvements. Surprisingly little attention has been given to the end user—what constitutes efficient GIS information for a decision maker, how maps should look, and so on. We hope that more research will also be carried out in this field in the future.

REFERENCES

Andersen, Ø., 1991, *Geografisk Informasjonssystem: Presisering og Problemstillinger.* Aas, Norway: Compendium.

Aronoff, S., 1989, *Geographic Information Systems: A Management Perspective.* Ottawa, Ontario, Canada: WDL.

Aslesen, L., 1996, *Paper on Copyright.* Hoenefoss, Norway: IHO.

Atenucci, J. C. (ed.), 1995, *Geographic Information Systems: A Guide to Technology.* New York: Van Nostrand Reinhold.

Balstrøm, T., Jacobi, O., Munk Sørensen, E. (eds.), 1994, *GIS i Danmark.* Copenhagen, Denmark: Teknisk Forlag, 9–21, 45–50, 61–71, 97–98.

Baudouin, A., and Anker, P., 1984, *Kartpresentasjon og EDB-Assistert Kartografi.* Oslo, Norway: Norwegian Computer Center.

Beck, R., 1997, Satellite imagery from earth watch and space imaging. *GIM Geomatics Info Magazine,* 10(3):60–63.

Bernhardsen, T., 1985, *Samfunnsmessig Nytteverdi Av Tekniske Kart.* Arendal, Norway: Asplan Viak

Bernhardsen, T., 1986, *A Cost/Benefit Study of GIS Methodology and Results.* Toronto, Ontario, Canada: FIG International Congress.

Berry, J., 1987, Fundamental operations in computer-assisted map analysis. *International Journal of Geographical Information Systems,* 2.

Bertin, J., 1983, *Semiology of Graphics.* Madison, Wis.: University of Wisconsin Press.

Beser, J., and Haunschild, M., 1995, Advantages of integrated GPS/GLONASS: operation GIM. *Geomatics Info Magazine,* 9(11):63–69.

Bjørke, J. T., 1988a, *Grunnleggende Datastrukturer: Eksempler på Deres Kartografiske Anvendelser.* Grimstad, Norway: paper presented at AID.

Bjørke, J. T., 1988b, *Kartinformasjon.* Grimstad, Norway: paper presented at AID.

Brande-Lavridsen, H., 1994, *Cartography.* Aalborg, Denmark: ICA paper.

Burrough, P. A., 1986, *Principles of Geographical Information Systems for Land Resource Assessment.* Oxford: Science Publications.

Carosio, A., and Zanini, M., 1996, Landscape modelling and visualization: from 2-D surface to 3-D volumes. *GIM Geomatics Info Magazine,* 10(5):6–8.

Castle, G. H., 1993*(ca), Profiting from a geographic information system. GIS World.*

Cullis, B. J., 1994, *A strategy for assessing organizational GIS adoption success. GIS/LIS* 1994: 208–17.

Dale, P. L., and Clarc, S. R., 1988, *Land Information Management: An Introduction with Special Reference to Cadastral Problems in Third World Countries.* Oxford: Clarendon Press.

Dangermond, J., 1997, *Address to NYC GIS Steering Committee.* Paper available on Internet, e-mail *dlashell@esri.com.*

Ericsson, G. (ed.), 1987, Community Benefit of Digital Spatial Information. Arendal, Norway: Nordic Kvantif/Asplan Viak.

ESRI, 1995, *Understanding GIS: The ArcInfo method.* Redlands, Calif.: ESRI.

Fotheringham, S. and Rogerson, P., 1994, *Spatial Analysis and GIS.* Bristol, Pa.: Taylor & Francis.

Gallardo, M., 1995, *Microsoft Windows NT.* Paper available on Internet, *http://ata.princeton.edu/falks/winnet/winnet/html*

Goodchild, M. F., and Kemp, K. K., 1990, *Introduction to GIS: Core Curriculum:* Berkeley, Calif.: National Center for Geographic Information and Analysis.

Gralla, P., 1997, *How the Internet Works—All-New Edition.* London: Prentice Hall.

Hofmann-Wellenhof, B., Lichtenegge, H., and Collins, J., 1994, *Global Positioning Systems: Theory and Practice.* New York: Springer-Verlag.

Huxhold, W. E., and Levinsohn, A. G., 1995, *Managing Geographic Information Systems Projects.* New York: Oxford University Press.

Ireland, P., 1996, *Inkjet versus laser in the output game.* GIS Europe, 11(5):38.

Jones, A., 1997, The new commercial satellites. *Mapping Awareness,* 11(3): 14–17.

Kofler, M., and Grüber, M., 1997, Towards a 3-D GIS Database. *GIM Geomatics Info Magazine,* 11(5):55–57.

Korte, P. (ed.), 1997, *The GIS Book: The Smart Manager's Guide to Purchasing, Implementation and Running a Geographic Information System.* Entrada: OnWord Press

Kraak, M. J., and Ormeling, F. J., 1996, *Cartography: Visualization of spatial data,* Harlow, Essex, England: Addison Wesley Longman.

Kylen, B., and Hekland, J., 1990, *Economics of Geographic Information: Organizational Impact of Technology Changes in the Road GIS Case.* Arendal, Norway: Nordic Kvantif/Asplan Viak.

Langran, G., 1992, *Time in Geographic Information Systems.* London: Taylor & Francis.

Larsen, J. M., 1991, *Brukerbehov.* Arendal, Norway: Geodatasenteret.

Laurini, R., and Thompson, D., 1992, *Fundamentals of Spatial Information Systems.* London: Academic Press: 102–106.

Lee, Y., and Plunkett, G., 1997, Desktop GIS implementation. *GIM Geomatics Info Magazine,* 11(3):33–35.

MacEachren, A. M., and Fraser, D. R., 1994, *Visualization in Modern Cartography.* Oxford: Taylor Pergamon.

Maguire, D. J., Goodchild, M. F., and Rhind, D. W., 1991, *Land applications.* In Geographical Information Systems, Vol. 1, *Principles.* Harlow, Essex, England: Longman Scientific and Technical; 251–67, 269–97, 319–35.

Medycykyj-Scott, D., and Hearnshaw, H. M., (eds.), 1993, *Human Factors in Geographical Information Systems.* London: Belhaven Press.

Newton, A. W., Zwart, P. R., and Cavill, M. E. (eds.), 1992, *Networking Spatial Information Systems.* London: Belhaven Press, 15–27.

Nielsen, F., 1994, *GIS Database Design and Organisation.* Paper available on Internet, e-mail *fofdat@inet.uni.c.dk.*

Nieuwenhuijs, S., 1995, Oracle MultiDimension: new frontiers in spatial data management. GIS Europe, 4(5):40–42.

Nix, M., 1997, *One-person mobile office for surveying.* GIM Geomatics Info Magazine, 11(4):32–35.

Onsrud, J. H., and Rushtom, G., 1995, *Sharing Geographic Information.* New Brunswick, N.J.: Center for Urban Policy Research.

Østensen, O., 1996, OSO TC/211: *ISO standards for GI infrastructures.* GIM Geomatics Info Magazine, 10(3):24–25.

Palm, C., 1991, *Karttecken—Kartsymbol.* Hoenefoss, Norway: Nordic Cartographic Society Summer School.

Sahai, R., 1996, *Inside MicroSation 95.* Entrada: OnWord Press

Sanchez, R., 1996, *Satellite image integration with spatial data.* GIM International Journal for Geomatics, 10(10):56–59.

Schulte, V., 1996, Conversion kit. *GIS Europe,* 5(6):36–38.

Seaborn, D., 1995, *Database management in GIS: is your system a poor relation?* GIS Europe, 4(5):34–38.

Somers, R., 1994a, *Alternative GIS development strategies.* GIS/LIS, 1994: 706–15.

Somers, R., 1994b, *GIS organization and staffing.* URISA, 1994:41–52.

Steinfort, A., 1996, *Is there life after GIS?* GIM Geomatics Info Magazine, 10(11):24–27.

Strand, G. H., 1991a, *Linear Combination of Models in Geographical Information Systems.* Oslo, Norway: NITO Summer School.

Strand, G. H., 1991b, *Sentrale Teknikker i Geografiske Informasjonssytemer.* Oslo, Norway: NITO Summer School.

Strand, G. H., and Kvenild, L., 1984, *Supermap, software for Tematisk Kartografi.* Trondheim, Norway: Institute of Geography.

Tomlin, C. D., 1990, *Geographic Information Systems and Cartographic Modelling.* Upper Saddle River, N.J.: Prentice Hall.

Tveitdal, S., (ed.), 1987, *Community Benefit of Digital Spatial Information: Digital Maps, Data Bases, Economics and User Experience in North America.* Norway: Arendal, Nordic Kvantif/Asplan Viak.

van der Schans, R., 1995, Differences and similarities of GIS and CAD. *GIM International Journal for Geomatics,* 9(8):43–47.

van Genderen, J. L., 1991, *Guidelines for Education and Training in Environmental Information Systems in Sub-Saharan Africa: Some Key Issues.* Delft, The Netherlands: ITC.

von Rimscha, S., and Wolf, W., 1997, Modelling in an uncertain world. *GIS Europe,* 6(1):18–20.

Worboys, M. F., 1995, *GIS: A Computing Perspective.* London: Taylor & Francis, 61–64, 309–14.

Worrall, L. (ed.), 1990, *Geographic Information Systems: Developments and Applications.* London: Belhaven Press.

INDEX